Sustainability & Green Polymer Chemistry
Volume 2: Biocatalysis and Biobased Polymers

ACS SYMPOSIUM SERIES **1373**

Sustainability & Green Polymer Chemistry Volume 2: Biocatalysis and Biobased Polymers

H. N. Cheng, Editor
USDA – Agricultural Research Service
New Orleans, Louisiana, USA

Richard A. Gross, Editor
Rensselaer Polytechnic Institute
Troy, New York, USA

Sponsored by the
ACS Division of Polymer Chemistry, Inc.

American Chemical Society, Washington, DC

Library of Congress Cataloging-in-Publication Data

Names: Cheng, H. N., editor. | Gross, Richard A., 1957- editor. | American Chemical Society. Division of Polymer Chemistry.
Title: Sustainability & green polymer chemistry / H.N. Cheng, editor, USDA -- Agricultural Research Service New Orleans, Louisiana, USA, Richard A. Gross, editor, Rensselaer Polytechnic Institute, Troy, New York, USA ; sponsored by the ACS Division of Polymer Chemistry, Inc.
Description: Washington, DC : American Chemical Society, [2021] | Series: ACS symposium series; 1373 | Includes bibliographical references and index. | Contents: volume 2. Biocatalysis and biobased polymers.
Identifiers: LCCN 2020049541 (print) | LCCN 2020049542 (ebook) | ISBN 9780841298521 (hardcover) | ISBN 9780841298514 (ebook other)
Subjects: LCSH: Polymers. | Polymerization. | Green chemistry. | Sustainability.
Classification: LCC QD381 .S88 2021 (print) | LCC QD381 (ebook) | DDC 668.9028/6--dc23
LC record available at https://lccn.loc.gov/2020049541
LC ebook record available at https://lccn.loc.gov/2020049542

The paper used in this publication meets the minimum requirements of American National Standard for Information Sciences—Permanence of Paper for Printed Library Materials, ANSI Z39.48n1984.

Foreword

The ACS Symposium Series is an established program that publishes high-quality volumes of thematic manuscripts. For over 40 years, the ACS Symposium Series has been delivering essential research from world leading scientists, including 36 Chemistry Nobel Laureates, to audiences spanning disciplines and applications.

Books are developed from successful symposia sponsored by the ACS or other organizations. Topics span the entirety of chemistry, including applications, basic research, and interdisciplinary reviews.

Before agreeing to publish a book, prospective editors submit a proposal, including a table of contents. The proposal is reviewed for originality, coverage, and interest to the audience. Some manuscripts may be excluded to better focus the book; others may be added to aid comprehensiveness. All chapters are peer reviewed prior to final acceptance or rejection.

As a rule, only original research papers and original review papers are included in the volumes. Verbatim reproductions of previous published papers are not accepted.

ACS Books

Contents

Indexes

Preface

According to the U.S. Environmental Protection Agency, "sustainability" entails the creation and maintenance of the conditions under which humans and nature can exist in productive harmony to support present and future generations. "Green Chemistry" is part of sustainability and it focuses on the design of chemical products and processes that reduce or eliminate the generation of hazardous substances. Currently sustainability and green chemistry are hot research topics, and there are numerous worldwide research and development (R&D) activities revolving around these topics. "Green *polymer* chemistry" places a particular emphasis on polymer science and its contributions to sustainability and green chemistry. As polymers are almost ubiquitous in our daily lives, they can play a major role in our pursuit of sustainability.

The two volumes of this book are based on an international symposium on sustainability and green polymer chemistry that was organized for the ACS Spring National Meeting in Philadelphia in March 2020. We had assembled 79 outstanding speakers and 25 poster presenters for this symposium. Unfortunately, owing to the Covid-19 pandemic, the entire ACS Philadelphia meeting was cancelled. In discussions with many of the symposium speakers, we felt that this field is highly relevant to the scientific enterprise, and it would be very useful to report in a timely fashion the current R&D in this field in a single compilation. Thus, we have undertaken to edit this book, which provides a convenient medium for the symposium speakers to share their latest research developments.

It may be noted that green (and sustainable) polymer chemistry has become popular among polymer scientists. In fact, for many years we have been organizing "green polymer chemistry" symposia (or variations of this theme) at ACS National Meetings at 3-year intervals. We have noticed that this field is highly multidisciplinary, where many researchers use techniques and tools from polymer chemistry, organic synthesis, material science, microbiology, molecular biology, catalysis, enzymology, environmental science, analytical chemistry, chemical engineering, and other disciplines. This field is also global, with active research being done in the Americas, Europe, Asia, and other parts of the world. Communication, collaboration, and coordination are therefore useful as we continue to promote, support, and advance this field. Hopefully this book will represent a useful step in this direction.

A total of 23 chapters (in two volumes) are included in this book. In this volume (entitled "*Sustainability & Green Polymer Chemistry, Vol. 2. Biocatalysis and Biobased Polymers*"), 11 chapters are included. These are grouped into three sections:

1) Use of Enzymes for Polymers,
2) Green Sustainable Building Blocks,
3) Green Sustainable Films and Bioplastics.

In addition, the first volume of this book (*Sustainability & Green Polymer Chemistry, Vol. 1. Green Products and Processes*) consists of 12 chapters, including an overview chapter (Chapter 1), two chapters on Green Sustainable Products from Plant Oils, four chapters on Green Processes and

Materials, three chapters on Green Materials for Medicine, and two chapters on Green Polymer Additives.

The two volumes of this book are targeted for scientists, engineers, and students, who are involved or interested in green polymer chemistry and sustainability. Hopefully the range of the topics covered and the depth of the information will be helpful to the readers. This book can also be used as a textbook for a course on green polymer chemistry and a reference book for libraries or for people who need information on specific polymer topics involving green chemistry and sustainability.

We thank the authors for their contributions and for their patience during the peer review process. Thanks are also due to the many reviewers for their time and effort to read the manuscripts and make suggestions. We also gratefully acknowledge the ACS Books staff (particularly Arlene Furman and Amanda Koenig) for their expert handling of the manuscripts and their editorial efforts. The 2020 Green Polymer Chemistry symposium in Philadelphia was sponsored by ACS Division of Polymer Chemistry (POLY), with the nominal co-sponsorships of the ACS Division of Agricultural and Food Chemistry (AGFD) and 2020 ACS President-Elect.

H. N. Cheng
Southern Regional Research Center
USDA – Agricultural Research Service
1100 Robert E. Lee Boulevard
New Orleans, Louisiana 70124, United States

Richard A. Gross
Department of Chemistry and Chemical Biology
Rensselaer Polytechnic Institute
110 8th Street
Troy, New York 12180, United States

Use of Enzymes in Polymers

Chapter 1

Green Pathways for the Enzymatic Synthesis of Furan-Based Polyesters and Polyamides

Fitrilia Silvianti,[1,2] Dina Maniar,[1] Laura Boetje,[1] and Katja Loos*[,1]

[1]Macromolecular Chemistry & New Polymeric Materials, Zernike Institute for Advanced Materials, University of Groningen, Nijenborgh 4, 9747 AG, Groningen, The Netherlands

[2]Politeknik ATK Yogyakarta, the Ministry of Industry of the Republic of Indonesia, Jl. Prof. Wirjono Projodikoro, DIY, Indonesia

***Email: k.u.loos@rug.nl**

The attention towards the utilization of sustainable feedstocks for polymer synthesis has grown exponentially in recent years. One of the spotlighted monomers derived from renewable resources is 2,5-furandicarboxylic acid (FDCA), one of the most promising bio-based monomers, due to its resemblance to petroleum-based terephthalic acid. Very interesting synthetic routes using this monomer have been reported in the last two decades. Combining the use of bio-based monomers and non-toxic chemicals via enzymatic polymerizations can lead to a robust and favorable approach towards a greener technology of bio-based polymer production. In this chapter, a brief introduction to FDCA-based monomers and enzymatic polymerizations is given, particularly focusing on furan-based polymers and their polymerization. In addition, an outline of the recent developments in the field of enzymatic polymerizations is discussed.

Introduction

Polymers are macromolecules that consist of many small molecular fragments known as repeating units (*1*). Allocated by their origin, polymers can be classified as natural or synthetic polymers. Natural polymers can be found in living organisms, for example, lignocellulose, starch, proteins, DNA, and RNA. Synthetic polymers are conventionally produced via the polymerization of simple building blocks and can be tailored to meet specific demands. Driven by the establishment and advancement of common polymerization techniques, the growth of synthetic polymers was prominent in the early twentieth century (*2*), which led to a significant rise in the consumption of this class of materials.

Polymers are known as ubiquitous and valued materials that play an important role in our daily lives due to their broad range of applications, such as in packaging, coatings, fibers, adhesives, foams, and specialty polymers. However, the massive demand for polymeric materials was accompanied by a severe environmental problem. The polymer industry is mostly dependent on the fossil feedstock and petroleum-based engineering processes (*3, 4*) . This is of great concern as fossil fuels are limited and are expected to be depleted within a few centuries (*5–7*). As a consequence, the polymer industry faces rising pressure to develop environmentally friendly and sustainable production processes and materials. To address these challenges, it is necessary to develop sustainable polymer materials. One approach that has attracted considerable research effort is the polymerization of monomers or platform chemicals from renewable resources. Hence, the synthesis and application of high-performance, sustainable polymers have increasingly attracted the attention of scientists. In addition, sustainable polymers have been developed in recent years due to their eco-friendly advantages and abundance in nature.

For a polymer to be sustainable, it should obey the following four criteria (*8–11*) (illustrated in Figure 1):

(1) renewable raw materials, catalysts, and solvents are used;
(2) green synthesis routes that are clean and prevent waste or toxic chemicals exposure to the environment should be used;
(3) high energy efficiency of the manufacturing process and applications are required;
(4) polymers should have a low carbon trace, for example, decomposable polymers or polymers that can be recycled or discarded with a low environmental effect.

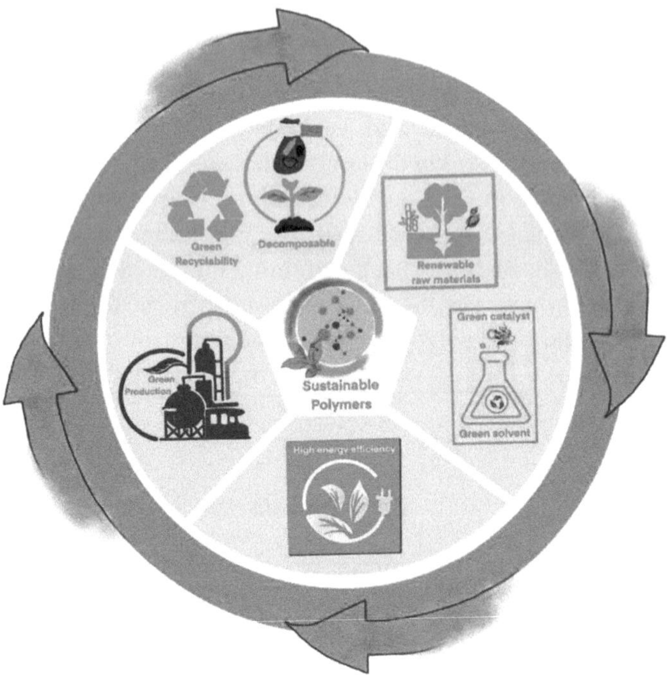

Figure 1. Entangle elements for sustainable polymers.

By utilizing bio-based monomers in enzymatic polymerizations, many requirements of the above-mentioned list can be fulfilled since bio-based monomers, and enzymic catalysts are derived from biomass feedstock, which consists of sustainable materials. Furthermore, the synthesis processes of enzymatic polymerizations are clean and energy-efficient without toxic residuals contaminating the final products and environment (*9, 12*).

2,5-Furandicarboxylic acid (FDCA) is one of the interesting bio-based furan monomers derived from biomass feedstock for polyesters and polyamides synthesis (*13–15*). As shown in Scheme 1, FDCA has two dicarboxylic acid groups, that can be used in polycondensation reactions, by esterification it can form dimethyl 2,5-furandicarboxylate (DMFDCA), which is generally used in polymerizations due to the better solubility compared to FDCA. FDCA-based polymers show analogous or even superior thermal and mechanical properties than petrochemical-based polymers, i.e., terephthalic acid (TPA) and isophthalic acid (IPA) based polymers (*6, 16, 17*). Hence, it is interesting to discover a green approach towards various furan-based polymers and to have the opportunity to explore their potential applications.

Scheme 1. Chemical structure of FDCA and its ester DMFDCA.

Bio-Based Polymers

Recently, there has been growing interest in polymers derived from renewable resources because of their environmentally friendly character and their abundance in nature (*8, 17, 18*). Many bio-based polymers are commercially available nowadays, i.e., starch, polylactic acid (PLA), poly(butylene succinate) (PBS), polyhydroxybutyrate, polyhydroxyalkanoates (PHAs), poly(ε-caprolactone) (PCL), poly(butylene adipate-co-butylene terephthalate) (PBAT), bio-poly(ethylene terephthalate), bio-poly(trimethylene terephthalate) (Bio-PTT), bio-polyamide (Bio-PA), and many others. One can attain polymers from renewable resources by modification of bio-based polymers, via conversion of biomass into monomers along with polymerization or using microorganisms and plants (*8, 19*).

Bio-based polymers are polymers made from renewable resources and are considered potential replacements for petroleum-based polymers (*20–23*). These polymers are classified into many types. For example, based on the availability of raw materials, the bio-based polymers can primarily be divided into three groups containing natural polymers, bio-synthetic polymers, and chemical-synthetic polymers (*3, 8, 19, 20, 24, 25*). Generally, a natural polymer can be directly used or produced by chemical or physical modification, while a bio-synthetic polymer is formed by synthesizing a polymer using organisms and plants, such as PLA, and a copolymer that is synthesized by direct fermentation of metabolically engineered *Escherichia coli* (*26*). Hereafter, chemical-synthetic polymers are bio-based polymers created from natural resources derived molecules or by the breakdown of bio-based macromolecules via the arrangement of chemical and biochemical methods (*3*). For example, polyethylene furanoate (PEF) is a bio-based polymer that classifies as a chemical-synthetic polymer. They can be made by 3 steps: (1) catalytic dehydration of fructose in an alcohol medium to form alkoxy-methylfurfural (RMF); (2) catalytic oxidation using acetic acid to produce furandicarboxylic acid (FDCA); and (3) catalytic transesterification polymerization of bis(hydroxyethyl)-2,5-furandicarboxylate in ethylene glycol (EG) (*27*).

One of the advantages of utilizing bio-based polymers is the lower emissions of carbon dioxide in contrast to conventional petrol-based polymers. The carbon atoms in bio-based polymers result from the atmospheric carbon dioxide; additionally, their incineration does not release extra carbon dioxide to the atmosphere (*21, 28*). Furthermore, bio-based polymers can have superior characteristics over traditional polymers, e.g., they are eco-friendly, sustainable, decomposable, and so on (*29, 30*).

Furan-Based Monomers

Shifting from petro-based feedstock to renewable resources can rectify some environmental problems associated with petrochemical extraction and render a sustainable polymer production. A nearly infinite supply of bio-based monomers can be generated from biomass as an origin of raw materials (*31*). Biomass is defined by the International Union of Pure and Applied Chemistry (IUPAC) as a "material produced by the growth of microorganisms, plants, or animals" (*32*). Among the various types of biomass, starch, vegetable oil, cellulose, lignin, and proteins are the most widely used sources to generate diverse building blocks for bio-based polymers. Each type of biomass gives access to different types of building blocks, and some of them are already industrially produced (*33*), as illustrated in Figure 2.

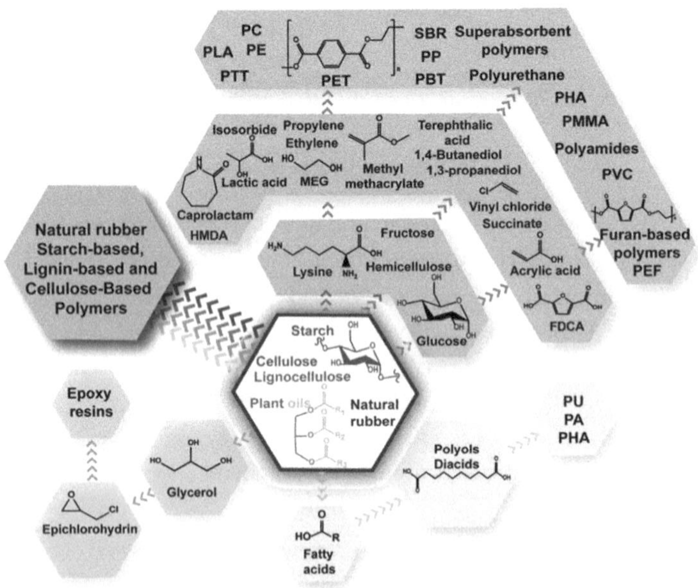

Figure 2. Major conversion pathways from biomass to building blocks and polymers.

Cellulose is one of the main elements of biomass, wood, cotton, and other plant fibers and represents the most abundant organic compound produced by living organisms (*34*). Hence, cellulose has been the primary raw material of bio-based polymers for nearly the last 150 years (*34*). The first thermoplastic polymer—celluloid—was developed from cellulose nitrate and nitric acid by Hyatt Manufacturing Company in 1870 and lead to the development of an industrial scale of new materials based on cellulose modification (*34, 35*). 2,5-furandicarboxylic acid (FDCA) is one of the interesting bio-based monomers derived from cellulosic biomass and is used in polyester, polyamide, and poly(ester amide) synthesis (*13, 14, 17, 36–38*). As already mentioned above, FDCA-based polymers have comparable properties to petrochemical-based polymers, making FDCA valuable as a bio-based monomer (*17, 39–42*). Within a large variety of renewable building blocks, furan derivatives and furan chemistry occupy a special position in polymer chemistry. The similarity between furan and phenyl rings opens an opportunity for a bio-based alternative for phenyl-based polymers. A whole new chapter in polymer science can be created from furan compounds that are readily prepared from sugars and/or polysaccharides.

Generally, FDCA can be obtained from polysaccharides or sugars by biocatalytic oxidation of 5-(hydroxymethyl)furfural (HMF) (*6, 9, 41, 43–45*). The biocatalytic oxidation route of HMF to

FDCA using HMF oxidase can occur by two routes (*44*), as can be seen on Scheme 2. The first presented oxidation of the aldehyde group of HMF into the carboxylic acid, followed by forming 5-(hydroxymethyl)furan-2-carboxylic acid (*route I*), the second oxidation yields 5-formylfuran-2-carboxylic acid, and this will ultimately be transformed into FDCA. In the second place (*route II*), the alcohol group of HMF is oxidized to the analogous aldehyde, forming furan-2,5-dicarbaldehyde, followed by additional oxidations similar to route I yielding FDCA. Another alternative route of FDCA derivation was reported by Fu et al. and van Es et al., they presented the synthesis of 2,5-FDCA using a catalytic conversion of furfural into 2,5-FDCA (*46, 47*). As can be seen in Scheme 2, after furfural oxidation forming furoic acid, dicarboxylic acids (FDCA) and furan unsubstituted compounds are obtained. Interestingly, not only the 2,5-isomer (70%) but also the 2,4-isomer (30%) and to a smaller amount extent the 3,4-isomer (<5%) are formed (*47*). Polyesters containing the three different isomers of the furandicarboxylic acid groups, i.e., 2,5-, 2,4- or 3,4-FDCA, and a series of industrially applicable aliphatic diols were already reported using titanium(IV) isopropoxide as a catalyst (*37, 48, 49*). Currently, as one of the promising monomers for polyesters and polyamides, FDCA is commercially produced by several chemical companies.

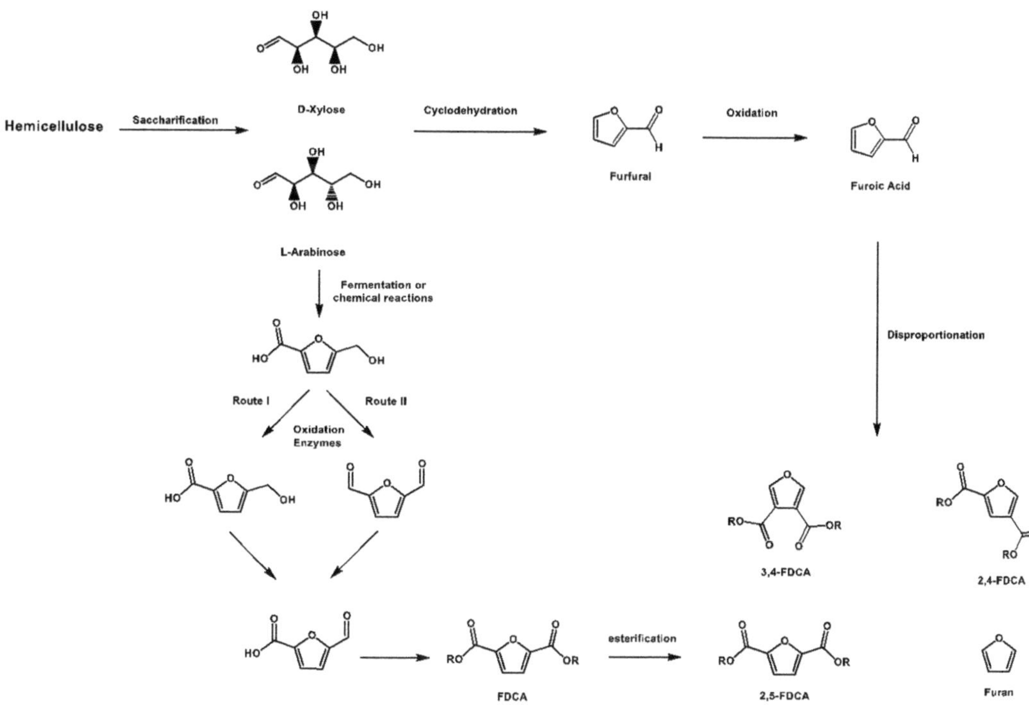

Scheme 2. FDCA derivation routes.

Enzymatic Polymerization

Biocatalytic synthesis of polymeric materials is of interest due to its advantages, e.g., selectivity and considerable environmental impact, etc. Currently, enzymes are already used in the production of food products, i.e., wine, vinegar, cheese, as well as in the manufacturing of products like leather and linen (*50, 51*). As a catalyst, enzymes are highly efficient, reaction rates can be increased up to 10 billion times than the normal reactions, and enzymes have a high selectivity for specific substrates (*50*).

Enzymatic polymerization is expressed as an in vitro polymer synthesis (*9, 12*). Non-toxic enzymes from natural resources can act under mild reaction conditions; hence, enzymatic polymerizations are regarded as an effective technique for developing sustainable polymers and for reducing dependence on fossil fuel resources (*9, 52*). Recently, many biocatalyzed synthetic routes towards polymers have been explored, including polyesters, polyamides, polylactate, vinyl polymers, etc. (*53, 54*).

Lipases are popular enzymes in the biocatalytic synthesis of polyesters. Lipases are enzymes that catalyze the hydrolysis reaction of ester bonds in lipids in nature. Water-insoluble long fatty acids from a triglyceride can be hydrolyzed with lipase to form diglycerides, mono-glycerides, and glycerol in aqueous media. As reported by Casas-Godoy et al. (*55*), a general catalytic reaction cycle of lipases begins with acylation, which involves transferring a proton among the serine, the histidine, and the aspartate residues of the active site of the lipase, resulting in a hydroxyl group activation of the catalytic serine. This serine hydroxyl residue, with consequently enhanced nucleophilicity, reacts with the carbonyl group's substrate. The initial tetrahedral intermediate is produced by a negative charge on the carbonyl group's oxygen. The hole of the oxyanion stabilizes the charge scattering and decreases the state energy of the tetrahedral intermediate by creating no less than two hydrogen-bonds. The next step is deacylation and recovering the active site of the enzyme. In terms of the nucleophile, it depends on the type of reactions; it can be water leading to hydrolysis while an alcohol leads to an alcoholysis reaction. The schematic cycle of lipase catalysis can be seen in Scheme 3.

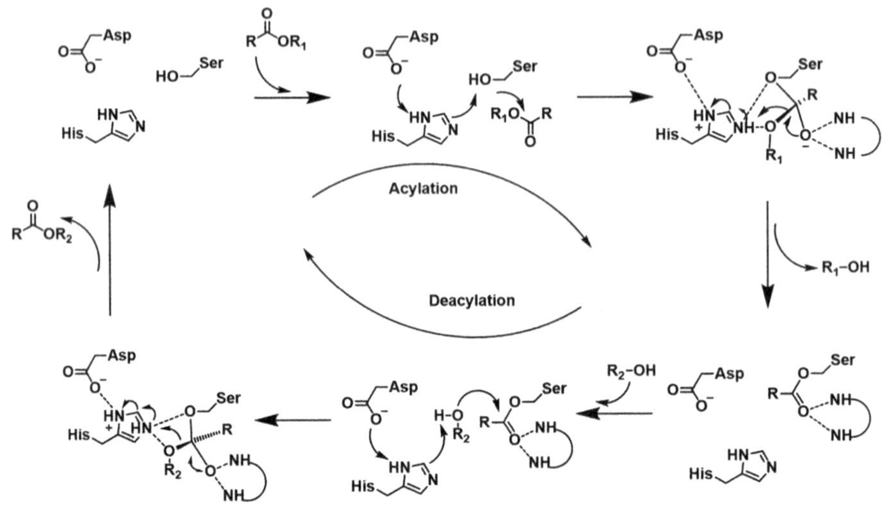

Scheme 3. General mechanism of Lipases catalysis.

Currently, *Candida antarctica* lipase B (CALB) is a very popular lipase for polyester and polyamide polymerization, because of its high selectivity, wide substrate availability, high thermal stability, and high catalytic performance (*9*). As mentioned by its name, CALB is isolated from the yeast species *Candida antartica* and exists in two forms: A and B, entitled as CALA and CALB. The specificity towards substrates of these two forms are different; the A-lipase is working in an unspecified way towards triglycerides and has a low activity regarding simple ester molecules, while the B-lipase by contrast is less working towards big triglycerides but very good towards a wide scope of esters, amides, and thiol molecules (*56, 57*).

Enzymatic synthesis of polyesters can be performed through three different approaches (Scheme 4): (a) step-growth polycondensation; (b) ring-opening polymerization; (c) an integrating of ring-opening polymerization and polycondensation (9). In general polycondensation and ring-opening are the preferred methods.

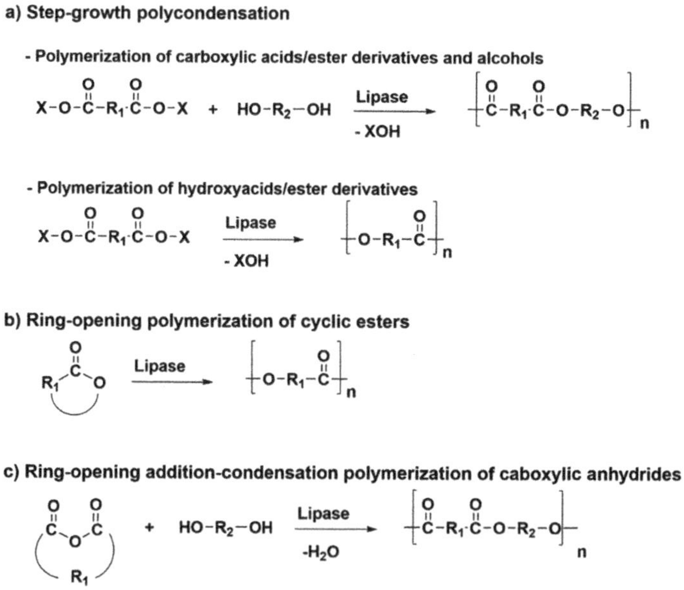

Scheme 4. Main lipase-catalyzed synthesis methods of polyesters.

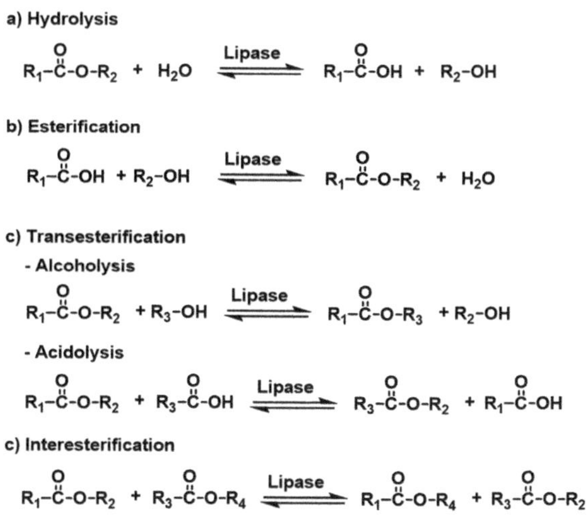

Scheme 5. Basic elemental modes lipase-catalyzed reaction on polyesters synthesis.

Lipases can have high catalytic reaction rates when the reactions are performed in nonpolar organic solvents, depending on the solvent hydrophobicity (9, 58–62). Recently, some solvents, which are generally claimed as "green," for example, ionic liquids (ILs), deep eutectic solvents (DESs), hydrofluorocarbon solvents (HFCs), and supercritical carbon dioxide ($scCO_2$), were also utilized as a medium for enzymatic polymerization. Some publications reported the enzymatic synthesis of polymers in green solvents, i.e., ring-opening polymerization of ε-caprolactone (PCL) using Novozyme-435 lipase, this reaction was successfully conducted in a hydrophobic ILs, namely 1-butyl-3-methylimidazolium hexafluorophosphate ($[BMIM]PF_6$), which successfully produced PCL in 85% monomer conversion with a high number-average molecular weight (M_n) of 5942 g mol-1 (58). While polylactide was obtained via a CALB-catalyzed ring-opening polymerization of l-lactide in 1-butyl-3-methylimidazolium tetrafluoroborate ([BMIM]BF_4), achieving M_n 55000 g mol-1 (56). Furthermore, based on the literature review, lipases can catalyze the ring-opening polymerization of caprolactone in DESs (*59*, *60*) which are considered to be green and can be simply arranged from low-cost renewable raw materials (*61–63*).

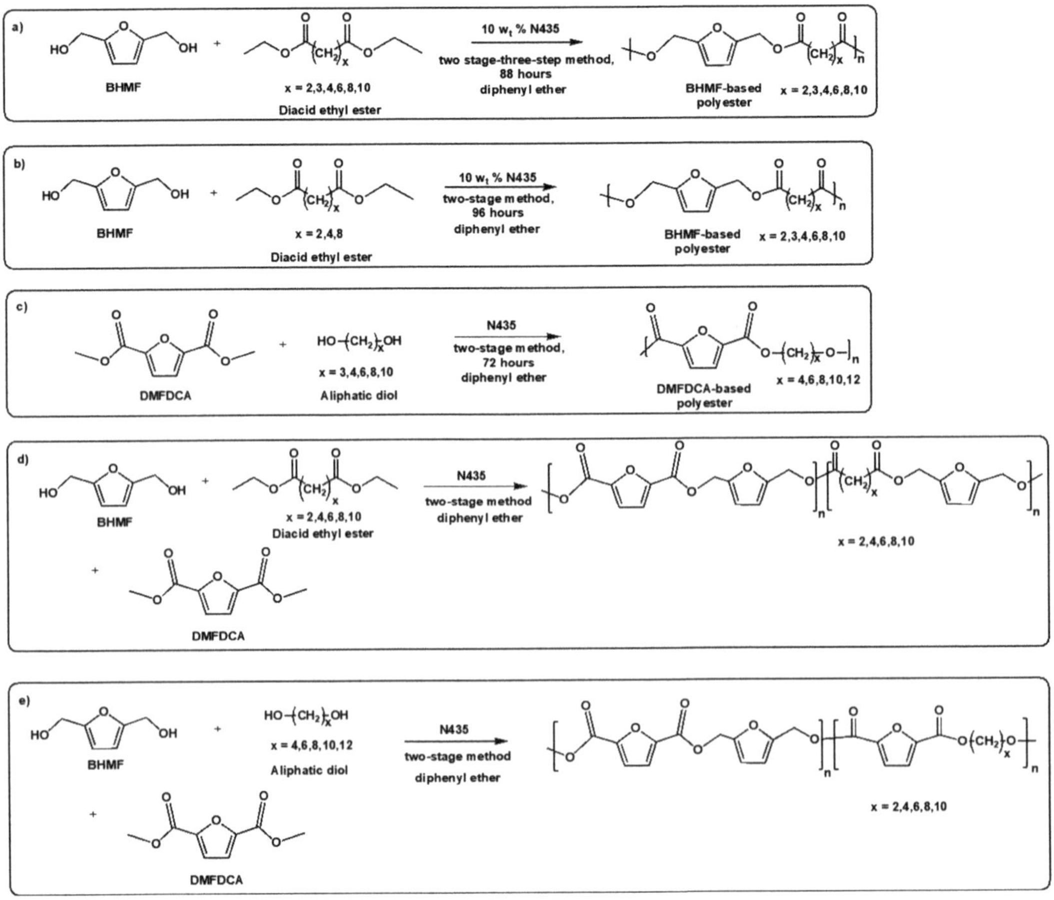

Scheme 6. Lipase-catalyzed polymerizations of furan-based polyesters.

Furan-Based Polyesters

One of the most popular 100% bio-based polyesters, poly(ethylene furanoate) (PEF), has been produced and commercialized by Avantium (NL). This polymer is synthesized via the copolymerization of FDCA with monoethylene glycol (MEG). Since both monomers can be derived from initial biomass material, and the obtained PEF is 100% recyclable, PEF is regarded as a bio-based analog to polyethylene terephthalate (PET) (*64*). In the late 1970s, Moore and Kelly (*65*, *66*) started their studies in the synthesis of furanic-aromatic polyesters. Consisting of furanic and benzenic or furanic rings only in their backbone, this family of furan polyesters became attractive due to their improved mechanical and thermal properties and/or their liquid crystalline properties. For instance, Gandini et al. (*67*, *68*) reported the synthesis of poly(1,4-phenylbismethylene 2,5-furandicarboxylate) (PHBMF), poly(2,5-furandimethylene 2,5-furandicarboxylate) (PBHMF), and poly(1,4-phenylene 2,5-furandicarboxylate) (PHQF). All described polyesters presented high thermal stability, especially PHQF with a very high decomposition temperature of almost 490 °C. PHQF was revealed to have high crystallinity and no thermal transition appeared below 400 °C, indicating that the melting temperature (T_m) was above its degradation temperature.

Currently, PET is the most produced polyester and is considered as the fourth largest produced polymer in the world, up to 70% of polymers fiber production (*2*). The total European plastics demand in the packaging market in 2018 reached the highest account at 39.9% per 51.2 million tons, which is predominantly composed of PE, PP, and PET (*69*). Therefore, it is demanding to develop bio-based alternatives such as PEF.

Table 1. Comparison of Furan-Based Polyesters Obtained by Enzymatic Polymerization of BHMF and Diacid Ethyl Ester (C_x, PE-BHMF-ac); DMFDCA and Aliphatic Diols (C_x, PE-DMFDCA-ol); BHMF, Diacid Ethyl Ester, and DMFDCA (C_x, PE-BHMF-DMFDCA-ac); BHMF, Aliphatic Diols, and DMFDCA (C_x, PE-BHMF-DMFDCA-ol)

Polymer	T_g[a] (°C)	T_m[a] (°C)	$T_{d\text{-}max}$[b] (°C)	M_n[c] (g/mol)	M_w[c] (g/mol)	X_c[d] (%)	*Ref.*
C_4, PE-BHMF-ac	-19	-	300/446	2200	3400	43	80
	-15	-	268	2800	25400	16	82
C_4, PE-DMFDCA-ol	36	169	396	1200	1700	42	79
C_4, PE-BHMF-DMFDCA-ac	-8[e]	-	240[e]	8900[e]	16850[e]	-	81
	-	-	-	1800[f]	4100[f]	-	81
C_4, PE-BHMF-DMFDCA-ol	15[g]	142[g]	370[g]	1975[g]	2820[g]	-	81
	-	118[h]	370[h]	1400[h]	1500[h]	-	81
C_6, PE-BHMF-ac	-38	51/66	319/453	2400	3300	55	80
C_6, PE-DMFDCA-ol	16	140	397	2500	3800	37	79
C_6, PE-BHMF-DMFDCA-ac	-16[e]	57[e,i]	250/310/460[e]	5400[e]	9300[e]	-	81
	-	-	-	-	-	-	
C_6, PE-BHMF-DMFDCA-ol	12[g]	120[g]	390[g]	5600[g]	18000[g]	-	81
	-0.7[h]	108[h]	380[h]	1620[h]	2110[h]	-	
C_8, PE-BHMF-ac	-29	69/79	327/452	2200	3600	55	80
	-22	81	291/443	3000	20400	19	82
C_8, PE-DMFDCA-ol	6	141	399	3300	4800	26	79
C_8, PE-BHMF-DMFDCA-ac	-19[e]	62[e,i]	250/310/450[e]	6350[e]	11650[e]	-	81
C_8, PE-BHMF-DMFDCA-ol	2[g]	123[g]	390[g]	16050[g]	35350[g]	-	81
	7[h]	88[h]	390[h]	3100[h]	5650[h]	-	81
C_{10}, PE-BHMF-ac	-8	80/85	332/452	2200	3900	65	80
C_{10}, PE-DMFDCA-ol	-5	106	402	23700	48700	21	79
C_{10}, PE-BHMF-DMFDCA-ac	-19[e]	79[e]	280/320/460[e]	2500[e]	3450[e]	-	81
	-	-	-	670[f]	1800[f]	-	81
C_{10}, PE-BHMF-DMFDCA-ol	6[g]	90[g,i]	390[g]	13900[g]	28650[g]	-	81
	-	92[h]	390[h]	1500[h]	1600[h]	-	81

[a] Glass transition and melting temperature measured from the second heating scan of differential scanning calorimetry (DSC). [b] temperature at a maximum rate of decomposition measured by thermogravimetric analysis (TGA). [c] M_n and M_w were determined by size-exclusion chromatography (SEC). [d] Xc = degree of crystallinity, calculated from wide-angle X-ray diffraction (WAXD) profile. [e] feed ratio of DMFDCA, BHMF, diacid ethyl ester = 12.5 %: 50 %: 37.5 %. [f] feed ratio of DMFDCA, BHMF, diacid ethyl ester = 50 %: 25 %: 25 %. [g] feed ratio of DMFDCA, BHMF, aliphatic diol = 50 %: 12.5 %: 37.5 %. [h] feed ratio of DMFDCA, BHMF, aliphatic diol = 50 %: 25 %: 25 %. [i] Measured from the first DSC heating scan.

Several synthetic routes towards PEF have been developed with polycondensation, and polytransesterification is mainly conducted using conventional two-stage melt-processes utilizing catalyst and reduced pressure (*48, 64, 70, 71*). However, speaking of eco-friendly polymer production, we have to consider not only the bio-based polymer itself but also the need to manage the whole production process. Enzymatic polymerizations offer the best solution to overcome the limitations of the conventional methods. They have been proven to be a robust route for polymer production in a sustainable manner with mild reaction conditions, mainly in view of temperature and the use of renewable non-toxic enzyme catalyst (*9*, *36*, *52*).

In terms of polymerization methods, several studies revealed useful method guidelines to achieve high molecular weight polyesters, which is necessary for high-performance properties. During the polymerization, four reactions may occur, namely: hydrolysis, esterification, transesterification (alcoholysis and acidolysis), and interesterification (Scheme 5). These reactions are all reversible; hence, it is crucial to maintain the right reaction conditions to obtain high molecular weight polymers. Generally, side reactions and monomer evaporation will lead to a stoichiometric disproportion of reactants; therefore, it is essential to keep a reaction system in balance. Limiting temperature and vacuum during the first stage of polymerization can prevent the evaporation of volatile monomers (*72*), and the side products, such as water or alcohol, can be removed by vacuum mostly during the second stage of the polymerization (*48*, *72*, *73*). Besides vacuum, molecular sieve also can be used to remove side products via adsorption (*72*). The molecular weight can also be improved by properly increasing the polymerization temperature, extending the time of polymerization, and using the exact amount of catalyst. Based on several studies, the molecular weight of the polymer product via enzymatic polymerization can be significantly influenced by the applied polymerization method, compared to polymerization in bulk, the enzymatic polymerization in the presence of a preferred solvent can be the best method to achieve high molecular weight polymers (*74–77*). This phenomenon is explained by a lower viscosity of the reaction system, which decreases diffusion constrictions among polymer chain reactants, monomers, and oligomers, while in bulk, the reaction media has a higher viscosity, thereby reducing the diffusion of the reactants (*74*, *77*, *78*).

We performed a series of enzymatic polymerizations towards furanic-aliphatic polyesters (*79–81*). The first study of furan-based polyesters was utilizing 2,5-Bis(hydroxymethyl)furan (BHMF) and several diacid ester (see Scheme 6a) (*80*), we found that methylene units of dicarboxylic acid significantly influence the thermal stability and crystallinity of the products, both increasing in line with the number of methylene units assembled into the polyesters chain. The recent study by Pellis et al. reports the enzymatic polyester synthesis utilizing BHMF (*82*). They used three different ester derivatives as co-monomer (see Scheme 6b). The results of thermal stability and M_n were shown to be in accordance with our previous study (*80*), despite the higher weight average molecular weight (M_w) achieved. A further study was published in 2015 in which we performed an enzymic reaction using DMFDCA and aliphatic diols in order to investigate the effect of diol structures (see Scheme 6c) (*79*). Besides, to produce high molecular weight polymers, the optimization of reaction conditions such as varied temperature and pressures were studied. Furthermore, recently we reported the enzymatic synthesis of furan-based copolyesters by applying polycondensation methods on DMFDCA and BHMF as the main precursors (see Scheme 6d-e) (*81*). In this study, we focused on using different furan monomers and changing the incorporated aliphatic chains from diols to diacid ethyl esters. A significant difference in the degree of polymerization (*DP*) was observed, and it becomes lower when diacid ethyl esters are used (see Figure 3). Besides, the thermal stability and crystallinity were also affected, furan-based copolyesters from diacid ethyl esters were less thermally stable, and some were detected as amorphous materials as can be seen in Figure 4. In Table 1, a comparison of several properties of these polymers is listed.

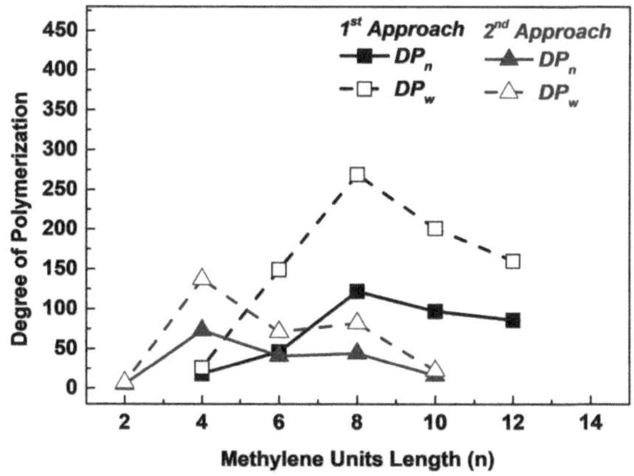

Figure 3. Comparison of degree of polymerization of the furan-based copolyesters using first approach obtained from feed ratio of DMFDCA, BHMF, and aliphatic diols = 50%: 12.5%: 37.5% and from the second approach the feed ratio of DMFDCA, BHMF and diacid ethyl esters = 12.5%: 50%: 37.5%. Reproduced with permission from reference (81). Copyright 2019 John Wiley & Sons.

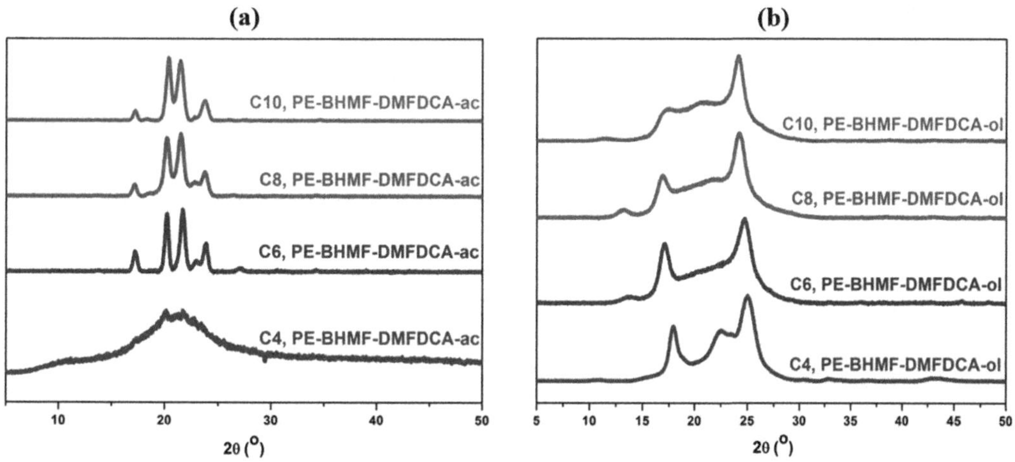

Figure 4. Comparison of Wide-Angle X-Ray Diffraction (WAXD) spectra of the furan-based polyesters from (a) DMFDCA, BHMF, diacid ethyl esters with the feed ratio 12.5 %: 50 % : 37.5 % and (b) DMFDCA, BHMF, aliphatic diol with the feed ratio of 50 %: 12.5 % : 37.5 %. Reproduced with permission from ref. (81). Copyright 2019 John Wiley & Sons.

Furan-Based Polyamides and Poly(ester amides)s

Besides polyester, polyamides are an important family of polymers as they typically exhibit high thermal stability, enhanced mechanical performance, impact strength, and abrasion resistance. Polyamides are generally used as fibers and films for consumer goods, the automotive industry, and electronics. All these appealing characteristics certainly have encouraged studies on furanic polyamides (*83–86*). FDCA-based polyamides display rather high glass transition temperature (T_g) values and good mechanical performance. The first synthesis of FDCA-based polyamides dates back to 1961 when Hopff and Krieger (*87, 88*) found that heating of a hexamethylene diammonium FDCA salt resulted in a strong decarboxylation and no discrete polymers were obtained from their attempts. Subsequently, Heertjes and Kok (*89*) reported that the decarboxylation of FDCA occurred at about 195 °C. They reported polymerization of FDCA, dimethyl 2,5-furandicarboxylate

(DMFDCA), or 2,5-furandicarbonyl dichloride (FDCDCL) with C4, C6, and C8 linear aliphatic diamines and obtained clear, fragile, and light-yellow to brown color FDCA-based polyamides. The T_m of PA-4F, PA-6F, and PA-8F were reported to be 250, 175, and 125 °C, respectively. In contrast, Grosshardt et al (*83*) reported several amorphous FDCA-based polyamides (PA-6F, PA-8F, PA-10F, and PA-12F) with a T_g ranging from 70 to 110 °C, obtained from the melt-polycondensation using Sn- or Ti-derived catalysts. The companies Avantium and Solvay reported an enhanced method to create higher molecular weight FDCA-based polyamides in two patents (*84, 85*). Here, Gruter et al. proposed a two-stage polymerization strategy to prevent the occurrence of N-methylation of the poly(amides). In the first stage of the proposed polymerization, DMFDCA and diamines are oligomerized; afterward, the polyamide oligomers are combined by a bifunctional linker.

Poly(ester amide)s are interesting materials as they generally display both characteristics of both polyesters and polyamides. Polyesters freely degrade within hydrolysis of the ester linkage; they are more flexible, and, in general, show better solubility in organic solvents. Due to their hydrogen bonding, polyamides show higher thermal and mechanical robustness. With this combination of features, poly(ester amide)s are mainly suggested for functions that oblige good thermo-mechanical properties, in line with biocompatibility and biodegradation, for example, they would be of high interest for biomedical purposes or as high-performance polymers with lower environmental impact. Surprisingly, compared to the furan-based polyesters and polyamides, recent literature dealing with furan poly(ester amide)s is scarce. Abid et al. (*38, 90*) developed several furan poly(ester amide)s via bulk polycondensation. They found that the obtained poly(ester amide)s are amorphous polymers, and the increase of the amide fraction is increasing the T_g and decreasing the degradation rate. In a more recent paper, they published the synthesis of sulfonated furanic poly(ester amide)s by coupled simultaneous polyesterification and polyamidation of difuranic diesters and dimethyl 5-sodiosulfoisophthalate (*91*). Rastogi and Noordover et al. (*92*) have successfully synthesized novel poly(ester amide)s containing 2,5-furandicarboxylic acid moieties and compared them to the terephthalic acid and isophthalic acid analogs. In a more recent publication, they reported the synthesis of FDCA-based cross-linked poly(ester amide)s via the polymerization of 2,5-bis(4,5-dihydrooxazol-2-yl)furan (2,5-FDCAox) with sebacic acid (*93*). In addition, the importance of this furan polymer family was also highlighted in a patent on the preparation of biodegradable furan poly(ester amide)s (*94*). Specifically on FDCA-based poly(ester amide)s, Robert et al. (*95*) successfully produced it using Titanium(IV) isopropoxide (TIS) as catalyst by melt polycondensation in three stages, i.e. transesterification of DMFDCA to bis(hydroxy-decamethylene)-2,5-furan dicarboxilate (BHDF), polyesterification and polycondensation of the co-monomers of BHDF, amido diol and DMFDCA to create poly(ester amide).

As lipases can catalyze the creation of amide bonds, their use as biocatalysts for the in vitro synthesis of polyamides and poly(ester amide)s is clearly attractive. Yet, the lipase-catalyzed polymerization of polyamides has not been studied extensively due to the high T_m and poor solubility of polyamides (*9*). At temperatures above the T_m of the polyamides, the catalytic activity of lipases decreases significantly because of the protein deactivation and denaturation. In addition, many polyamides can only be dissolved in destructive solvents (concentrated H_2SO_4, formic acid, and trifluoroacetic acid), in which lipases lose their activity.

Nevertheless, some polyamides and oligoamides have been synthesized via an enzymatic pathway (*96, 97*). In analogy to polyesters, the basic modes of elemental reactions during the polyamide syntheses consist of direct amidation and transamidation (aminolysis) (Scheme 7). The application of vacuum is still necessary for the enzymatic syntheses of polyamides. However, a higher water amount can be tolerated due to the higher reaction equilibrium constant (hundreds of times larger than for polyesters) (*98*). Hence, the applied vacuum in the lipase-catalyzed synthesis of polyamides is relatively high in comparison to the polyester analogs (100 to 2 mmHg). Besides that, molecular sieves are also used merged with a vacuum step (*99–101*).

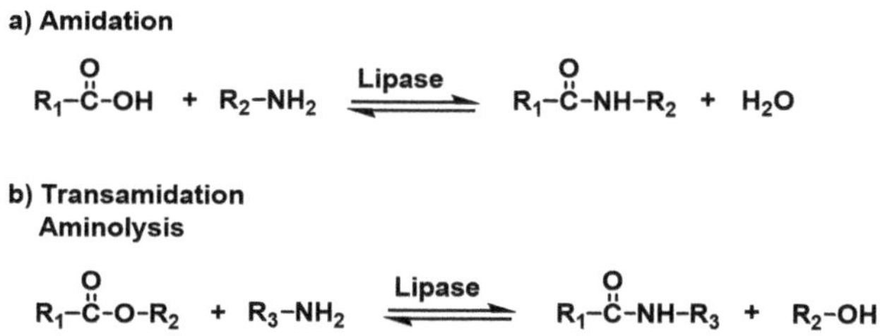

Scheme 7. Basic elemental modes of lipase-catalyzed reactions in polyamide synthesis.

We recently reported various enzymatic pathways towards aliphatic and semi-aromatic polyamides or oligoamides (*99*, *102–105*), including FDCA-based furanic aliphatic polyamides (*106*). We successfully investigated the enzymatic synthesis of FDCA-based polyamides by focusing on synthesizing poly(octamethylene furanamide) (PA8F) (*106*) (see Scheme 8a). In this study, the obtained PA8F shows a comparable crystalline structure and thermal stability to poly(octamethylene terephtalamide), which is known as petroleum-based polyamide. We optimized the reaction conditions and found that using 20 wt% of N435 gave the highest yields and M_n on both methods of one stage and two stage. To continue our study on the one-stage method of enzymatic polymerization, we performed the enzymatic synthesis of FDCA-based polyamides by using several bio-based aliphatic diamines with different chain lengths (see Scheme 8b) (*107*). The enzymatic polymerization kinetics was studied using DMFDCA and 1,8-ODA as the precursors. This study highlighted that the phase separation of FDCA-based oligoamides/polyamides happened in the early stages of reaction and followed to a subsequent enzymatic solid-state polymerization. Furthermore, enzymatic polymerization time could be shortened from 72 ~ 24 hours, which means enzymatic polymerization can be comparable to conventional methods based on the reaction time and the resulting molecular weights (*107*).

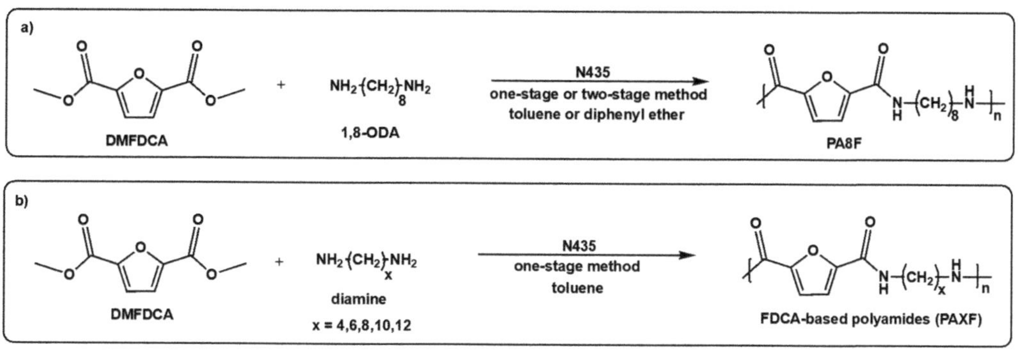

Scheme 8. Lipase-catalyzed polymerization of FDCA-based polyamides.

To the best of our knowledge, despite the extensive studies on enzymatic synthesis of furan polyesters and polyamides, similar studies on furan poly(ester amide)s are very limited up to now, especially on FDCA-based poly(ester amide)s. So far, as we mentioned above, Abid et al. have reported several studies on the synthesis of furan-based poly(ester amide)s utilizing $Zn(AcO)_2$ as catalyst (*38*, *90*, *91*), Noordover et al. utilized triphenyl phosphite as a catalyst for the poly(ester amide)s synthesis (*93*), and Robert et al. reported FDCA-based poly(ester amide)s using TIS as catalyst (*95*).

Recently, we successfully performed an enzymatic polymerization of FDCA-based poly(ester amide)s using two different approaches and reaction media, including ILs, i.e., [BMIM]PF_6 and 1-ethyl-3-methylimidazolium tetrafluoroborate ([EMIM]BF_4) (see Scheme 9) (*108*). The purpose of using ILs is to improve the sustainability of the whole synthetic process. ILs are reported as green solvents due to their potential of low volatility and low toxicity (*109–111*), besides they have outstanding catalytic activity and tunabilities (*112–114*). According to this study, the first novel FDCA-based poly(ester amide)s reported via enzymatic polymerization using N435, achieved high M_n up to 13000 g mol^{-1} for poly(dodecamethylene furanoate-co-dodecamethylene furanamide) (PEAF12) and high T_m of 92 °C, while the $T_{d\text{-}max}$ of this semicrystalline material presented at 395 °C is comparable to PEF, which is around 389 °C (*108*, *115*), see Table 2. Furthermore, we also revealed in this study, the selectivity of CALB towards the substrate, even though characteristically the reactivity of aminoalcohols are more reactive than their diols. On the first approach, using aliphatic diols and aliphatics diamines as substrates successfully obtained the higher molecular weight than the second one using aliphatic aminoalcohols. However, in general, CALB have the same preference towards ester and amide formation.

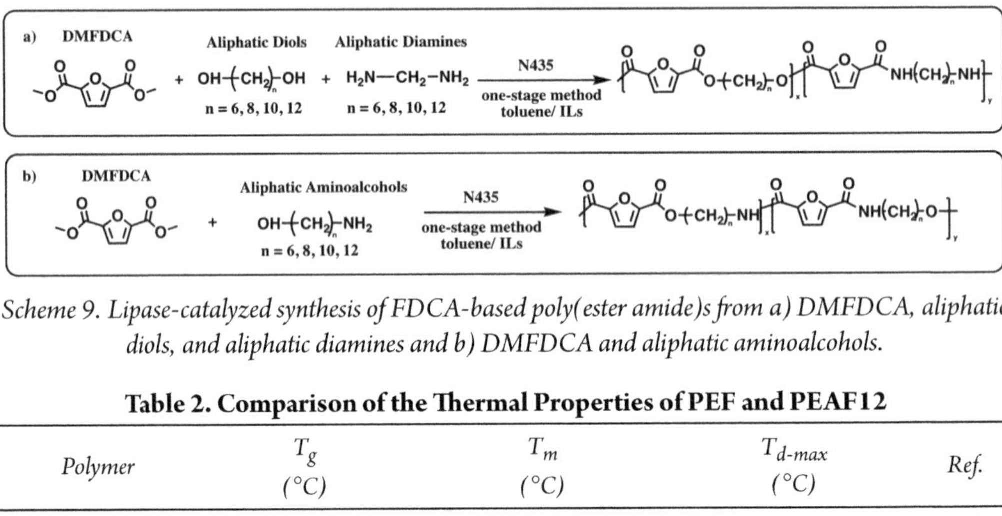

Scheme 9. Lipase-catalyzed synthesis of FDCA-based poly(ester amide)s from a) DMFDCA, aliphatic diols, and aliphatic diamines and b) DMFDCA and aliphatic aminoalcohols.

Table 2. Comparison of the Thermal Properties of PEF and PEAF12

Polymer	T_g (°C)	T_m (°C)	$T_{d\text{-}max}$ (°C)	*Ref.*
PEF	80.7/88	210/213.8/230	398/567.5	(3, *115*)
PEAF12	25	92	395	(*108*)

ILs and DESs as Alternative Green Reaction Medium

Solvent is one of the important parts in enzymatic polymerization as mentioned above, a preferred solvent can enhance the catalytic activity of lipases and tends to keep a good equilibrium reaction which is related to high molecular polymer product. Improvement and implementation of sustainable solvent media have been a trending topic in the last two decades. ILs are considered to be "green solvents" mostly due to the benefit of their nonvolatile properties as compared to the volatile organic compounds (VOCs) in traditional industrial solvents. Using ILs would avoid the emission of VOCs, a primary source of environmental contamination (*116–118*). Moreover, they have very worthy properties as reaction media, due to their high recyclability, low toxicity, low volatility, being less-flammable, and unique solubility for organic and inorganic materials (*109–112, 117, 119, 120*). In addition, ILs are suitable for enzymatic reactions, as they can stabilize lipases (*121, 122*). However, for several reasons, ILs are not always considered as "green," specifically, they can cause water contamination as commonly used chemicals. Furthermore, some ILs syntheses and purifications employed toxic chemicals, yet they can be aimed to be eco-friendly with large possible profits for sustainable chemical (*123, 124*).

ILs are generally defined as salts with a melting temperature below 100 °C, developed from conventional high-temperature melted salts (*116, 117, 120, 122, 125–127*). They are liquids that consist of a mixture of salts ionic species, an anion, and a cation. Although both individual species have extremely high melting points, their proper incorporation leads to the creation of a liquid phase that has a lower melting point temperature. This occurs due to an asymmetrical size between the anion and cation, which lean to decrease the lattice energy of salts crystalline form (*121, 122*). Standard ILs are created from organic cations, for example, 1,3-dialkylimidazolium, tetraalkylammonium, paired with various anions that have a strongly delocalized negative charge (e.g., BF_4, PF_6, NO_3, $MeOSO_3$, bis((trifluoromethyl)sulfonyl)imide, or NTf_2) (*118, 122*). The physical properties of ILs (e.g., density, viscosity, melting points, and polarity) can be easily tuned by the appropriate selection of anions and/or cations (*112, 118, 121*) – see Figure 5.

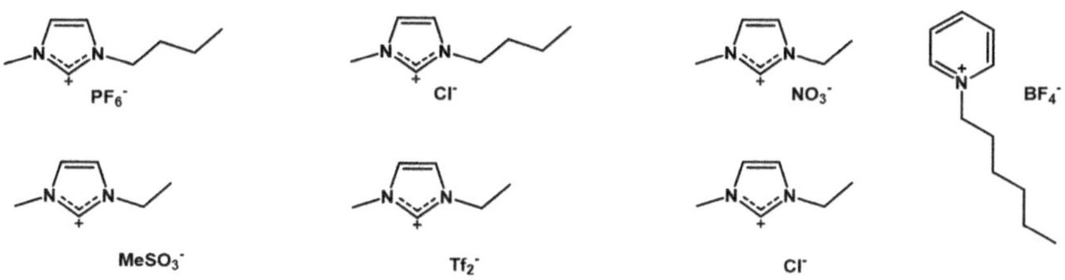

Figure 5. Examples of ILs involving of simple salts.

Another sustainable alternative solvent developed from ILs, called DESs, is emerging recently. Same way to ILs, they consist of a combination of compounds, hydrogen-bond donors and acceptors. DESs are generally obtained by combining two elements, namely, hydrogen-bond acceptor (HBA) and hydrogen-bond donor (HBD), which can be related with each other via hydrogen-bond interactions (*61, 128–132*). Due to this formation of intermolecular hydrogen-bonds, forming asymmetrical ions with low lattice energy, making DESs generally have a much lower melting point than the individual elements (*120, 128, 132*). Therefore, as like ILs, it is possible to perform reactions in DESs with a temperature below 100 °C. Moreover, comparing to VOCs, DESs are not measured as volatile and flammable, making them "green solvents." Currently, the most preferred DESs are created by choline chloride (ChCl), carboxylic acids, and other hydrogen-bond donors, i.e., glycerol, urea, succinic acid, and citric acid (*128–130, 133, 134*). These are known as cheap chemicals. Therefore, DESs possess advantages of low-cost reaction media compared to ILs. This is valuable since the low production cost was mainly considered for commercial manufacturing. In addition, a mixture of ChCl and glycerol is the most used DESs and is non-toxic, because both individual chemicals are commonly used as food additives (*135–138*). Some structures of the chemical that can be used to form DESs are shown in Figure 6.

Based on the literature, several applications of DESs have been reported, i.e., separation processes (*120, 139, 140*), carbon dioxide adsorption (*141, 142*), biomass processing (*129, 131, 143*), electrochemical processes (*132*), bio-catalysis (*59, 144*), polymerization reaction (*62, 128,*

145), and so on. Related to our interest, some papers reported good catalytic activity of enzymes in DESs, including enzymatic polymerization with a lipase (*60*, *144*, *146–148*). It was discovered that CALB effectively produces polymers by ring-opening polymerization of caprolactone using DESs (*60*). Lipase-catalyzed polymerization of furan-based polyester(amide)s using ILs have been successfully produced in our group, as explained above (see Scheme 9). This shows that alternative green solvents can be a good medium for the enzymatic polymerization of furan-based polymers.

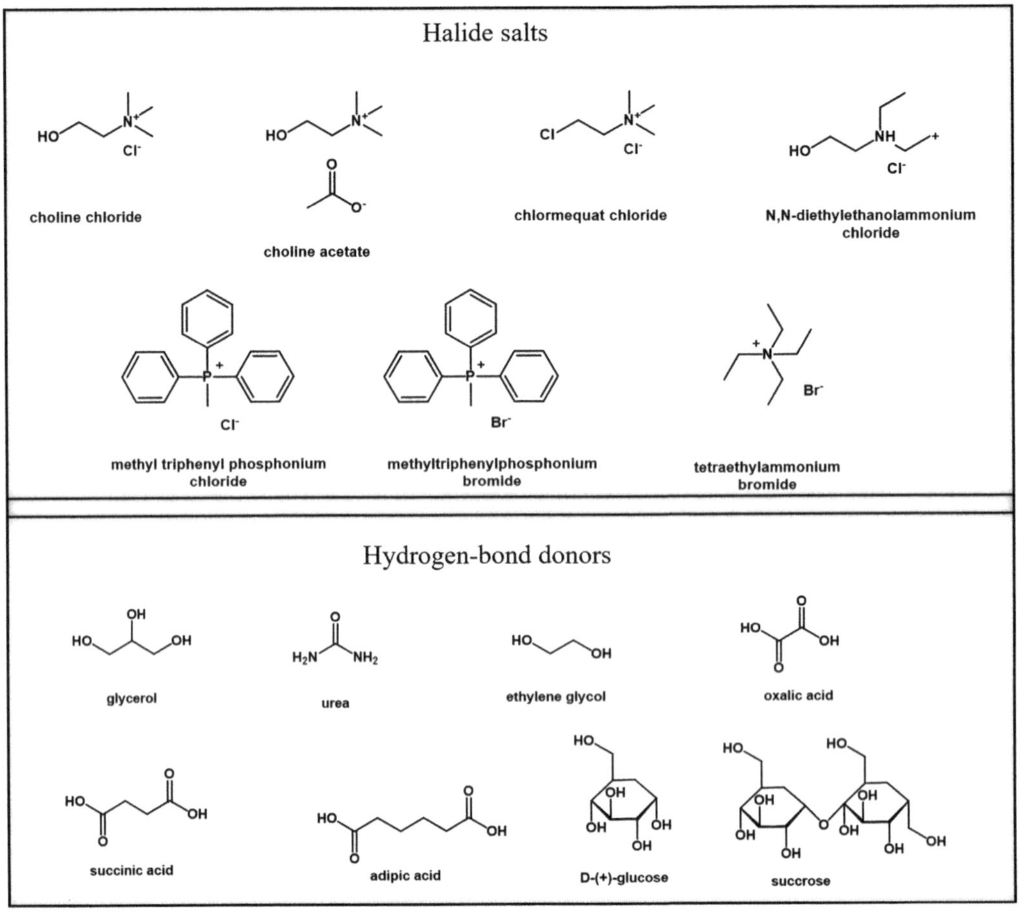

Figure 6. Structures of some halide salts and hydrogen-bond donors used in the formation of DESs.

Conclusion

The enzymatic polymerization of furan-based polymers is highlighted as a favorable green technology towards sustainable polymers. Based on several articles, enzymatic polymerization is recognized to be a robust green technique for the production of bio-based polyesters and polyamides, especially for FDCA-based polyesters, that are considered as an alternative for petroleum-based PET.

The enzymatic polymerization of furanic polyesters and polyamides was extensively reported in the past few years. In terms of the synthetic methods, decades of research have provided a useful guide about which parameters are affecting the enzymatic reaction systems for polyester or polyamides synthesis, even broadened to poly(ester amide)s. The effect of polymerization conditions such as the polymerization parameters including solvents, oligomeration time, vacuum, catalyst dosage, and the usage of several co-monomers was investigated in regard to reaction optimization to achieved high molecular weight materials.

The core significance of this method as a green process obviously offers a solution to issues in furan-based polymers synthesis in general. However, in specific terms, for sustainable polymer production, it is essential that the entire procedure be "green," from the starting materials, synthetic process, disposal, and energy consumption. Hence, future studies should focus on a green method of the whole process in enzymatic polymerization, for instance, by using green media such as ILs and DESs. As presented, several combinations of molecules can be chosen to form ILs or DESs. Choosing the appropriate medium is essential because it can affect the yields and especially the molecular weight of polymers. On the other hand, besides pursuing the optimization of the green synthetic process, economic considerations should be taken into account for the competitive scale up in the manufacturing industry, for example, by focusing on low-priced chemicals and performing the synthetic process at low energy consumption.

Currently, the development of enzymatic polycondensations of FDCA-based polymers using several ILs and DESs is conducted in our group to accomplish this goal of greener enzymatic polymerization routes.

Acknowledgments

Fitrilia Silvianti gratefully acknowledges the financial support from the Industrial Human Resource Development Agency of the Ministry of Industry of the Republic of Indonesia (BPSDMI, Kemenperin) and Politeknik ATK Yogyakarta.

References

1. Odian, G. *Principles of Polymerization*; John Wiley & Sons, 2004.
2. Geyer, R.; Jambeck, J. R.; Law, K. L. Production, Use, and Fate of All Plastics Ever Made. *Sci. Adv.* **2017**, *3* (7), e1700782. https://doi.org/10.1126/sciadv.1700782.
3. Nakajima, H.; Dijkstra, P.; Loos, K. The Recent Developments in Biobased Polymers toward General and Engineering Applications: Polymers That Are Upgraded from Biodegradable Polymers, Analogous to Petroleum-Derived Polymers, and Newly Developed. *Polymers*. MDPI AG October 18, 2017. https://doi.org/10.3390/polym9100523.
4. Isikgor, F. H.; Becer, C. R. Lignocellulosic Biomass: A Sustainable Platform for the Production of Bio-Based Chemicals and Polymers. *Polym. Chem.* **2015**, *6* (25), 4497–4559. https://doi.org/10.1039/c5py00263j.
5. Williams, C.; Hillmyer, M. Polymers from Renewable Resources: A Perspective for a Special Issue of Polymer Reviews. *Polym. Rev.* **2008**, *48* (1), 1–10. https://doi.org/10.1080/15583720701834133.
6. van Putten, R. J.; van der Waal, J. C.; de Jong, E.; Rasrendra, C. B.; Heeres, H. J.; de Vries, J. G. Hydroxymethylfurfural, a Versatile Platform Chemical Made from Renewable Resources. *Chem. Rev.*. **2013**, *113*, 1499–1597. https://doi.org/10.1021/cr300182k.
7. Rass-Hansen, J.; Falsig, H.; Jørgensen, B.; Christensen, C. H. Bioethanol: Fuel or Feedstock? *J. Chem. Technol. Biotechnol.* **2007**, *82* (4), 329–333. https://doi.org/10.1002/jctb.1665.
8. Mülhaupt, R. Green Polymer Chemistry and Bio-Based Plastics: Dreams and Reality. *Macromol. Chem. Phys.* **2013**, *214* (2), 159–174. https://doi.org/10.1002/macp.201200439.

9. Jiang, Y.; Loos, K. Enzymatic Synthesis of Biobased Polyesters and Polyamides. *Polymers* **2016**, *8* (7), 243. https://doi.org/10.3390/polym8070243.
10. RameshKumar, S.; Shaiju, P.; O'Connor, K. E.; P, R. B. Bio-Based and Biodegradable Polymers - State-of-the-Art, Challenges and Emerging Trends. *Curr. Opin. Green Sustainable Chem.* **2020**, *21*, 75–81. https://doi.org/10.1016/j.cogsc.2019.12.005.
11. Papageorgiou, G. Thinking Green: Sustainable Polymers from Renewable Resources. *Polymers* **2018**, *10* (9), 952. https://doi.org/10.3390/polym10090952.
12. Kobayashi, S. Enzymatic Polymerization. In *Encyclopedia of Polymer Science and Technology*; John Wiley & Sons, Inc.: Hoboken, NJ, USA, 2011. https://doi.org/10.1002/0471440264.pst118.pub2.
13. Chernyshev, V. M.; Kravchenko, O. A.; Ananikov, V. P. Conversion of Plant Biomass to Furan Derivatives and Sustainable Access to the New Generation of Polymers, Functional Materials and Fuels. *Russ. Chem. Rev.* **2017**, *86* (5), 357–387. https://doi.org/10.1070/rcr4700.
14. Gharbi, S.; Afli, A.; Gharbi, R. el; Gandini, A. Polyamides Incorporating Furan Moieties: 4. Synthesis, Characterisation and Properties of a Homologous Series. *Polym. Int.* **2001**, *50* (5), 509–514. https://doi.org/10.1002/pi.648.
15. Muthusamy, K.; Lalitha, K.; Prasad, Y. S.; Thamizhanban, A.; Sridharan, V.; Maheswari, C. U.; Nagarajan, S. Lipase-Catalyzed Synthesis of Furan-Based Oligoesters and Their Self-Assembly-Assisted Polymerization. *ChemSusChem* **2018**, *11* (14), 2453–2463. https://doi.org/10.1002/cssc.201800446.
16. Rosatella, A. A.; Simeonov, S. P.; Frade, R. F. M.; Afonso, C. A. M. 5-Hydroxymethylfurfural (HMF) as a Building Block Platform: Biological Properties, Synthesis and Synthetic Applications. *Green Chem.* **2011**, *13* (4), 754–793. https://doi.org/10.1039/c0gc00401d.
17. Sousa, A. F.; Vilela, C.; Fonseca, A. C.; Matos, M.; Freire, C. S. R.; Gruter, G. J. M.; Coelho, J. F. J.; Silvestre, A. J. D. Biobased Polyesters and Other Polymers from 2,5-Furandicarboxylic Acid: A Tribute to Furan Excellency. *Polym. Chem.* **2015**, *6*, 5961–5983. https://doi.org/10.1039/c5py00686d.
18. Zhang, X.; Fevre, M.; Jones, G. O.; Waymouth, R. M. Catalysis as an Enabling Science for Sustainable Polymers. *Chem. Rev.* **2018**, *118*, 839–885. https://doi.org/10.1021/acs.chemrev.7b00329.
19. Poulopoulou, N.; Kasmi, N.; Bikiaris, D. N.; Papageorgiou, D. G.; Floudas, G.; Papageorgiou, G. Z. Sustainable Polymers from Renewable Resources: Polymer Blends of Furan-Based Polyesters. *Macromol. Mater. Eng.* **2018**, *303* (8), 1800153. https://doi.org/10.1002/mame.201800153.
20. Babu, R. P.; O'Connor, K.; Seeram, R. Current Progress on Bio-Based Polymers and Their Future Trends. *Prog. Biomater.* **2013**, *2* (1), 8. https://doi.org/10.1186/2194-0517-2-8.
21. Fukushima, K.; Kimura, Y. Stereocomplexed Polylactides (Neo-PLA) as High-Performance Bio-Based Polymers: Their Formation, Properties, and Application. *Polym. Int.*. **2006**, *55*, 626–642. https://doi.org/10.1002/pi.2010.
22. Gao, S.; Tang, G.; Hua, D.; Xiong, R.; Han, J.; Jiang, S.; Zhang, Q.; Huang, C. Stimuli-Responsive Bio-Based Polymeric Systems and Their Applications. *J. Mater. Chem. B* **2019**, 709–729. https://doi.org/10.1039/c8tb02491j.

23. Kawaguchi, H.; Ogino, C.; Kondo, A. Microbial Conversion of Biomass into Bio-Based Polymers. *Bioresource Technol.*. **2017**, *245*, 1664–1673. https://doi.org/10.1016/j.biortech.2017.06.135.
24. Meraldo, A. Introduction to Bio-Based Polymers. In *Multilayer Flexible Packaging: Second Edition*; Elsevier Inc., 2016; pp 47–52. https://doi.org/10.1016/B978-0-323-37100-1.00004-1.
25. Gao, M.; Leng, X.; Zhang, W.; Wei, Z.; Li, Y. A Biobased Aliphatic Polyester Derived from 10-Hydroxydecanoic Acid: Molecular Weight Dependence of Physical Properties. *Polym. Test.* **2020**, *82*, 106295. https://doi.org/10.1016/j.polymertesting.2019.106295.
26. Jung, Y. K.; Kim, T. Y.; Park, S. J.; Lee, S. Y. Metabolic Engineering of *Escherichia Coli* for the Production of Polylactic Acid and Its Copolymers. *Biotechnol. Bioeng.* **2010**, *105* (1), 161–171. https://doi.org/10.1002/bit.22548.
27. Nguyen, H. T. H.; Qi, P.; Rostagno, M.; Feteha, A.; Miller, S. A. The Quest for High Glass Transition Temperature Bioplastics. *J. Mater. Chem. A* **2018**, 9298–9331. https://doi.org/10.1039/c8ta00377g.
28. Yoshiharu, K. SPSJ Mitsubishi Chemical Award Accounts Molecular, Structural, and Material Design of Bio-Based Polymers. *Polym. J.* **2009**, *41* (10), 797–807. https://doi.org/10.1295/polymj.PJ2009154.
29. Iles, A.; Martin, A. N. Expanding Bioplastics Production: Sustainable Business Innovation in the Chemical Industry. *J. Cleaner Prod.* **2013**, *45*, 38–49. https://doi.org/10.1016/j.jclepro.2012.05.008.
30. Lambert, S.; Wagner, M. Environmental Performance of Bio-Based and Biodegradable Plastics: The Road Ahead. *Chem. Soc. Rev* **2017**, *46*, 6855. https://doi.org/10.1039/c7cs00149e.
31. Gandini, A.; Lacerda, T. M. From Monomers to Polymers from Renewable Resources: Recent Advances. *Prog. Polym. Sci.* **2015**, *48*, 1–39. https://doi.org/10.1016/j.progpolymsci.2014.11.002.
32. McNaught, A. D. (comp.); International Union of Pure and Applied Chemistry, O. (United K. eng; Wilkinson, A. (comp.). *Compendium of Chemical Terminology. IUPAC Recommendations.* Oxford (United Kingdom) Blackwell 1997.
33. Carus, M.; Aeschelmann, F.; Essel, R.; Rascka, A. *Market Study on Bio-Based Building Blocks and Polymers in the World-Capacities, Production and Applications: Status Quo and Trends toward 2020 & Environmental Aspects & CO 2-Based Polymers.*
34. Klemm, D.; Heublein, B.; Fink, H. P.; Bohn, A. Cellulose: Fascinating Biopolymer and Sustainable Raw Material. *Angew. Chem., Int. Ed.* **2005**, *44*, 3358–3393. https://doi.org/10.1002/anie.200460587.
35. Ullmann, F.; Gerhartz, W.; Yamamoto, Y.; Campbell, F. *Ullmann's Encyclopedia of Industrial Chemistry*; 1985.
36. Maniar, D.; Hohmann, K. F.; Jiang, Y.; Woortman, A. J. J.; van Dijken, J.; Loos, K. Enzymatic Polymerization of Dimethyl 2,5-Furandicarboxylate and Heteroatom Diamines. *ACS Omega* **2018**, *3* (6), 7077–7085. https://doi.org/10.1021/acsomega.8b01106.
37. Nolasco, M. M.; Araujo, C. F.; Thiyagarajan, S.; Rudić, S.; Vaz, P. D.; Silvestre, A. J. D.; Ribeiro-Claro, P. J. A.; Sousa, A. F. Asymmetric Monomer, Amorphous Polymer?

Structure–Property Relationships in 2,4-FDCA and 2,4-PEF. *Macromolecules* **2020**. https://doi.org/10.1021/acs.macromol.9b02449.

38. Triki, R.; Bougarech, A.; Tessier, M.; Abid, S.; Fradet, A.; Abid, M. Furanic–Aliphatic Polyesteramides by Bulk Polycondensation Between Furan-Based Diamine, Aliphatic Diester and Diol. *J. Polym. Environ.* **2018**, *26* (3), 1272–1278. https://doi.org/10.1007/s10924-017-1037-y.
39. Delidovich, I.; Hausoul, P. J. C.; Deng, L.; Pfützenreuter, R.; Rose, M.; Palkovits, R. Alternative Monomers Based on Lignocellulose and Their Use for Polymer Production. *Chem. Rev.* **2016**, *116* (3), 1540–1599. https://doi.org/10.1021/acs.chemrev.5b00354.
40. Joshi, A. S.; Lawrence, J. G.; Coleman, M. R. Effect of Biaxial Orientation on Microstructure and Properties of Renewable Copolyesters of Poly(Ethylene Terephthalate) with 2,5-Furandicarboxylic Acid for Packaging Application. *ACS Appl. Polym. Mater.* **2019**, *1* (7), 1798–1810. https://doi.org/10.1021/acsapm.9b00330.
41. Sun, L.; Zhang, Y.; Wang, J.; Liu, F.; Jia, Z.; Liu, X.; Zhu, J. 2,5-Furandicarboxylic Acid as a Sustainable Alternative to Isophthalic Acid for Synthesis of Amorphous Poly(Ethylene Terephthalate) Copolyester with Enhanced Performance. *J. Appl. Polym. Sci.* **2019**, *136* (11), 47186. https://doi.org/10.1002/app.47186.
42. Wang, J. G.; Liu, X. Q.; Zhu, J. From Furan to High Quality Bio-Based Poly(Ethylene Furandicarboxylate). *Chin. J. Polym. Sci. (Engl. Ed.)* **2018**, *36* (6), 720–727. https://doi.org/10.1007/s10118-018-2092-0.
43. Werpy, T.; Petersen, G. *Top Value Added Chemicals from Biomass: Volume I -- Results of Screening for Potential Candidates from Sugars and Synthesis Gas*; Golden, CO (United States), 2004. https://doi.org/10.2172/15008859.
44. Dijkman, W. P.; Groothuis, D. E.; Fraaije, M. W. Enzyme-Catalyzed Oxidation of 5-Hydroxymethylfurfural to Furan-2,5-Dicarboxylic Acid. *Angew. Chem., Int. Ed.* **2014**, *53* (25), 6515–6518. https://doi.org/10.1002/anie.201402904.
45. Wang, J. G.; Liu, X. Q.; Zhu, J. From Furan to High Quality Bio-Based Poly(Ethylene Furandicarboxylate). *Chin. J. Polym. Sci. (Engl. Ed.)* **2018**, *36* (6), 720–727. https://doi.org/10.1007/s10118-018-2092-0.
46. Pan, T.; Deng, J.; Xu, Q.; Zuo, Y.; Guo, Q.-X.; Fu, Y. Catalytic Conversion of Furfural into a 2,5-Furandicarboxylic Acid-Based Polyester with Total Carbon Utilization. *ChemSusChem* **2013**, *6* (1), 47–50. https://doi.org/10.1002/cssc.201200652.
47. Thiyagarajan, S.; Pukin, A.; van Haveren, J.; Lutz, M.; van Es, D. S. Concurrent Formation of Furan-2,5- and Furan-2,4-Dicarboxylic Acid: Unexpected Aspects of the Henkel Reaction. *RSC Adv.* **2013**, *3* (36), 15678–15686. https://doi.org/10.1039/c3ra42457j.
48. Thiyagarajan, S.; Vogelzang, W.; J. I. Knoop, R.; Frissen, A. E.; van Haveren, J.; van Es, D. S. Biobased Furandicarboxylic Acids (FDCAs): Effects of Isomeric Substitution on Polyester Synthesis and Properties. *Green Chem.* **2014**, *16* (4), 1957–1966. https://doi.org/10.1039/c3gc42184h.
49. Zaidi, S.; Thiyagarajan, S.; Bougarech, A.; Sebti, F.; Abid, S.; Majdi, A.; Silvestre, A. J. D.; Sousa, A. F. Highly Transparent Films of New Copolyesters Derived from Terephthalic and 2,4-Furandicarboxylic Acids. *Polym. Chem.* **2019**, *10* (39), 5324–5332. https://doi.org/10.1039/c9py00844f.

50. Gurung, N.; Ray, S.; Bose, S.; Rai, V. A Broader View: Microbial Enzymes and Their Relevance in Industries, Medicine, and Beyond. *BioMed Res. Int.* **2013**, *2013*. https://doi.org/10.1155/2013/329121.
51. Sheldon, R. A.; Pereira, P. C. Biocatalysis Engineering: The Big Picture. *Chem. Soc. Rev.* **2017**, *46*, 2678–2691. https://doi.org/10.1039/c6cs00854b.
52. Douka, A.; Vouyiouka, S.; Papaspyridi, L.-M.; Papaspyrides, C. D. A Review on Enzymatic Polymerization to Produce Polycondensation Polymers: The Case of Aliphatic Polyesters, Polyamides and Polyesteramides. *Prog. Polym. Sci.* **2018**, *79*, 1–25. https://doi.org/10.1016/j.progpolymsci.2017.10.001.
53. Kobayashi, S., Uyama, H., Kadokawa, J., Eds. *Enzymatic Polymerization towards Green Polymer Chemistry*; Springer: Singapore, 2019.
54. Bruns, N.; Loos, K. *Enzymatic Polymerizations*; Academic Press, 2019; Vol. 627.
55. Casas-Godoy, L.; Duquesne, S.; Bordes, F.; Sandoval, G.; Marty, A. Lipases: An Overview. *Methods in Molecular Biology*; 2012; pp 3–30. https://doi.org/10.1007/978-1-61779-600-5_1.
56. Anderson, E. M.; Larsson, K. M.; Kirk, O. One Biocatalyst-Many Applications: The Use of Candida Antarctica B-Lipase in Organic Synthesis. *Biocatal. Biotransform.* **1998**, *16* (3), 181–204. https://doi.org/10.3109/10242429809003198.
57. Rogalska, E.; Cudrey, C.; Ferrato, F.; Verger, R. Stereoselective Hydrolysis of Triglycerides by Animal and Microbial Lipases. *Chirality* **1993**, *5* (1), 24–30. https://doi.org/10.1002/chir.530050106.
58. Yuan, J.; Dai, Y.; Yu, Y.; Wang, P.; Wang, Q.; Fan, X. Biocatalytic Synthesis of Poly(ε-Caprolactone) Using Modified Lipase in Ionic Liquid Media. *Eng. Life Sci.* **2016**, *16* (4), 371–378. https://doi.org/10.1002/elsc.201500081.
59. Gorke, J. T.; Srienc, F.; Kazlauskas, R. J. Deep Eutectic Solvents for Candida Antarctica Lipase B-Catalyzed Reactions. In *ACS Symposium Series*; American Chemical Society, 2010; Vol. 1038, pp 169–180. https://doi.org/10.1021/bk-2010-1038.ch014.
60. Gorke, J. T.; Kazlauskas, R. J.; Srienc, F. Enzymatic Processing in Deep Eutectic Solvents. US8247198B2.
61. Durand, E.; Lecomte, J.; Villeneuve, P. Deep Eutectic Solvents: Synthesis, Application, and Focus on Lipase-Catalyzed Reactions. *Eur. J. Lipid Sci. Technol.* **2013**, *115* (4), 379–385. https://doi.org/10.1002/ejlt.201200416.
62. Jablonský, M.; Škulcová, A.; Šima, J. Use of Deep Eutectic Solvents in Polymer Chemistry–a Review. *Molecules* **2019**, *24*, 3978. https://doi.org/10.3390/molecules24213978.
63. Tan, J. N.; Dou, Y. Deep Eutectic Solvents for Biocatalytic Transformations: Focused Lipase-Catalyzed Organic Reactions. *Appl. Microbiol. Biotechnol.* **2020**, *104*, 1481–1496. https://doi.org/10.1007/s00253-019-10342-y.
64. *Technology Profile: Polyethylene Furanoate Production (PEF) – Bioplastics News* https://bioplasticsnews.com/2019/05/03/technology-profile-polyethylene-furanoate-production-pef/ (accessed 2020-03-27).
65. Kelly, J. E. *Polyesters Based on 2,5-Disubstituted Furans in Various States Reduction*, 1975.
66. Moore, J. A.; Kelly, J. E. Polyesters Derived from Furan and Tetrahydrofuran Nuclei. *Macromolecules* **1978**, *11* (3), 568–573.

67. Gandini, A.; Coelho, D.; Gomes, M.; Reis, B.; Silvestre, A. Materials from Renewable Resources Based on Furan Monomers and Furan Chemistry: Work in Progress. *J. Mater. Chem.* **2009**, *19* (45), 8656–8664. https://doi.org/10.1039/b909377j.
68. Gomes, M.; Gandini, A.; Silvestre, A. J. D.; Reis, B. Synthesis and Characterization of Poly(2,5-Furan Dicarboxylate)s Based on a Variety of Diols. *J. Polym. Sci., Part A: Polym. Chem.* **2011**, *49* (17), 3759–3768. https://doi.org/10.1002/pola.24812.
69. *Plastics-the Facts 2019 An Analysis of European Plastics Production, Demand and Waste Data.*
70. Haernvall, K.; Zitzenbacher, S.; Amer, H.; Zumstein, M. T.; Sander, M.; McNeill, K.; Yamamoto, M.; Schick, M. B.; Ribitsch, D.; Guebitz, G. M. Polyol Structure Influences Enzymatic Hydrolysis of Bio-Based 2,5-Furandicarboxylic Acid (FDCA) Polyesters. *Biotechnol. J.* **2017**, *12* (9), 1600741. https://doi.org/10.1002/biot.201600741.
71. Jia, Z.; Wang, J.; Sun, L.; Zhu, J.; Liu, X. Fully Bio-Based Polyesters Derived from 2,5-Furandicarboxylic Acid (2,5-FDCA) and Dodecanedioic Acid (DDCA): From Semicrystalline Thermoplastic to Amorphous Elastomer. *J. Appl. Polym. Sci.* **2018**, *135* (14), 46076. https://doi.org/10.1002/app.46076.
72. Vouyiouka, S. N.; Topakas, E.; Katsini, A.; Papaspyrides, C. D.; Christakopoulos, P. A Green Route for the Preparation of Aliphatic Polyesters via Lipase-Catalyzed Prepolymerization and Low-Temperature Postpolymerization. *Macromol. Mater. Eng.* **2013**, *298* (6), 679–689. https://doi.org/10.1002/mame.201200188.
73. Kasmi, N.; Papageorgiou, G.; Achilias, D.; Bikiaris, D. Solid-State Polymerization of Poly(Ethylene Furanoate) Biobased Polyester, II: An Efficient and Facile Method to Synthesize High Molecular Weight Polyester Appropriate for Food Packaging Applications. *Polymers* **2018**, *10* (5), 471. https://doi.org/10.3390/polym10050471.
74. Jiang, Y.; van Ekenstein, G. O. R. A.; Woortman, A. J. J.; Loos, K. Fully Biobased Unsaturated Aliphatic Polyesters from Renewable Resources: Enzymatic Synthesis, Characterization, and Properties. *Macromol. Chem. Phys.* **2014**, *215* (22), 2185–2197. https://doi.org/10.1002/macp.201400164.
75. Azim, H.; Dekhterman, A.; Jiang, Z.; Gross, R. A. Candida Antarctica Lipase B-Catalyzed Synthesis of Poly(Butylene Succinate): Shorter Chain Building Blocks Also Work. *Biomacromolecules* **2006**, *7* (11), 3093–3097. https://doi.org/10.1021/bm060574h.
76. Pellis, A.; Comerford, J. W.; Weinberger, S.; Guebitz, G. M.; Clark, J. H.; Farmer, T. J. Enzymatic Synthesis of Lignin Derivable Pyridine Based Polyesters for the Substitution of Petroleum Derived Plastics. *Nat. Commun.* **2019**, *10* (1), 1–9. https://doi.org/10.1038/s41467-019-09817-3.
77. Yang, Y.; Lu, W.; Zhang, X.; Xie, W.; Cai, M.; Gross, R. A. Two-Step Biocatalytic Route to Biobased Functional Polyesters from ω-Carboxy Fatty Acids and Diols. *Biomacromolecules* **2010**, *11* (1), 259–268. https://doi.org/10.1021/bm901112m.
78. Zhaozhong, J.; Chen, L.; Wenchun, X.; Gross, R. A. Controlled Lipase-Catalyzed Synthesis of Poly(Hexamethylene Carbonate). *Macromolecules* **2007**, *40* (22), 7934–7943. https://doi.org/10.1021/ma070665m.
79. Jiang, Y.; Woortman, A. J. J.; Alberda Van Ekenstein, G. O. R.; Loos, K. A Biocatalytic Approach towards Sustainable Furanic-Aliphatic Polyesters. *Polym. Chem.* **2015**, *6* (29), 5198–5211. https://doi.org/10.1039/c5py00629e.

80. Jiang, Y.; Woortman, A. J. J.; Alberda Van Ekenstein, G. O. R.; Petrović, D. M.; Loos, K. Enzymatic Synthesis of Biobased Polyesters Using 2,5-Bis(Hydroxymethyl) Furan as the Building Block. *Biomacromolecules* **2014**, *15* (7), 2482–2493. https://doi.org/10.1021/bm500340w.
81. Maniar, D.; Jiang, Y.; Woortman, A. J. J.; van Dijken, J.; Loos, K. Furan-Based Copolyesters from Renewable Resources: Enzymatic Synthesis and Properties. *ChemSusChem* **2019**, *12* (5), 990–999. https://doi.org/10.1002/cssc.201802867.
82. Pellis, A.; Weinberger, S.; Gigli, M.; Guebitz, G. M.; Farmer, T. J. Enzymatic Synthesis of Biobased Polyesters Utilizing Aromatic Diols as the Rigid Component. *Eur. Polym. J.* **2020**, *130*, 109680. https://doi.org/10.1016/j.eurpolymj.2020.109680.
83. Fehrenbacher, U.; Grosshardt, O.; Kowollik, K.; Tübke, B.; Dingenouts, N.; Wilhelm, M. Synthese Und Charakterisierung von Polyestern Und Polyamiden Auf Der Basis von Furan-2,5-Dicarbonsäure. *Chem. Ing. Tech.* **2009**, *81* (11), 1829–1835. https://doi.org/10.1002/cite.200900090.
84. Duursma, A.; Aberson, R.; Smith, D. D.; Flores, J.; Dam, M. A.; Gruter, G. J. M. Process for Preparing a Furan-Based Polyamide, a Furan-Based Oligomer and Compositions and Articles Comprising the Furan-Based Polyamide. WO2015060718 A1, 2015.
85. Smith, D. D.; Flores, J.; Aberson, R.; Dam, M. A.; Duursma, A.; Gruter, G. J. M. Polyamides Containing the Bio-Based 2,5-Furandicarboxylic Acid. WO2015059047 A1, 2015.
86. Mitiakoudis, A.; Gandini, A. *Synthesis and Characterization of Furanic Polyamides*; 1991; Vol. 24.
87. Hopff, H.; Krieger, A. Über Decarboxylierung Und Dissoziation Heterocyclischer Dicarbonsäuren. *Helv. Chim. Acta* **1961**, *44* (4), 1058–1063. https://doi.org/10.1002/hlca.19610440425.
88. Hopff, V. H.; Krieger, A. Über Polyamide Aus Heterocyclischen Dicarbonsäuren. *Makromol. Chem.* **1961**, *47* (1), 93–113. https://doi.org/10.1002/macp.1961.020470109.
89. Heertjes, P.; Kok, G. Polycondensation Products of 2, 5-Furandicarboxylic Acid. *Delft Prog. Rep.* **1974** (1), 59–63Ser. A.
90. Triki, R.; Abid, M.; Tessier, M.; Abid, S.; el Gharbi, R.; Fradet, A. Furan-Based Poly(Esteramide)s by Bulk Copolycondensation. *Eur. Polym. J.* **2013**, *49* (7), 1852–1860. https://doi.org/10.1016/j.eurpolymj.2013.04.014.
91. Abid, M.; Mhiri, S.; Bougarech, A.; Triki, R.; Abid, S. Preparation, Characterization and Degradation Study of Novel Sulfonated Furanic Poly(Ester-Amide)s. *Des. Monomers Polym.* **2020**, *23* (1), 16–24. https://doi.org/10.1080/15685551.2020.1727171.
92. Wilsens, C. H. R. M.; Deshmukh, Y. S.; Noordover, B. A. J.; Rastogi, S. Influence of the 2, 5-Furandicarboxamide Moiety on Hydrogen Bonding in Aliphatic-Aromatic Poly(Ester Amide)s. *Macromolecules* **2014**, *47* (18), 6196–6206. https://doi.org/10.1021/ma501310f.
93. Wilsens, C. H. R. M.; Wullems, N. J. M.; Gubbels, E.; Yao, Y.; Rastogi, S.; Noordover, B. A. J. Synthesis, Kinetics, and Characterization of Bio-Based Thermosets Obtained through Polymerization of a 2,5-Furandicarboxylic Acid-Based Bis(2-Oxazoline) with Sebacic Acid. *Polym. Chem.* **2015**, *6* (14), 2707–2716. https://doi.org/10.1039/c4py01609b.
94. Alfonso Alegre, M. J.; Zagalaz Lasierra, P.; Caballero López, M. Á. Biodegradable Polyesteretheramide. *EP 2 628 761 A1, 21.08.2013 Bulletin 2013/34*. 2013.
95. Kluge, M.; Papadopoulos, L.; Magaziotis, A.; Tzetzis, D.; Zamboulis, A.; Bikiaris, D.; Robert, T. A Facile Method to Synthesize Semicrystalline Poly(Ester Amide)s from 2,5-

Furandicarboxylic Acid, 1,10-Decanediol, and Crystallizable Amido Diols. *ACS Sustainable Chem. Eng.* **2020**, *8*, 10812–10821. https://doi.org/10.1021/acssuschemeng.0c02730.

96. Cheng, H. N. *Enzyme-Catalyzed Synthesis of Polyamides and Polypeptides*; Wiley-VCH, 2010.
97. Stavila, E.; Loos, K. Synthesis of Polyamides and Their Copolymers via Enzymatic Polymerization. *J. Renewable Mater.* **2015**, *3* (4), 268–280. https://doi.org/10.7569/JRM.2015.634102.
98. Vouyiouka, S. N.; Karakatsani, E. K.; Papaspyrides, C. D. Solid State Polymerization. *Prog. Polym. Sci. (Oxford)* **2005**, *30*, 10–37. https://doi.org/10.1016/j.progpolymsci.2004.11.001.
99. Stavila, E.; Arsyi, R. Z.; Petrovic, D. M.; Loos, K. Fusarium Solani Pisi Cutinase-Catalyzed Synthesis of Polyamides. *Eur. Polym. J.* **2013**, *49* (4), 834–842. https://doi.org/10.1016/j.eurpolymj.2012.12.010.
100. Poulhès, F.; Mouysset, D.; Gil, G.; Bertrand, M. P.; Gastaldi, S. Speeding-up Enzyme-Catalyzed Synthesis of Polyamides Using ω-Amino-α-Alkoxy-Acetate as Monomer. *Polymer* **2013**, *54* (14), 3467–3471. https://doi.org/10.1016/j.polymer.2013.05.011.
101. Poulhès, F.; Mouysset, D.; Gil, G.; Bertrand, M. P.; Gastaldi, S. CAL-B Catalyzed Synthesis of Chiral Polyamides. *Tetrahedron Asymmetry* **2012**, *23* (11–12), 867–875. https://doi.org/10.1016/j.tetasy.2012.05.021.
102. Stavila, E.; Alberda Van Ekenstein, G. O. R.; Woortman, A. J. J.; Loos, K. Lipase-Catalyzed Ring-Opening Copolymerization of ε-Caprolactone and β-Lactam. *Biomacromolecules* **2014**, *15* (1), 234–241. https://doi.org/10.1021/bm401514k.
103. Stavila, E.; Alberda Van Ekenstein, G. O. R.; Loos, K. Enzyme-Catalyzed Synthesis of Aliphatic-Aromatic Oligoamides. *Biomacromolecules* **2013**, *14* (5), 1600–1606. https://doi.org/10.1021/bm400243a.
104. Baum, I.; Elsässer, B.; Schwab, L. W.; Loos, K.; Fels, G. Atomistic Model for the Polyamide Formation from β-Lactam Catalyzed by Candida Antarctica Lipase B. *ACS Catal.* **2011**, *1* (4), 323–336. https://doi.org/10.1021/cs1000398.
105. Schwab, L. W.; Kroon, R.; Schouten, A. J.; Loos, K. Enzyme-Catalyzed Ring-Opening Polymerization of Unsubstituted β-Lactam. *Macromol. Rapid Commun.* **2008**, *29* (10), 794–797. https://doi.org/10.1002/marc.200800117.
106. Jiang, Y.; Maniar, D.; Woortman, A. J. J.; Alberda Van Ekenstein, G. O. R.; Loos, K. Enzymatic Polymerization of Furan-2,5-Dicarboxylic Acid-Based Furanic-Aliphatic Polyamides as Sustainable Alternatives to Polyphthalamides. *Biomacromolecules* **2015**, *16* (11), 3674–3685. https://doi.org/10.1021/acs.biomac.5b01172.
107. Jiang, Y.; Maniar, D.; Woortman, A. J. J.; Loos, K. Enzymatic Synthesis of 2,5-Furandicarboxylic Acid-Based Semi-Aromatic Polyamides: Enzymatic Polymerization Kinetics, Effect of Diamine Chain Length and Thermal Properties. *RSC Adv.* **2016**, *6* (72), 67941–67953. https://doi.org/10.1039/c6ra14585j.
108. Maniar, D.; Silvianti, F.; Ospina, V.; Woortman, A.; van Dijken, J.; Loos, K. On the Way to Greener Furanic-Aliphatic Poly(Ester Amide)s: Enzymatic Polymerization in Ionic Liquid. *Polymer* **2020**, *205*, 122662. https://doi.org/10.1016/j.polymer.2020.122662.
109. Dong, F. X.; Zhang, L.; Tong, X. Z.; Chen, H. B.; Wang, X. L.; Wang, Y. Z. Ionic Liquid Coated Lipase: Green Synthesis of High Molecular Weight Poly(1,4-Dioxan-2-One). *J. Mol. Catal. B: Enzym.* **2012**, *77*, 46–52. https://doi.org/10.1016/j.molcatb.2012.01.006.

110. Haiß, A.; Jordan, A.; Westphal, J.; Logunova, E.; Gathergood, N.; Kümmerer, K. On the Way to Greener Ionic Liquids: Identification of a Fully Mineralizable Phenylalanine-Based Ionic Liquid. *Green Chem.* **2016**, *18* (16), 4361–4373. https://doi.org/10.1039/c6gc00417b.
111. Maton, C.; de Vos, N.; Stevens, C. v. Ionic Liquid Thermal Stabilities: Decomposition Mechanisms and Analysis Tools. *Chem. Soc. Rev.* **2013**, *42* (13), 5963–5977. https://doi.org/10.1039/c3cs60071h.
112. Hallett, J. P.; Welton, T. Room-Temperature Ionic Liquids: Solvents for Synthesis and Catalysis. 2. *Chem. Rev.* **2011**, *111*, 3508–3576. https://doi.org/10.1021/cr1003248.
113. Myles, L.; Gore, R. G.; Gathergood, N.; Connon, S. J. A New Generation of Aprotic yet Brønsted Acidic Imidazolium Salts: Low Toxicity, High Recyclability and Greatly Improved Activity. *Green Chem.* **2013**, *15* (10), 2740–2746. https://doi.org/10.1039/c3gc40975a.
114. Zhang, Q.; Zhang, S.; Deng, Y. Recent Advances in Ionic Liquid Catalysis. *Green Chem.* **2011**, *13* (10), 2619–2637. https://doi.org/10.1039/c1gc15334j.
115. Matos, M.; Sousa, A. F.; Fonseca, A. C.; Freire, C. S. R.; Coelho, J. F. J.; Silvestre, A. J. D. A New Generation of Furanic Copolyesters with Enhanced Degradability: Poly(Ethylene 2,5-Furandicarboxylate)- *Co* -Poly(Lactic Acid) Copolyesters. *Macromolecular Chem. Phys.* **2014**, *215* (22), 2175–2184. https://doi.org/10.1002/macp.201400175.
116. Rogers, R. D.; Seddon, K. R. Ionic Liquids - Solvents of the Future? *Science* **2003**, 792–793. https://doi.org/10.1126/science.1090313.
117. Zhu, S.; Chen, R.; Wu, Y.; Chen, Q.; Zhang, X.; Yu, Z. A Mini-Review on Greenness of Ionic Liquids. *Chem. Biochem. Eng. Q.* **2009**, *23* (2), 207–211.
118. Lozano, P.; García-Verdugo, E. Ionic Liquids and Other Nonconventional Solvents in Biotransformations: Medium Engineering and Process Development. In *Ionic Liquids in Biotransformations and Organocatalysis*; John Wiley & Sons, Inc.: Hoboken, NJ, USA, 2012; pp 103–149. https://doi.org/10.1002/9781118158753.ch4.
119. Philippi, F.; Rauber, D.; Springborg, M.; Hempelmann, R. Density Functional Theory Descriptors for Ionic Liquids and the Charge-Transfer Interpretation of the Haven Ratio. *J. Phys. Chem. A* **2019**, *123* (4), 851–861. https://doi.org/10.1021/acs.jpca.8b10827.
120. Pena-Pereira, F.; Namieśnik, J. Ionic Liquids and Deep Eutectic Mixtures: Sustainable Solvents for Extraction Processes. *ChemSusChem* **2014**, *7* (7), 1784–1800. https://doi.org/10.1002/cssc.201301192.
121. Earle, M. J.; Seddon, K. R. Ionic Liquids. Green Solvents for the Future. *Pure Appl. Chem.* **2000**, *72* (7), 1391–1398. https://doi.org/10.1351/pac200072071391.
122. Gorke, J.; Srienc, F.; Kazlauskas, R. Toward Advanced Ionic Liquids. Polar, Enzyme-Friendly Solvents for Biocatalysis. *Biotechnol. Bioprocess Eng.* **2010**, *15*, 40–53. https://doi.org/10.1007/s12257-009-3079-z.
123. Ranke, J.; Stolte, S.; Störmann, R.; Aming, J.; Jastorff, B. Design of Sustainable Chemical Products - The Example of Ionic Liquids. *Chem. Rev.* **2007**, *107*, 2183–2206. https://doi.org/10.1021/cr050942s.
124. Deetlefs, M.; Seddon, K. R. Assessing the Greenness of Some Typical Laboratory Ionic Liquid Preparations. *Green Chem.* **2010**, *12*, 17–30. https://doi.org/10.1039/b915049h.
125. Wilkes, J. S. A Short History of Ionic Liquids - From Molten Salts to Neoteric Solvents. *Green Chem.* **2002**, *4*, 73–80. https://doi.org/10.1039/b110838g.

126. Kunz, W.; Maurer, E.; Klein, R.; Touraud, D.; Rengstl, D.; Harrar, A.; Dengler, S.; Zech, O. Low Toxic Ionic Liquids, Liquid Catanionics, and Ionic Liquid Microemulsions. *J. Dispersion Sci. Technol.* **2011**, *32* (12), 1694–1699. https://doi.org/10.1080/01932691.2011.616109.
127. Hejazifar, M.; Lanaridi, O.; Bica-Schröder, K. Ionic Liquid Based Microemulsions: A Review. *J. Mol. Liquids* **2020**, 112264. https://doi.org/10.1016/j.molliq.2019.112264.
128. Carriazo, D.; Serrano, M. C.; Gutiérrez, M. C.; Ferrer, M. L.; del Monte, F. Deep-Eutectic Solvents Playing Multiple Roles in the Synthesis of Polymers and Related Materials. *Chem. Soc. Rev.* **2012**, *41*, 4996–5014. https://doi.org/10.1039/c2cs15353j.
129. Zhang, Q.; de Oliveira Vigier, K.; Royer, S.; Jérôme, F. Deep Eutectic Solvents: Syntheses, Properties and Applications. *Chem. Soc. Rev.* **2012**, *41*, 7108–7146. https://doi.org/10.1039/c2cs35178a.
130. Paiva, A.; Craveiro, R.; Aroso, I.; Martins, M.; Reis, R. L.; Duarte, A. R. C. Natural Deep Eutectic Solvents - Solvents for the 21st Century. *ACS Sustainable Chem. Eng.* **2014**, *2*, 1063–1071. https://doi.org/10.1021/sc500096j.
131. Loow, Y. L.; New, E. K.; Yang, G. H.; Ang, L. Y.; Foo, L. Y. W.; Wu, T. Y. Potential Use of Deep Eutectic Solvents to Facilitate Lignocellulosic Biomass Utilization and Conversion. *Cellulose* **2017**, *24*, 3591–3618. https://doi.org/10.1007/s10570-017-1358-y.
132. Smith, E. L.; Abbott, A. P.; Ryder, K. S. Deep Eutectic Solvents (DESs) and Their Applications. *Chem. Rev.* **2014**, *114*, 11060–11082. https://doi.org/10.1021/cr300162p.
133. Rodriguez Rodriguez, N.; van den Bruinhorst, A.; Kollau, L. J. B. M.; Kroon, M. C.; Binnemans, K. Degradation of Deep-Eutectic Solvents Based on Choline Chloride and Carboxylic Acids. *ACS Sustainable Chem. Eng.* **2019**, *7* (13), 11521–11528. https://doi.org/10.1021/acssuschemeng.9b01378.
134. Lee, C. B. T. L.; Wu, T. Y.; Ting, C. H.; Tan, J. K.; Siow, L. F.; Cheng, C. K.; Md. Jahim, J.; Mohammad, A. W. One-Pot Furfural Production Using Choline Chloride-Dicarboxylic Acid Based Deep Eutectic Solvents under Mild Conditions. *Bioresource Technol.* **2019**, *278*, 486–489. https://doi.org/10.1016/j.biortech.2018.12.034.
135. Liu, W.; Zhang, K.; Chen, J.; Yu, J. Ascorbic Acid and Choline Chloride: A New Natural Deep Eutectic Solvent for Extracting Tert-Butylhydroquinone Antioxidant. *J. Mol. Liq.* **2018**, *260*, 173–179. https://doi.org/10.1016/j.molliq.2018.03.092.
136. Zeisel, S. H. Choline: An Important Nutrient in Brain Development, Liver Function and Carcinogenesis. *J. Am. Coll. Nutr.* **1992**, *11* (5), 473–481. https://doi.org/10.1080/07315724.1992.10718251.
137. Gaba, R.; Kaur, N.; Pal, A.; Sharma, D.; Kumar, H. Molecular Interactions and Taste Perception of an Artificial Sweetener Saccharin Sodium in Aqueous and in Aqueous Solutions of Choline Chloride at Different Temperatures. *J. Mol. Liq.* **2020**, *305*, 112855. https://doi.org/10.1016/j.molliq.2020.112855.
138. Taladrid, D.; Marín, D.; Alemán, A.; Álvarez-Acero, I.; Montero, P.; Gómez-Guillén, M. C. Effect of Chemical Composition and Sonication Procedure on Properties of Food-Grade Soy Lecithin Liposomes with Added Glycerol. *Food Res. Int.* **2017**, *100*, 541–550. https://doi.org/10.1016/j.foodres.2017.07.052.

139. Bi, W.; Tian, M.; Row, K. H. Evaluation of Alcohol-Based Deep Eutectic Solvent in Extraction and Determination of Flavonoids with Response Surface Methodology Optimization. *J. Chromatogr. A* **2013**, *1285*, 22–30. https://doi.org/10.1016/j.chroma.2013.02.041.
140. Abbott, A. P.; Cullis, P. M.; Gibson, M. J.; Harris, R. C.; Raven, E. Extraction of Glycerol from Biodiesel into a Eutectic Based Ionic Liquid. *Green Chem.* **2007**, *9* (8), 868–872. https://doi.org/10.1039/b702833d.
141. Leron, R. B.; Li, M. H. Solubility of Carbon Dioxide in a Choline Chloride-Ethylene Glycol Based Deep Eutectic Solvent. *Thermochim. Acta* **2013**, *551*, 14–19. https://doi.org/10.1016/j.tca.2012.09.041.
142. Francisco, M.; van den Bruinhorst, A.; Zubeir, L. F.; Peters, C. J.; Kroon, M. C. A New Low Transition Temperature Mixture (LTTM) Formed by Choline Chloride+lactic Acid: Characterization as Solvent for CO2 Capture. *Fluid Phase Equilib.* **2013**, *340*, 77–84. https://doi.org/10.1016/j.fluid.2012.12.001.
143. Francisco, M.; van den Bruinhorst, A.; Kroon, M. C. New Natural and Renewable Low Transition Temperature Mixtures (LTTMs): Screening as Solvents for Lignocellulosic Biomass Processing. *Green Chem.* **2012**, *14* (8), 2153–2157. https://doi.org/10.1039/c2gc35660k.
144. Gorke, J. T.; Srienc, F.; Kazlauskas, R. J. Hydrolase-Catalyzed Biotransformations in Deep Eutectic Solvents. *Chem. Commun.* **2008**, *0* (10), 1235–1237. https://doi.org/10.1039/b716317g.
145. Gómez, A. v.; Biswas, A.; Tadini, C. C.; Furtado, R. F.; Alves, C. R.; Cheng, H. N. Use of Natural Deep Eutectic Solvents for Polymerization and Polymer Reactions. *J. Braz. Chem. Soc.* **2019**, 717–726. https://doi.org/10.21577/0103-5053.20190001.
146. Zhao, H.; Zhang, C.; Crittle, T. D. Choline-Based Deep Eutectic Solvents for Enzymatic Preparation of Biodiesel from Soybean Oil. *J. Mol. Catal. B: Enzym.* **2013**, *85–86*, 243–247. https://doi.org/10.1016/j.molcatb.2012.09.003.
147. Zhao, H.; Baker, G. A.; Holmes, S. New Eutectic Ionic Liquids for Lipase Activation and Enzymatic Preparation of Biodiesel. *Org. Biomol. Chem.* **2011**, *9* (6), 1908–1916. https://doi.org/10.1039/c0ob01011a.
148. Durand, E.; Lecomte, J.; Baréa, B.; Piombo, G.; Dubreucq, E.; Villeneuve, P. Evaluation of Deep Eutectic Solvents as New Media for Candida Antarctica B Lipase Catalyzed Reactions. *Process Biochem.* **2012**, *47* (12), 2081–2089. https://doi.org/10.1016/j.procbio.2012.07.027.

Chapter 2

α-Glucan Phosphorylase-Catalyzed Enzymatic Reactions to Precisely Synthesize Non-natural Polysaccharides

Jun-ichi Kadokawa

Graduate School of Science and Engineering, Kagoshima University, 1-21-40 Korimoto, Kagoshima 890-0065, Japan
Email: kadokawa@eng.kagoshima-u.ac.jp

This chapter presents the development of precision synthesis of non-natural polysaccharides by α-glucan phosphorylase (GP)-catalyzed enzymatic reactions. GP is one of the enzymes that has, been used as a catalyst to practically synthesize polysaccharides. It catalyzes enzymatic polymerization of α-D-glucose 1-phosphate to produce an α(1→4)-glucan with a well-defined structure—that is, amylose—with liberating inorganic phosphate. To initiate the polymerization, the nonreducing ends of α(1→4)-glucan chains, such as maltooligosaccharides, should be present as the initiating sites (primers) in the reaction system. As GP recognizes substrate analogs of α-D-glucose 1-phosphate (monosaccharide 1-phosphates) as glycosyl donors to catalyze glycosylations, the corresponding monosaccharide units can be introduced to the nonreducing ends of glycosyl acceptors. As thermostable GP shows more tolerance for the recognition of substrates than the most extensively studied GPs isolated from a potato, the former enzyme catalyzes consecutive glycosylations using some monosaccharide 1-phosphates, which could be extended to the enzymatic polymerization under the conditions for removal of inorganic phosphate. For example, this enzyme catalyzes the enzymatic polymerization of α-D-glucosamine 1-phosphate as a monomer to produce an aminopolysaccharide composed of an α(1→4)-linked α-D-glucosamine chain. The thermostable GP-catalyzed enzymatic polymerization by the same operation has also produced non-natural heteropolysaccharides. Through the enzymatic glycosylation catalyzed by thermostable GP, carboxylate-terminated maltooligosaccharides were prepared and employed as a cross-linker with water-soluble chitin to obtain network chitin. The thermostable GP-catalyzed enzymatic reactions have also produced amphoteric polysaccharide materials.

Introduction

Polysaccharides are widely distributed in nature and show specific important functions in a wide range of biological systems (*1*). Such versatile functions of polysaccharides come from their structural variety, owing to many kinds of monosaccharide units and regio- and stereoarrangements of glycosidic linkages among them in polysaccharides (2). Due to the complicated structures, chemical synthesis of polysaccharides by repeating reactions for the formation of glycosidic linkages (chemical glycosylations) has not been efficiently developed (3). Compared to the approach by general organic reactions, the enzymatic method has been identified as a useful approach to synthesizing polysaccharides with well-defined structures. This is because they move along in a highly controlled fashion in regio- and stereoarrangements of glycosidic linkages, in accordance with the specificity of each enzyme, even without the protection of hydroxy groups in sugar substrates (*3–6*). Of the six main classes of enzymes, glycosyl transferase, and glycosyl hydrolase have been used successfully as catalysts in practical synthesis of well-defined polysaccharides via the formation of glycosidic linkages (*7–9*).

Enzymes generally recognize native substrates, strictly according to a 1:1 ratio, in enzyme–substrate complexes, like a lock-and-key relationship. The enzymatic approach is useful to synthesizing natural polysaccharides. For example, cellulase belonging to glycosyl hydrolase catalyzes enzymatic polymerization of β-cellobiosyl fluoride as a monomer via consecutive glycosylations to produce the synthetic cellulose (Figure 1a) (*10*). However, some enzymes show weak specificities for the recognition of substrates, and accordingly can be employed for the synthesis of non-natural polysaccharides (3). Cellulase has been also employed to synthesize non-natural cellulose derivatives (Figures 1b and 1c) (*11*, *12*).

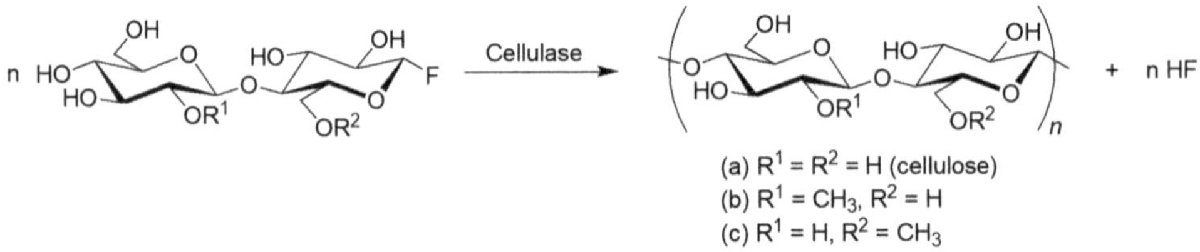

Figure 1. Cellulase-catalyzed enzymatic polymerization to produce (a) cellulose and (b,c) its derivatives.

α-Glucan phosphorylase (GP), belonging to glycosyltransferase family 35 (enzyme commission number 2.4.1.1), is one of the enzymes that have been used as catalysts to practically synthesize oligo- and polysaccahrides (*13–16*). GP catalyzes enzymatic polymerization of α-D-glucose 1-phosphate (Glc-1-P) as a monomer to produce α(1→4)-glucan; that is, amylose (*17–20*). As GP shows weak specificity for the recognition of substrates, it catalyzes reactions for the formation of α(1→4) glycosidic linkages (i.e., glycosylations) using some substrate analogs of Glc-1-P to make non-natural oligo- and polysaccharides (*21–24*). Based on the aforementioned background and viewpoints, this chapter deals with the precise enzymatic synthesis of non-natural polysaccharides catalyzed by GP.

Characteristics of GP-Catalyzed Enzymatic Reactions

GP is the enzyme that catalyzes the reversible phosphorolysis of α(1→4)-glucans, such as amylose and glycogen, at the nonreducing end in the presence of inorganic phosphate (Pi) (*25–27*). Glc-1-P and α(1→4)-glucans with one smaller degree of polymerization (DP) are therefore obtained as the reaction's products (Figure 2a). Owing to reversibility of the reaction, GP also catalyzes the enzymatic chain elongation (glycosylation) using two substrates—that is, Glc-1-P and α(1→4)-glucan—under suitable conditions (Figure 2b). In such α(1→4) glycosylation, a glucose residue is transferred from Glc-1-P to the nonreducing end of α(1→4)-glucan, with the formation of an α(1→4) glycosidic linkage to produce a α(1→4)-glucan with one larger DP. Therefore, the former and latter substrates act as a glycosyl donor and acceptor, respectively, in the glycosylation; Pi is also produced from Glc-1-P. In particular, GPs have the lowest DP in α(1→4)-glucan chains to recognize at the nonreducing end, depending on their sources. Accordingly, α(1→4)-oligoglucans—that is, maltooligosaccharides—with DPs greater than the lowest one should be used as the glycosyl acceptors in the glycosylation. The smallest substrates for the phosphorolysis and glycosylation, recognized by the most extensively studied GP isolated from a potato, are maltopentaose and maltotetraose (Glc_4), respectively. The lowest DPs of substrates recognized by GPs isolated from thermophilic bacteria sources (thermostable GPs) for phosphorolysis and glycosylation are one less than those recognized by potato GPs—the substrates Glc_4 and maltotriose (Glc_3), respectively (*28–31*).

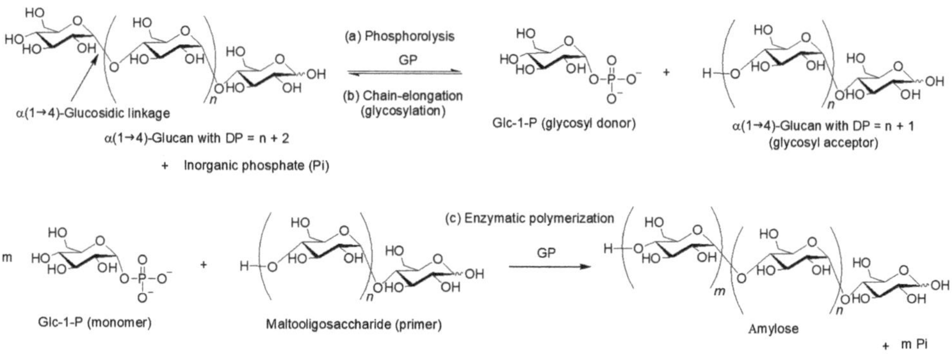

Figure 2. GP-catalyzed (a) phosphorolysis, (b) glycosylation, and (c) enzymatic polymerization.

When an excess donor/acceptor (Glc-1-P/maltooligosaccharide) molar ratio is employed in the reaction system, GP catalyzes consecutive glycosylations as the manner of polymerization to produce the α(1→4)-glucan polymer (i.e., amylose, Figure 2c) (*17–20*). As the enzymatic polymerization of Glc-1-P is initiated at the nonreducing end of the maltooligosaccharide acceptor, and propagation then occurs by consecutive reactions in the same manner, the GP-catalyzed enzymatic polymerization is categorized as chain-growth polymerization. Accordingly, the maltooligosaccharide acceptor is often referred to as a "primer" for polymerization. The enzymatically produced amylose spontaneously forms a double-helical assembly as the precipitate in the aqueous reaction media. Within the double helix, interstrand stabilization has been reported to be achieved without any steric conflict and through the occurrence of O(2)•••O(6) types of hydrogen bonds (*32, 33*). Amylose is a known component in starch with the other component, amylopectin. With the difficulty of complete separation of amylose from amylopectin in starch, the GP-catalyzed enzymatic polymerization is useful to obtain a pure amylose sample. As termination and chain transfer reactions do not occur, the GP-catalyzed enzymatic polymerization takes place in a manner similar to living polymerization. DPs of the amylose products are controlled by Glc-1-P monomer/primer feed ratios while retaining narrow polydispersity (typically <1.2) (*34*).

As the reducing ends of the maltooligosaccharide primers do not participate in the reaction, the GP-catalyzed enzymatic polymerization can be conducted using the modified maltooligosaccharide primers, which are covalently immobilized on other materials at the reducing ends, such as polymeric chains. Such modified primers are typically composed of the multiple nonreducing maltooligosaccharide chain ends on the substances. This results in the production of various amylose-grafted materials (Figure 3), representatively, amylose-grafted polysaccharides, and polypeptides as biopolymer conjugates (*35–39*).

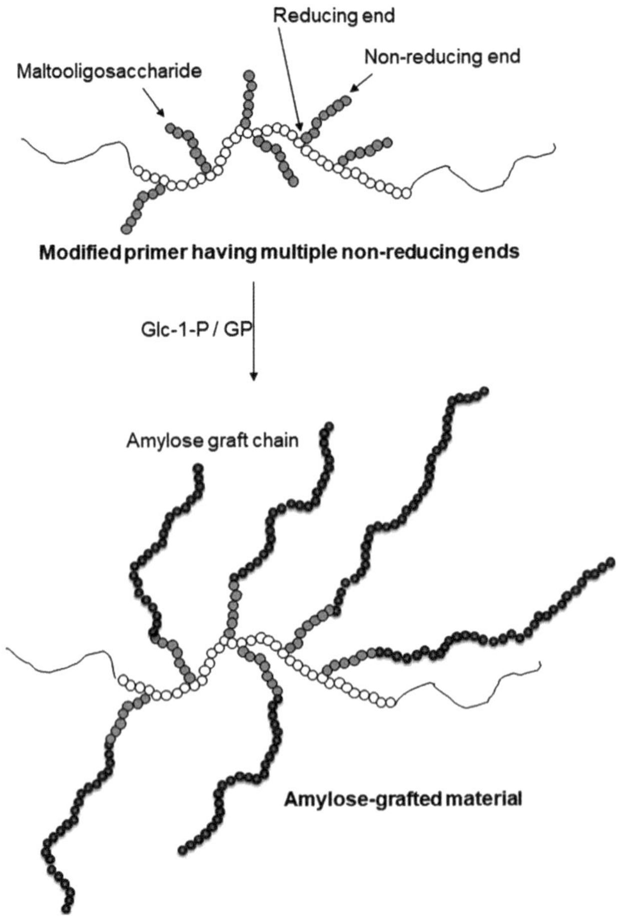

Figure 3. GP-catalyzed enzymatic polymerization using modified primer to produce amylose-grafted material.

Glycogen—which is a water-soluble, highly branched natural polysaccharide composed of α(1→4)-glucan chains further interlinked by α(1g6)branching points—can also be used as the primer for the GP-catalyzed enzymatic polymerization, owing to the presence of a number of nonreducing α(1→4)-glucan chain ends (*40*). When the GP-catalyzed enzymatic polymerization of Glc-1-P from the nonreducing ends of glycogen was conducted in acetate buffer—and the reaction mixture was subsequently left standing at room temperature for 24 h—it turned into a hydrogel form. Hydrogelation was reasonably explained by the double-helix formation of the elongated amylose chains among glycogen molecules, which acted as cross-linking points for construction of polymeric networks (Figure 4a).

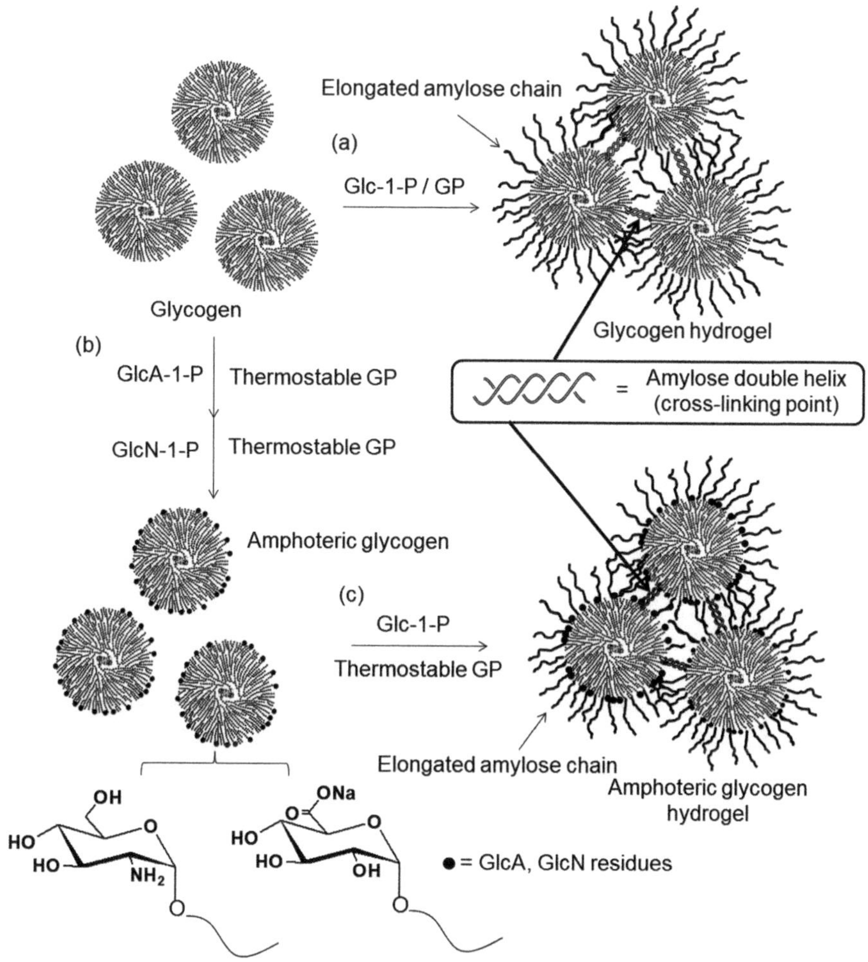

Figure 4. GP-catalyzed (a) polymerization to produce glycogen hydrogel, (b) glucuronylation and subsequent glucosaminylation to produce amphoteric glycogen, and (c) subsequent polymerization to produce amphoteric glycogen hydrogel.

Synthesis of Non-natural Oligo- and Polysaccharides by GP-Catalyzed Enzymatic Glycosylations

As previously mentioned, GPs have been known to exhibit weak specificity for the recognition of substrate structures. The extension of the GP-catalyzed enzymatic glycosylations using several monosaccharide 1-phosphates (i.e., substrate analogs of Glc-1-P) as the glycosyl donors has been investigated to obtain non-natural oligosaccharides comprising the corresponding monosaccharide residues at the nonreducing ends (*21–24*). For example, potato GP has shown recognition of α-D-mannose (Man), α-D-xylose, α-D-glucosamine (GlcN), and *N*-formyl-α-D- GlcN 1-phosphates (Man-1-P, Xyl-1-P, GlcN-1-P, and GlcNF-1-P, respectively) as the glycosyl donors in glycosylations. This is using Glc_4 as a glycosyl acceptor, to produce non-natural α-mannosylated, α-xylosylated, α-glucosaminylated, and *N*-formyl-α-glucosaminylated pentasaccharides, respectively (Figure 5) (*41–44*). Of the resulting non-natural pentasaccharides, the α-glucosaminylated pentasaccharide obtained from the glycosylation (glucosaminylation) using GlcN-1-P is particularly interesting—it is a basic oligosaccharide with an amino group at the C2 position of a GlcN unit at the nonreducing end.

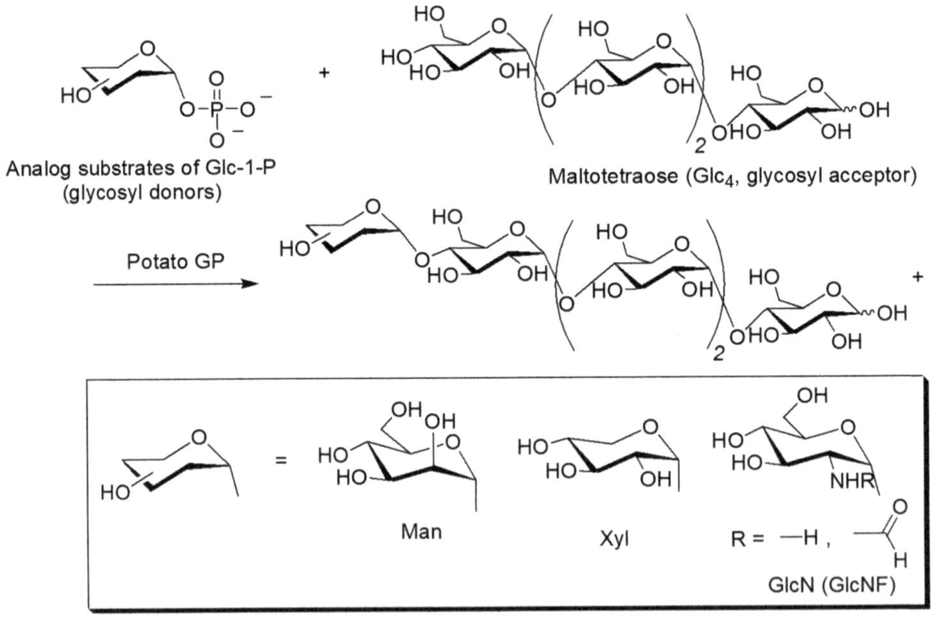

Figure 5. Potato GP–catalyzed enzymatic glycosylations using monosaccharide 1-phosphates (substrate analogs of Glc-1-P) as glycosyl donors, with Glc_4 as a glycosyl acceptor, to produce non-natural pentasaccharides.

An attempt was also made to synthesize an acidic oligosaccharide by the potato GP–catalyzed glycosylation using α-D-glucuronic acid 1-phosphate (GlcA-1-P), as an acidic source with a carboxylate group, with Glc_4. However, this enzyme did not recognize GlcA-1-P. Alternatively, thermostable GP isolated from *Aquifex aeolicus* VF5 recognized GlcA-1-P and catalyzed the glycosylation (glucuronylation) with Glc_3, the smallest glycosyl acceptor for this enzyme. This yielded acidic tetrasaccharide comprising a GlcA unit at the nonreducing end (Figure 6a) (*45*).

The enzymatic glucuronylation was applied to prepare maltooligosaccharides with carboxylate groups at both ends (carboxylate-terminated maltooligosaccharides) (Figure 6b) (*46*). First, the reducing end of maltoheptaose was oxidized by sodium hypoiodite to make a maltooligosaccharide with a carboxylate group (Glc_6-GlcCOONa). The thermostable GP-catalyzed glucuronylation of GlcA-1-P with the product was then conducted to obtain the desired carboxylate-terminated maltooligosaccharides (GlcA-Glc_n-GlcCOONa) with varying DP values. The occurrence of disproportionation on Glc_6-GlcCOONa, simultaneously with the glucuronylation, gave rise to the different DP products. The resulting GlcA-Glc_n-GlcCOONa was employed as cross-linker for condensation with a water-soluble chitin in the presence of a condensing agent—*N*-hydroxysuccinimide/1-ethyl-3-(3-dimethylaminopropyl)carbodiimide hydrochloride—to produce network chitins (Figure 7). Chitin with an approximately 50% degree of *N*-acetylation, derived by partial deacetylation of a native chitin, specifically showed water solubility and was called water-soluble chitin (*47*). With the progress of the condensation, the reaction mixture turned completely into hydrogel form. The hydrogel could be further converted into a film by pressing.

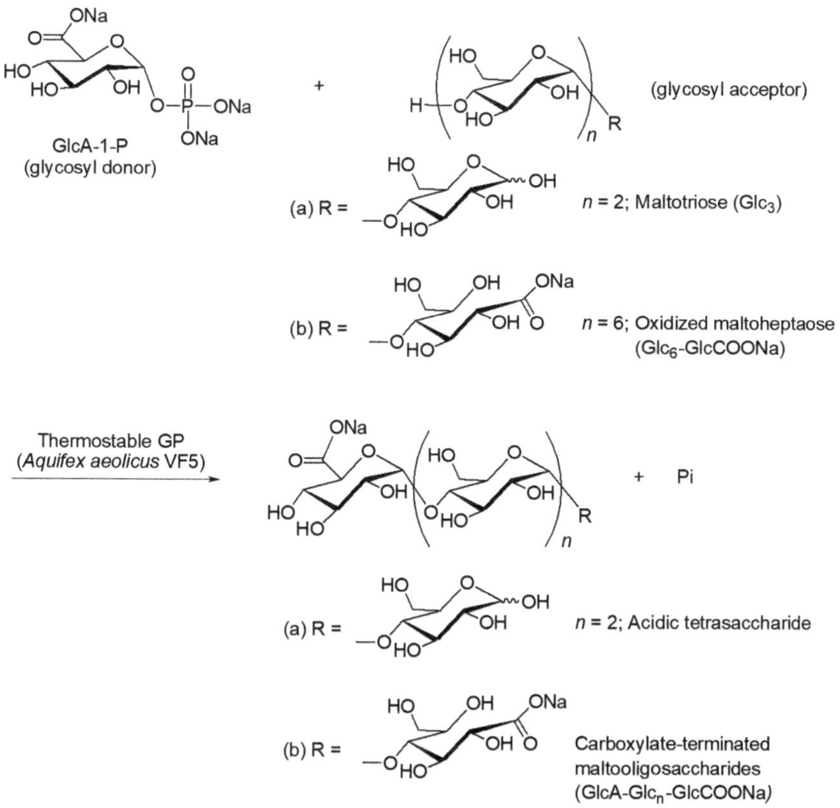

Figure 6. Thermostable GP-catalyzed enzymatic glucuronylation using GlcA-1-P as a glycosyl donor with glycosyl acceptor; (a) Glc_3 and (b) oxidized maltoheptaose (Glc_6-GlcCOONa) produce α-glucuronylated tetrasaccharides and carboxylate-terminated maltooligosaccharides (GlcA-Glc_n-GlcCOONa), respectively.

Figure 7. Cross-linking of water-soluble chitin with GlcA-Glc_n-GlcCOONa using a condensing agent to produce network chitin.

Through the previously mentioned GP-catalyzed glucosaminylation and glucuronylation, amphoteric polysaccharides with a highly branched structure (such as an amphoteric glycogen), with both the basic GlcN and acidic GlcA residues at a number of the nonreducing ends, have been synthesized (Figure 4b) (*48*, *49*). The thermostable GP-catalyzed glucuronylation using GlcA-1-P with highly branched polysaccharides was first conducted to produce acidic materials (*50*). The thermostable GP-catalyzed glucosaminylation using GlcN-1-P with the acidic products was then carried out to obtain amphoteric highly branched polysaccharides. The degree of functionality for GlcA/GlcN residues depended on the feed ratios employed in the reactions. The ζ-potential values of the amphoteric products changed from positive to negative as the pH of the solution increased. The values of the isoelectric points of the amphoteric products could be calculated by the pH values observed when the ζ-potential became zero, and accordingly, they were dependent on the GlcA/GlcN ratios present in the products. The produced amphoteric glycogen was converted into its hydrogel form by the thermostable GP-catalyzed enzymatic polymerization of Glc-1-P from the pure nonreducing α(1→4)-glucan chain ends, without the functionalization by GlcA or GlcN units (Figure 4c). The elongated amylose chains formed double-helical assemblies among the amphoteric glycogen molecules, which acted as cross-linking points for hydrogelation. The resulting hydrogels exhibited pH-responsive behavior, in which they were soluble in alkaline conditions and returned to hydrogels upon acidification of the system. The hydrogels exhibited pH-dependent shrinking/swelling behavior.

GP-Catalyzed Enzymatic Polymerization to Produce Amylose Analog Polysaccharides

A series of studies related to the previously mentioned potato GP–catalyzed glycosylations using substrate analogs has indicated if a monosaccharide residue is transferred from the substrate analogs to the nonreducing end of the maltooligosaccharide acceptor, subsequent glycosylations do not potentially take place. This is probably because the produced structure that is different from the Glc residue at the nonreducing end is no longer recognized by the enzyme. GP shows more tolerance at a C2 position in weak specificity for the recognition of monosaccharide 1-phosphates than that at other positions. The enzyme has been reported to induce the enzymatic polymerization of such a C2 substrate analog as the monomer, according to consecutive glycosylation processes, to produce non-natural α(1→4)-linked oligo- and polysaccharides (*8, 41*).

For example, a potato GP can catalyze the enzymatic polymerization of 2-deoxy-α-D-glucose 1-phosphate (dGlc-1-P), produced in situ from 1,2-dideoxy-D-glucose (D-glucal), in the presence of Pi, with maltooligosaccharide primers to produce α(1→4)-linked 2-deoxyglucose chains—that is, 2-deoxyamylose (DP = ~22) (*51–53*). The following mechanism has been reported to produce 2-deoxyamylose through the GP-catalyzed enzymatic polymerization using D-glucal (*41, 51*).

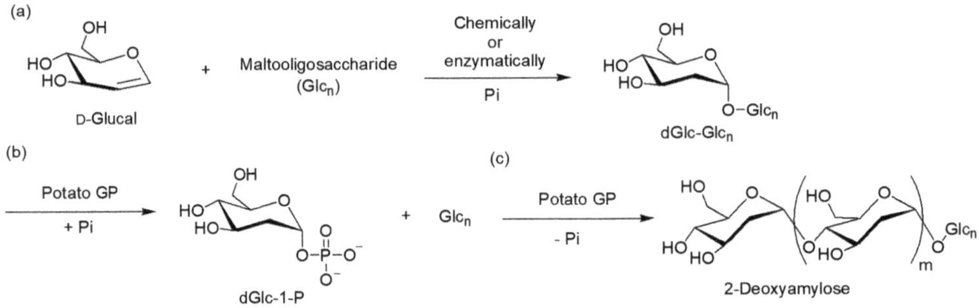

Figure 8. Two-step synthesis of dGlc-1-P in the presence of Pi and potato GP (a) and (b) and (c) potato GP–catalyzed enzymatic polymerization of resulting dGlc-1-P to produce 2-deoxyamylose.

A dGlc unit was chemically or enzymatically transferred from D-glucal to the nonreducing end of the maltooligosaccharide primer in an addition reaction manner, assisted by Pi (Figure 8a). In the second step, dGlc residue was released by GP-catalyzed enzymatic phosphorolysis to yield dGlc-1-

P in situ (Figure 8b), which acted as the actual substrate for the enzymatic polymerization (Figure 8c). The enzymatically produced 2-deoxyamylose was spontaneously precipitated in the reaction media. Powder X-ray diffraction analysis of the precipitate indicated the formation of a specific crystalline structure that was completely different from the well-known double helix of amylose (*54*). Molecular dynamics simulations showed that the isolated chains of 2-deoxyamylose spontaneously assembled to a novel double helix based on building blocks, with controlled hydrophobicity arising from pyranose ring stacking.

As previously mentioned, the potato GP–catalyzed glycosylation using GlcN-1-P or Man-1-P, which also has a different structure at the C2 position from Glc-1-P, with Glc_4, has produced the pentasaccharide with a GlcN or Man unit at the nonreducing end. However, this has not led to the enzymatic polymerization (*41*, *43*). There is a difference in the reaction manners using dGlc-1-P and GlcN-1-P/Man-1-P catalyzed by potato GP. This indicates that the absence of a hydroxy group at the C2 position in the glycosyl donor is more acceptable by a catalytic center of the enzyme than by the substitution to another group (an amino group) or the opposite stereoarrangement (axial) at that position. There is more tolerance for the recognition of substrates by thermostable GP (from *A. aeolicus* VF5) than by potato GP, as seen in the glucuronylation using GlcA-1-P. The former enzyme has been found to give rise to the different reaction manners from the latter enzyme when GlcN-1-P or Man-1-P is used as the glycosyl donor. When the thermostable GP-catalyzed glycosylations using Glc_3 and excess molar ratios of GlcN-1-P and Man-1-P were conducted in acetate buffer, consecutive glucosaminylations/mannosylations occurred, obtaining non-natural heterooligosaccharides composed of α(1→4)-linked oligo-GlcN/Man chains (Figure 9) (*55*). The matrix-assisted laser desorption ionization time-of-flight mass spectra of the products obtained by the donor/acceptor feed ratio of 10:1 observed several peaks, corresponding to the molecular masses of tetra- and octasaccharides with one and five GlcN or Man residues with Glc_3. However, further enzymatic chain elongation toward the higher DP products was inhibited by Pi produced from the glycosyl donors in these reaction systems, because it was the native substrate for the GP-catalyzed phosphorolysis (reverse reaction of chain elongation).

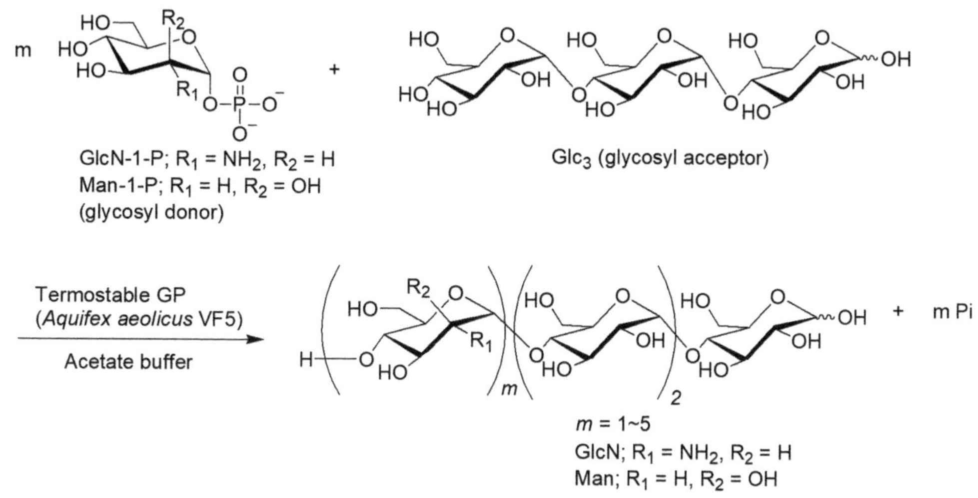

Figure 9. Thermostable GP-catalyzed consecutive glucosaminylations/mannosylations using GlcN-1-P/Man-1-P with Glc_3 in acetate buffer to produce non-natural heterooligosaccharides, composed of α(1→4)-linked GlcN/Man chains.

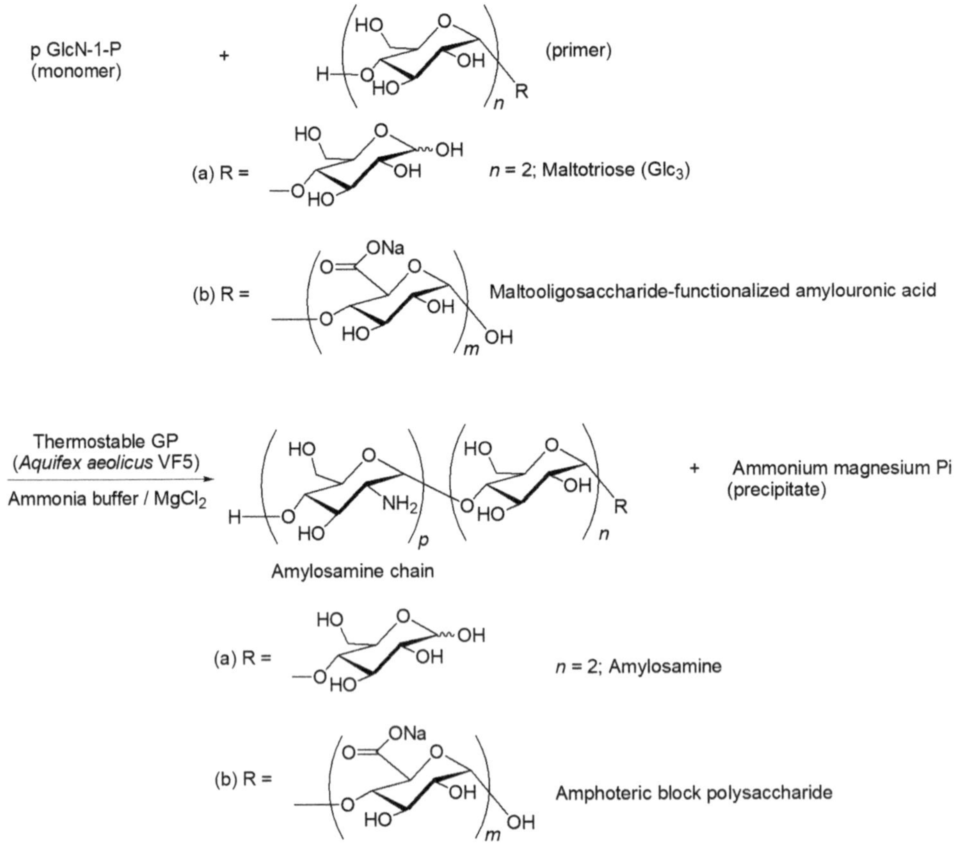

Figure 10. Thermostable GP-catalyzed enzymatic polymerization of GlcN-1-P using (a) Glc_3 and (b) maltooligosaccharide-functionalized amylouronic acid as primers in an ammonium buffer containing $MgCl_2$ to produce amylosamine and amphoteric block polysaccharides, composed of α(1→4)-linked GlcN and GlcA chains, respectively.

An attempt was made to remove Pi as a precipitate from the thermostable GP-catalyzed glucosaminylation system. This was achieved by conducting the reaction in an ammonium buffer (0.5 M, pH 8.6) containing magnesium chloride ($MgCl_2$), as it has been reported that Pi forms an insoluble salt with ammonium and magnesium ions (56). The thermostable GP-catalyzed enzymatic polymerization of GlcN-1-P as the monomer with the Glc_3 primer (30:1) successfully occurred in the buffer system at 40 °C for 7 days. This produced a non-natural aminopolysaccharide composed of α(1→4)-linked GlcN units with a DP of ~20 GlcN units ($M_n = 3760$) (Figure 10a) (57). The new amylose analog polysaccharide corresponded to a chitosan stereoisomer, which has been named "amylosamine." This product could be converted into a chitin stereoisomer by *N*-acetylation using acetic anhydride. Amylosamine showed water solubility because it had not formed a regular higher-order assembly in water, unlike amylose. However, the cationic amylosamine formed a double helix with anionic amylouronic acid, α(1→4)-linked GlcA polymer—another amylose analog polysaccharide—by electrostatic interaction between amino and carboxylate groups (58). Reductive amination of amylosamine formed hierarchically controlled assemblies to obtain nanoparticles, microaggregates, and macrohydrogels, depending on reaction conditions (59). The desired reductive amination of amylosamine was carried out in the presence of sodium cyanoborohydride as a reductant in 0.1 mol/L acetic acid aqueous solution at 60 °C for 1 h. When the lower feed ratios of the reductant/reducing end of amylosamine were employed for the reaction, nanoparticles formed. By increasing the feed ratios of the reductant/reducing end, the larger number of α(1→4)-linked GlcN chains assembled to further form the water-insoluble microaggregates and macrohydrogels.

The thermostable GP-catalyzed enzymatic copolymerizations of Glc-1-P with GlcN-1-P, and with Man-1-P as the comonomer using the Glc_3 primer, have successfully taken place under conditions for removal of Pi in an ammonia/$MgCl_2$ buffer. This obtained non-natural glucosaminoglucan composed of Glc/GlcN units and mannoglucan composed of Glc/Man units (Figure 11) (*60, 61*). The thermostable GP-catalyzed enzymatic polymerization of GlcN-1-P using maltooligosaccharide-functionalized amylouronic acid—amylouronic acid with a short α(1→4)-linked Glc chain at the nonreducing end—as the primer in the same operation produced an amylose analog amphoteric block polysaccharide composed of α(1→4)-linked GlcN and GlcA chains (Figure 10b) (*62*). The block polysaccharide showed a pH-responsive property.

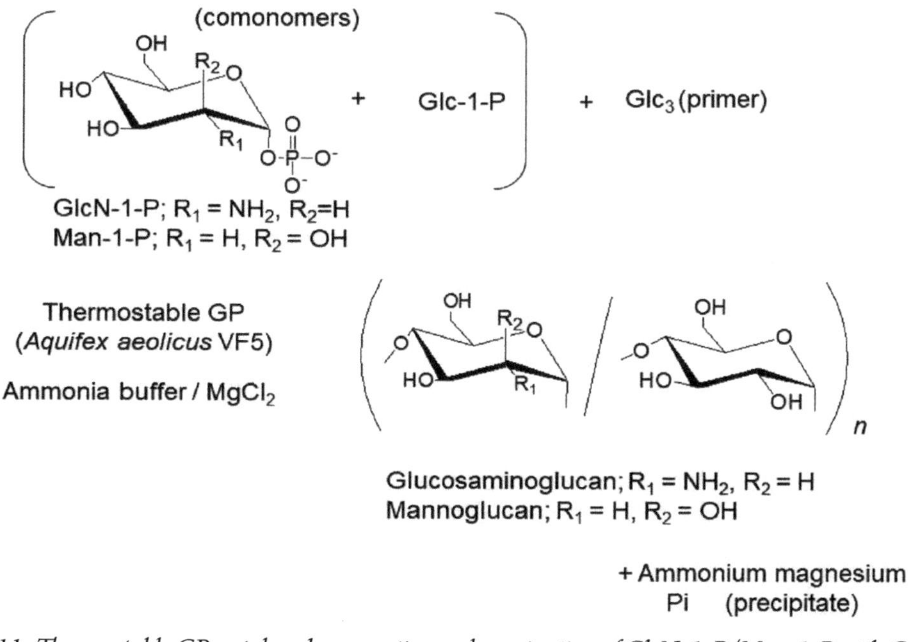

Figure 11. Thermostable GP-catalyzed enzymatic copolymerization of GlcN-1-P/Man-1-P with Glc-1-P, using Glc_3 as a primer in an ammonium buffer containing $MgCl_2$ to produce non-natural glucosaminoglucan and mannoglucan.

Conclusions

In this chapter, the author presented a theory that GP-catalyzed enzymatic reactions are a useful tool to synthesize a variety of non-natural polysaccharides. The enzymatic method has precisely produced basic (cationic), acidic (anionic), and amphoteric polysaccharides. Based on the viewpoint that non-natural materials can be expected to exhibit new functions and properties different from native ones, the produced non-natural polysaccharides have been representatively used as cross-linking, pH-responsive, gelling, and self-assembling materials. The present polysaccharides have the potential to be practically employed as new functional materials in further applications, such as in biomedicine, tissue engineering, and sustainable fields. As enzymatic reactions are the environmentally benign approach, the present reactions will contribute to advancing green aspects in organic, polymer, and materials chemistry in the future. As substrate analogs for the present enzymatic approaches have been prepared by organic synthetic methods, more active collaboration with synthetic organic researchers will lead to further progress in the research on enzymatically synthesized non-natural polysaccharides. Based on the viewpoint of green chemistry, alternatively, the development of new enzymatic methods for the synthesis of substrate analogs should be investigated as possibly more eco-friendly in the present research field.

References

1. Schuerch, C. Polysaccharides. In *Encyclopedia of Polymer Science and Engineering*, 2nd ed.; Mark, H. F., Bilkales, N., Overberger, C. G., Eds.; John Wiley & Sons: New York, 1986; Vol. 13, pp 87–162.
2. Yalpani, M. *Polysaccharides: Syntheses, Modifications, and Structure/Property Relations*; Elsevier: Amsterdam, New York, 1988.
3. Kadokawa, J. Precision polysaccharide synthesis catalyzed by enzymes. *Chem. Rev.* **2011**, *111*, 4308–4345.
4. Shoda, S.; Izumi, R.; Fujita, M. Green process in glycotechnology. *Bull. Chem. Soc. Jpn.* **2003**, *76*, 1–13.
5. Kobayashi, S.; Makino, A. Enzymatic polymer synthesis: An opportunity for green polymer chemistry. *Chem. Rev.* **2009**, *109*, 5288–5353.
6. Shoda, S.; Uyama, H.; Kadokawa, J.; Kimura, S.; Kobayashi, S. Enzymes as green catalysts for precision macromolecular synthesis. *Chem. Rev.* **2016**, *116*, 2307–2413.
7. Shoda, S.; Noguchi, M.; Li, G.; Kimura, S. Synthesis of Polysaccharides I: Hydrolase as Catalyst. In *Enzymatic Polymerization towards Green Polymer Chemistry*; Kohayasni, S., Uyama, H., Kadokawa, J., Eds.; Springer: Heidelberg, 2019; pp 15–46.
8. Loos, K.; Kadokawa, J. Synthesis of Polysaccharides II: Phosphorylase as Catalyst. In *Enzymatic Polymerization towards Green Polymer Chemistry*; Kobayashi, S., Uyama, H., Kadokawa, J., Eds.; Springer: Heidelberg, 2019; pp 47–87.
9. Kimura, S.; Iwata, T. Synthesis of Polysaccharides III: Sucrase as Catalyst. In *Enzymatic Polymerization towards Green Polymer Chemistry*; Kobayashi, S., Uyama, H., Kadokawa, J., Eds.; Springer: Heidelberg, 2019; pp 89–104.
10. Kobayashi, S.; Kashiwa, K.; Kawasaki, T.; Shoda, S. Novel method for polysaccharide synthesis using an enzyme - The 1st in vitro synthesis of cellulose via a nonbiosyntehtic path utilizing cellulase as catalyst. *J. Am. Chem. Soc.* **1991**, *113*, 3079–3084.
11. Okamoto, E.; Kiyosada, T.; Shoda, S.; Kobayashi, S. Synthesis of Alternatingly 6-*O*-Methylated Cellulose via Enzymatic Polymerization of a Substituted Cellobiosyl Fluoride Monomer Catalyzed by Cellulase. *Cellulose* **1997**, *4*, 161–172.
12. Izumi, R.; Suzuki, Y.; Shimizu, Y.; Fujita, M.; Ishihara, M.; Noguchi, M.; Kobayashi, A.; Shoda, S. Synthesis of Artificial Oligosaccharides by Polycondensation of 2'-*O*-Methyl Cellobiosyl Fluoride and Mannosyl-Glucosyl Fluoride Catalyzed by Cellulase. In *Interfacial Researches in Fundamental and Material Sciences of Oligo- and Polysaccharides*; Kadokawa, J., Ed.; Transworld Research Network: Trivandrum, India, 2009; pp 45–67.
13. Seibel, J.; Jordening, H. J.; Buchholz, K. Glycosylation with activated sugars using glycosyltransferases and transglycosidases. *Biocatal. Biotransform.* **2006**, *24*, 311–342.

14. O'Neill, E. C.; Field, R. A. Enzymatic synthesis using glycoside phosphorylases. *Carbohydr. Res.* **2015**, *403*, 23–37.
15. Kadokawa, J. Precision synthesis of functional polysaccharide materials by phosphorylase-catalyzed enzymatic reactions. *Polymers* **2016**, *8*, 138.
16. Kadokawa, J. α-Glucan phosphorylase: A useful catalyst for precision enzymatic synthesis of oligo- and polysaccharides. *Curr. Org. Chem.* **2017**, *21*, 1192–1204.
17. Ziegast, G.; Pfannemüller, B. Linear and star-shaped hybrid polymers. Phosphorolytic syntheses with di-functional, oligo-functional and multifunctional primers. *Carbohydr. Res.* **1987**, *160*, 185–204.
18. Fujii, K.; Takata, H.; Yanase, M.; Terada, Y.; Ohdan, K.; Takaha, T.; Okada, S.; Kuriki, T. Bioengineering and application of novel glucose polymers. *Biocatal. Biotransform.* **2003**, *21*, 167–172.
19. Yanase, M.; Takaha, T.; Kuriki, T. a-Glucan phosphorylase and its use in carbohydrate engineering. *J. Sci. Food Agric.* **2006**, *86*, 1631–1635.
20. Ohdan, K.; Fujii, K.; Yanase, M.; Takaha, T.; Kuriki, T. Enzymatic synthesis of amylose. *Biocatal. Biotransform.* **2006**, *24*, 77–81.
21. Kadokawa, J. Facile synthesis of unnatural oligosaccharides by phosphorylase-catalyzed enzymatic glycosylations using new glycosyl donors. In *Oligosaccharides: Sources, Properties and Applications*; Gordon, N. S., Ed.; Nova Science Publishers, Inc.: Hauppauge, NY, 2011; pp 269–281.
22. Kadokawa, J. Synthesis of non-natural oligosaccharides by a-glucan phosphorylase-catalyzed enzymatic glycosylations using analogue substrates of α-D-glucose 1-phosphate. *Trends Glycosci. Glycotechnol.* **2013**, *25*, 57–69.
23. Kadoakwa, J. Enzymatic synthesis of non-natural oligo- and polysaccharides by phosphorylase-catalyzed glycosylations using analogue substrates. In *Green Polymer Chemistry: Biobased Materials and Biocatalysis*; Cheng, H. N., Gross, R. A., Smith, P. B., Eds.; ACS Symposium Series 1192; American Chemical Society: Washington, DC, 2015; pp 87–99.
24. Kadokawa, J. I. α-Glucan phosphorylase-catalyzed enzymatic reactions using analog substrates to synthesize non-natural oligo-and polysaccharides. *Catalysts* **2018**, *8*, 473.
25. Kitaoka, M.; Hayashi, K. Carbohydrate-processing phosphorolytic enzymes. *Trends Glycosci. Glycotechnol.* **2002**, *14*, 35–50.
26. Nakai, H.; Kitaoka, M.; Svensson, B.; Ohtsubo, K. Recent development of phosphorylases possessing large potential for oligosaccharide synthesis. *Curr. Opin. Chem. Biol.* **2013**, *17*, 301–309.
27. Puchart, V. Glycoside phosphorylases: Structure, catalytic properties and biotechnological potential. *Biotechnol. Adv.* **2015**, *33*, 261–276.
28. Ubiparip, Z.; Beerens, K.; Franceus, J.; Vercauteren, R.; Desmet, T. Thermostable a-glucan phosphorylases: characteristics and industrial applications. *Appl. Microbiol. Biotechnol.* **2018**, *102*, 8187–8202.
29. Boeck, B.; Schinzel, R. Purification and characterisation of an α-glucan phosphorylase from the thermophilic bacterium *Thermus thermophilus*. *Eur. J. Biochem.* **1996**, *239*, 150–155.
30. Takaha, T.; Yanase, M.; Takata, H.; Okada, S. Structure and properties of *Thermus aquaticus* α-glucan phosphorylase expressed in *Escherichichia coli*. *J. Appl. Glycosci.* **2001**, *48*, 71–78.

31. Yanase, M.; Takata, H.; Fujii, K.; Takaha, T.; Kuriki, T. Cumulative effect of amino acid replacements results in enhanced thermostability of potato type L α-glucan phosphorylase. *Appl. Environ. Microbiol.* **2005**, *71*, 5433–5439.
32. Imberty, A.; Chanzy, H.; Perez, S.; Buleon, A.; Tran, V. The double-helical nature of the crystalline part of A-starch. *J. Mol. Biol.* **1988**, *201*, 365–378.
33. Imberty, A.; Perez, S. A revisit to the three-dimensional structure of B-type starch. *Biopolymers* **1988**, *27*, 1205–1221.
34. Kitamura, S. Starch polymers, natural and synthetic. In *The Polymeric Materials Encyclopedia, Synthesis, Properties and Applications*; Salamone, C., Ed.; CRC Press: New York, 1996; Vol. 10, pp 7915–7922.
35. Kadokawa, J. Synthesis of amylose-grafted polysaccharide materials by phosphorylase-catalyzed enzymatic polymerization. In *Biobased Monomers, Polymers, and Materials*; Smith, P. B., Gross, R. A., Eds.; ACS Symposium Series 1105; American Chemical Society: Washington, DC, 2012; pp 237–255.
36. Kadokawa, J. Synthesis of new polysaccharide materials by phosphorylase-catalyzed enzymatic α-glycosylations using polymeric glycosyl acceptors. In *Green Polymer Chemistry: Biocatalysis and Materials II*; Cheng, H. N., Gross, R. A., Smith, P. B., Eds.; ACS Symposium Series 1144; American Chemical Society: Washington, DC, 2013; pp 141–161.
37. Kadokawa, J. Chemoenzymatic synthesis of functional amylosic materials. *Pure Appl. Chem.* **2014**, *86*, 701–709.
38. Nishimura, T.; Akiyoshi, K. Amylose engineering: phosphorylase-catalyzed polymerization of functional saccharide primers for glycobiomaterials. *Wiley Interdiscip. Rev. Nanomed. Nanobiotechnol.* **2016**, *9*, e1423.
39. Kadokawa, J. Enzymatic preparation of functional polysaccharide hydrogels by phosphorylase catalysis. *Pure Appl. Chem.* **2018**, *90*, 1045–1054.
40. Izawa, H.; Nawaji, M.; Kaneko, Y.; Kadokawa, J. Preparation of glycogen-based polysaccharide materials by phosphorylase-catalyzed chain elongation of glycogen. *Macromol. Biosci.* **2009**, *9*, 1098–1104.
41. Evers, B.; Thiem, J. Further syntheses employing phosphorylase. *Bioorg. Med. Chem.* **1997**, *5*, 857–863.
42. Nawaji, M.; Izawa, H.; Kaneko, Y.; Kadokawa, J. Enzymatic synthesis of α-D-xylosylated maltooligosaccharides by phosphorylase-catalyzed xylosylation. *J. Carbohydr. Chem.* **2008**, *27*, 214–222.
43. Nawaji, M.; Izawa, H.; Kaneko, Y.; Kadokawa, J. Enzymatic α-glucosaminylation of maltooligosaccharides catalyzed by phosphorylase. *Carbohydr. Res.* **2008**, *343*, 2692–2696.
44. Kawazoe, S.; Izawa, H.; Nawaji, M.; Kaneko, Y.; Kadokawa, J. Phosphorylase-catalyzed *N*-formyl-α-glucosaminylation of maltooligosaccharides. *Carbohydr. Res.* **2010**, *345*, 631–636.
45. Umegatani, Y.; Izawa, H.; Nawaji, M.; Yamamoto, K.; Kubo, A.; Yanase, M.; Takaha, T.; Kadokawa, J. Enzymatic α-glucuronylation of maltooligosaccharides using α-glucuronic acid 1-phosphate as glycosyl donor catalyzed by a thermostable phosphorylase from *Aquifex aeolicus* vf5. *Carbohydr. Res.* **2012**, *350*, 81–85.

46. Kadokawa, J.; Chigita, H.; Yamamoto, K. Chemoenzymatic synthesis of carboxylate-terminated maltooligosaccharides and their use for cross-linking of chitin. *Int. J. Biol. Macromol.* **2020**, *159*, 510–516.
47. Kurita, K. Chitin and chitosan: Functional biopolymers from marine crustaceans. *Marine Biotechnol.* **2006**, *8*, 203–226.
48. Takata, Y.; Shimohigoshi, R.; Yamamoto, K.; Kadokawa, J. Enzymatic synthesis of dendritic amphoteric α-glucans by thermostable phosphorylase catalysis. *Macromol. Biosci.* **2014**, *14*, 1437–1443.
49. Takata, Y.; Yamamoto, K.; Kadokawa, J. Preparation of pH-responsive amphoteric glycogen hydrogels by α-glucan phosphorylase-catalyzed successive enzymatic reactions. *Macromol. Chem. Phys.* **2015**, *216*, 1415–1420.
50. Takemoto, Y.; Izawa, H.; Umegatani, Y.; Yamamoto, K.; Kubo, A.; Yanase, M.; Takaha, T.; Kadokawa, J. Synthesis of highly branched anionic α-glucans by thermostable phosphorylase-catalyzed α-glucuronylation. *Carbohydr. Res.* **2013**, *366*, 38–44.
51. Klein, H. W.; Palm, D.; Helmreich, E. J. M. General acid-base catalysis of α-glucan phosphorylases: Stereospecific glucosyl transfer from D-glucal is a pyridoxal 5′-phosphate and orthophosphate (arsenate) dependent reaction. *Biochemistry* **1982**, *21*, 6675–6684.
52. Evers, B.; Mischnick, P.; Thiem, J. Synthesis of 2-deoxy-α-D-arabino-hexopyranosyl phosphate and 2-deoxy-maltooligosaccharides with phosphorylase. *Carbohydr. Res.* **1994**, *262*, 335–341.
53. Evers, B.; Thiem, J. Synthesis of 2-Deoxy-maltooligosaccharides with Phosphorylase and Their Degradation with Amylases. *Starch - Stärke* **1995**, *47*, 434–439.
54. Uto, T.; Nakamura, S.; Yamamoto, K.; Kadokawa, J. Evaluation of artificial crystalline structure from amylose analog polysaccharide without hydroxy groups at C-2 position. *Carbohydr. Polym.* **2020**, *240*, 116347.
55. Shimohigoshi, R.; Takemoto, Y.; Yamamoto, K.; Kadokawa, J. Thermostable a-glucan phosphorylase-catalyzed successive a-mannosylations. *Chem. Lett.* **2013**, *42*, 822–824.
56. Borgerding, J. Phosphate deposits in digestion systems. *J. Water Pollut. Control Fed.* **1972**, *44*, 813–819.
57. Kadokawa, J.; Shimohigoshi, R.; Yamashita, K.; Yamamoto, K. Synthesis of chitin and chitosan stereoisomers by thermostable α-glucan phosphorylase-catalyzed enzymatic polymerization of α-D-glucosamine 1-phosphate. *Org. Bimol. Chem.* **2015**, *13*, 4336–4343.
58. Yui, T.; Uto, T.; Nakauchida, T.; Yamamoto, K.; Kadokawa, J. Double helix formation from non-natural amylose analog polysaccharides. *Carbohydr. Polym.* **2018**, *189*, 184–189.
59. Nakauchida, T.; Yamamoto, K.; Kadokawa, J. Hierarchically controlled assemblies from amylose analog aminopolysaccharides by reductive amination: From nano- to macrostructures. *J. Appl. Polym. Sci.* **2018**, *135*, 45890.
60. Yamashita, K.; Yamamoto, K.; Kadoakwa, J. Synthesis of non-natural heteroaminopolysaccharides by a-glucan phosphorylase-catalyzed enzymatic copolymerization: α(1->4)-linked glucosaminoglucans. *Biomacromolecules* **2015**, *16*, 3989–3994.
61. Baba, R.; Yamamoto, K.; Kadokawa, J. Synthesis of α(1-->4)-linked non-natural mannoglucans by alpha-glucan phosphorylase-catalyzed enzymatic copolymerization. *Carbohydr. Polym.* **2016**, *151*, 1034–1039.

62. Nakauchida, T.; Takata, Y.; Yamamoto, K.; Kadokawa, J. Chemoenzymatic synthesis and pH-responsive properties of amphoteric block polysaccharides. *Org. Biomol. Chem.* **2016**, *14*, 6449–6456.

Chapter 3

Enzyme Catalyzed Hydrolysis of Synthetic Polymers

Priya Katyal[1] and Jin Kim Montclare*[,1,2,3,4]

[1]Department of Chemical and Biomolecular Engineering, New York University, Tandon School of Engineering, Brooklyn, New York 11201, United States

[2]Department of Radiology, New York University Langone Health, New York, New York 10016, United States

[3]Department of Biomaterials, New York University College of Dentistry, New York, New York 10010, United States

[4]Department of Chemistry, New York University, New York, New York 10003, United States

***Email: montclare@nyu.edu**

Traditional plastics pose a grave threat to the environment with several hundred million tons of plastics being accumulated as waste. With conventional approaches requiring extreme temperatures and harsh conditions, degradation or recyclability achieved via enzymatic methods offer a meaningful alternative. In this chapter, we discuss enzymes that have been utilized for polymer degradation. We highlighted strategies to improve stability and catalytic efficiencies of these enzymes via protein engineering or anchoring peptides/domains to the catalytic domain or using additives that can enhance the binding of these enzymes to polymeric surfaces.

Introduction

Plastics are considered engineering marvels owing to their extraordinary performance and unparalleled versatility (*1*). They exhibit a range of unique properties including but not limited to their high strength and toughness, durability, and are light-weight and water-resistant (*2*). If a plastic undergoes an irreversible change in its structure and acquire its shape once upon molding, they are referred as thermosets. Examples include polyethylene, polypropylene (PP), polyvinyl chloride (PVC), polystyrene (PS) and polyethylene terephthalate (PET). However, if a plastic is capable of being repeatedly molded by heating and cooling, such polymers are referred to as thermoplastics, for example, polyurethanes (PUs) and polyesters (*3*). While thermoplastics account for 90% of commodity plastics, they are difficult to degrade and can impose significant environmental risks (*1*, *3*).

Since there is an exponential growth in the use of plastics with industries continuously pushing boundaries and achieving far-reaching applications, significant efforts are being directed towards plastic waste management and development of remediation technologies and recycling systems (*4*).

While traditional approaches like incineration and pyrolysis are energy intensive processes, degradation or recyclability achieved via chemical or biological methods offers meaningful alternatives with economic prospects. Although degradation via chemical means is continuously improving, for example, by employing ionic liquids, superacids or supercritical fluids, the conditions required to conduct such reactions need powerful reagents, such as strong acids or alkalis and high temperatures (3, 5).

Table 1. Bacterial and Fungal Enzymes That Are Used for Polymer Hydrolysis

Origin	*Name*	*Source*	*Reference*
Bacterial	BTA-1/BTA-2 Tfu_0882/Tfu_0883 Tf_Cut1/Tf_Cut2	*Thermobifida fusca*	(23, 24)
	Thc_Cut1/Thc_Cut2	*Thermobifida cellulosilytica*	(25)
	LCC	*Metagenome from leaf-branch compost*	(26)
	Est119	*Thermobifida alba*	(27, 28)
	Tcur1278/Tcur0390	*Thermomonospora curvata*	(29)
	Thh_Est	*Thermobifida halotolerans*	(14)
	Cut190/AHK190	*Saccharomonospora viridis*	(30, 31)
	PETase MHETase	*Ideonella sakaiensis*	(11, 12)
	Polyamidase	*Nocardia farcinica*	(32)
	Laccase	*Rhodococcus ruber*	(17)
	Peroxidase	*Azotobacter beijerinckii HM121*	(18)
Fungal	Lipase	*Candida antarctica*	(33)
	FsC	*Fusarium solani pisi*	(23, 34)
	AoC	*Aspergillus oryzae*	(35)
	AfC	*Aspergillus fumigatus*	(13)
	AbC	*Alternaria brassicicola*	(13)
	HiC	*Humicola insolens*	(36)
	Polyesterase	*Penicillium citrinum*	(37)
	Polyesterase	*Pseudomonas mendocina*	(36)

The alternative approach of using living microorganisms is much safer, requiring milder conditions to degrade plastics from the environment. Few bacteria and fungi have been identified that can breakdown and metabolize plastics (6). Recently, *Ideonella sakaiensis* has been identified as a bacterium that secretes two enzymes, PETase and mono(2-hydroxyethyl) terephthalate degrading enzyme (MHETase), capable of digesting PET (7). Microfungus *Fusarium solani*, isolated from plant surfaces and soil samples, is also known for its PET modifying abilities (8). However, this process of degradation is slow, with some polymers, such as highly crystalline plastics, not amenable to support microbial growth (6, 9). While it may be tedious to identify new microorganisms that directly interact with the polymeric surfaces, biodegradation using bacterial and fungal enzymes have generated widespread interest in plastics remediation and recycling efforts.

Enzymes including lipases, cutinases, PETases, carboxylesterases, and certain amidases have been investigated for plastic degradation (Table 1) (*10*). These enzymes are active against a wide range of polymers ranging from PET (*11*), MHET (*12*), poly(ε-caprolactone) (PCL) (*13*), polylactic acid (PLA) (*14*), polyhydroxyalkanoates (PHAs) (*15*), poly(1,4-butylene adipate) (PBA) (*16*), polypropylene (PE) (*17*), PS (*18*), PUs (*19*) and polyamides (*20*).

There are many examples demonstrating the utility of these enzymes in hydrolyzing several other polymers, and these have been reviewed extensively elsewhere (*10, 21, 22*). The focus here is to provide a brief overview of some of the polymer degrading enzymes, primarily for PET based systems. In this chapter, we highlight studies that compare different polymer active enzymes either from the structural or functional perspectives. In addition, we review approaches utilized to improve the binding of these enzymes to polymeric surfaces.

Structural and Functional analysis of Polymer Degrading Enzymes

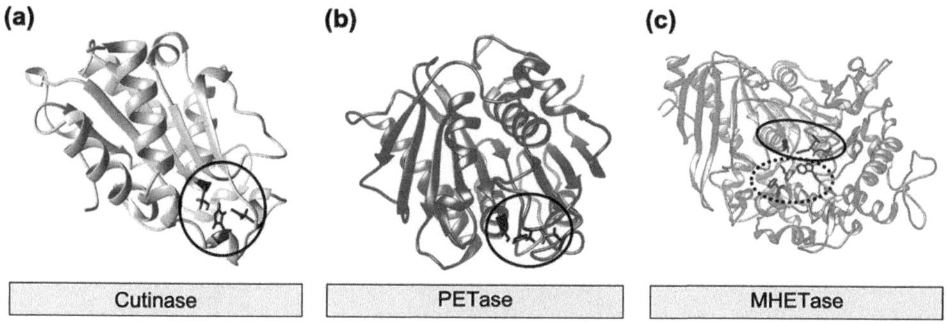

Figure 1. X-ray crystal structures of (a) Cutinase from Fusarium solani (PDB: 1CEX) (38), (b) PETase from Ideonella sakaiensis (PDB: 6EQE) (11) and (c) MHETase from Ideonella sakaiensis (PDB: 6QG9) (12). The catalytic triad comprising of serine, histidine, and aspartic acid is encircled with the solid line while the residues from the lid domain (Phe415, Leu254 and Trp397) encircled with the dotted line.

Fungal and Bacterial Cutinases

Cutinase, an enzyme produced by various bacteria and fungus (Table 1) (*23–37*), breaks down cutin, a major component of plant cuticle that prevents invasion of pathogens (*24*). It is a versatile enzyme used in textile, food and chemical industries (*34*). It is a serine hydrolase with an α/β-fold. The most well studied cutinase is derived from the fungus, *Fusarium solani pisi* (referred as FsC) (*34, 38*). It consists of five β-sheets with five α-helices, with its active site composed of a catalytic triad of Ser120, Asp175 and His188 (Figure 1) (*39*). While FsC favors short chain substrates, cutinase derived from another fungus, *Aspergillus oryzae* (AoC) prefers binding to longer chain substrates (*35*). A continuous groove in AoC, located across the catalytic site, is responsible for the substrate specificity. AoC is faster in degrading PCL films with 87% degradation achieved within 6 hours, compared to 30% degradation with FsC under similar reaction conditions (*35*).

Baker et al. have compared the thermal stability and activity of five different fungal cutinases derived from *Alternaria brassicicola* (AbC), *Aspergillus fumigatus* (AfC), *Humicola insolens* (HiC), AoC and FsC (*13*). Of all the cutinases, HiC is highly stable and exhibits the highest activity under all test conditions of pH 3, 5 and 8. At pH 8 and 40°C, PCL films are fully degraded within 3 hours of treatment with HiC. Under similar conditions, a longer incubation time of 6 hours is required for AoC and AfC to achieve complete degradation (*13*). Both FsC and AbC demonstrate partial degradation of PCL films even after 6 hours of incubation. Although a similar trend is noted at pH 5, complete degradation is only achieved after 6 hours of treatment with AbC showing no detectable activity (*13*). At pH 3, the residues near the active site (referred as 'crowning region') of HiC and AoC show near-neutral charge, enabling these two cutinases to maintain their activity even at lower pH. AbC, AfC and FsC, on the other hand, acquire a high positive charge, rendering the enzymes inactive (*13*).

HiC has also been compared to FsC and cutinase from *Pseudomonas mendocina* (PmC) for its ability to degrade low (7%) and higher (35%) crystallinity PET films (*36*). Of the three cutinases, HiC shows remarkable thermostability and exhibits 30-fold increase in activity upon increasing the temperature from 30 to 80°C (Figure 2a). Scanning electron microscopy (SEM) micrograph shows increased surface roughness of low-crystalline PET films (lcPET) with 95% weight loss within 96 hours of exposure of HiC (Figure 2b) (*36*). The crystallinity of the PET films influences the enzymatic activity with higher crystalline films impacting the activity of all the three cutinases. A 25-, 10- and 6- fold decrease in activities are observed for HiC, PmC and FsC on higher crystalline films, respectively (*36*).

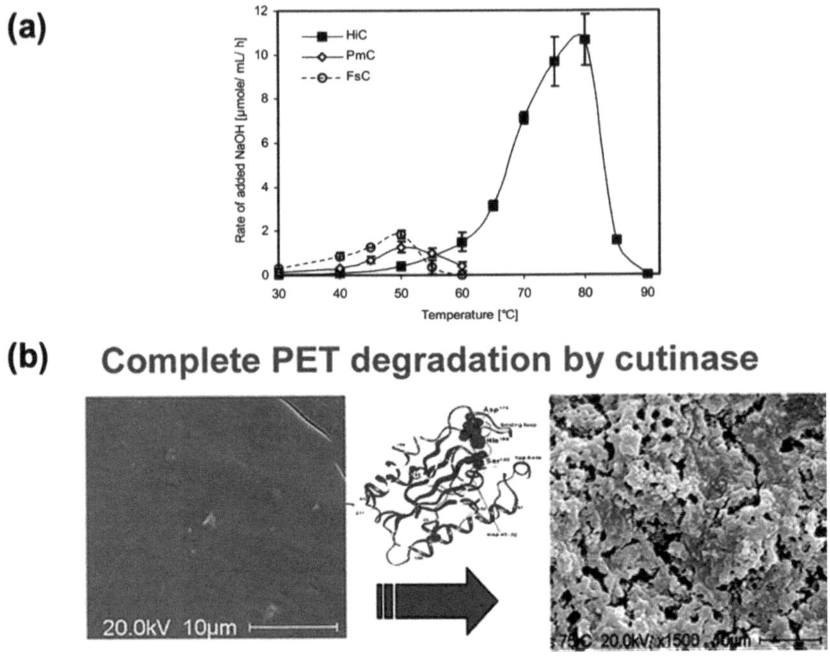

Figure 2. (a) Activity profile of HiC, PmC and FsC as a function of temperature. Low-crystalline PET films were incubated with enzymes at pH 8 and the activities were measured using a pH STAT titration system. (b) SEM micrographs showing HiC mediated hydrolysis of lcPET films (left). An increase in surface roughness with 95% weight loss was observed after 96 hours of enzyme exposure (right). Adapted with permission from reference (36). Copyright 2009 American Chemical Society.

Cutinases from bacteria, mainly from *Thermobifida* species have also been extensively studied (*22–24*). Although cutinases from *Thermobifida fusca* have long been known as cutin-degrading enzymes (*40*), Muller et al. for the first time have reported its ability to degrade commercial PET bottles (*23*). This *T. fusca* hydrolase (TfH) is reported to have a specific activity of 0.05 mg/h/mg of the enzyme at pH 7 and 55°C (*26*). Other cutinases obtained from *T. fusca* are BTA-1, BTA 2, Tfu_0882 and Tfu_0883 (*24, 41, 42*). Despite having high homology (93%), Thc_Cut1 and Thc_Cut2, two of the PET degrading cutinases from *T. cellulosilytica*, hydrolyze PET differently. While Thc_Cut1 possesses higher activity and releases a higher amount of terephthalic acid (TPA), Thc_Cut2 predominantly hydrolyzes PET to MHET (*25*). Another enzyme, having 57.4% sequence identity to cutinase from *T. fusca* (*24*), have been isolated from a metagenomic study of leaf-branch compost (*26*). The enzyme, referred as Leaf-branch compost cutinase (LCC), can efficiently degrade PET and PCL films at 50°C and pH 8 (*26*). While the specific PET degrading activity was found to be 12 mg/h/mg of enzyme, LCC has been more effective in degrading PCL films with a specific activity of 300 mg/h/mg of enzyme (*26*). Interestingly, LCC is capable of completely degrading PET to its monomeric units, ethylene glycol (EG) and TPA.

Tournier et al. compared five different enzymes, BTA1, BTA2, LCC, FsC and PETase, for their ability to depolymerize amorphous PET (*43*). Of the five enzymes, LCC exhibited the highest thermostability with a melting temperature of 84.7°C and was also found to be 33 times more efficient than all the other enzymes tested. The catalytic efficiency of LCC was slightly reduced when bottle grade PET was used. To improve the thermal stability and catalytic efficiency of LCC, computer-aided design was utilized. While variants Phe243Ile and Phe243Trp improved catalytic activity, variants with Thr96Met, Tyr127Gly, Asn246Asp or Asn246Met mutations improved thermostability relative to wild-type LCC. Introduction of a disulfide bridge between residues 238 and 283 (Asp238Cys/Ser283Cys), increased the melting temperature by 9.8°C; however, it adversely impacted its activity, with a decrease of 28% compared to wild-type LCC (*43*). To further generate the variants of LCC that exhibit both higher thermostability and improved activity, ICC and WCC variants were constructed (Table 2) (*43*). ICC and WCC revealed 122% and 98% increase in activity accompanied by a 6.2 and 10.1°C rise in their T_m, respectively. Addition of Thr96Met, Tyr127Gly, Asn246Asp or Asn246Met mutations, further improved the thermostability with variants ICCG, ICCM, WCCG and WCCM (Table 2), exhibiting improved melting temperatures ranging from +9.3 °C to +13.4 °C. Both WCCG and ICCG showed 90% depolymerization of PET, particularly the post-consumer colored flakes PET waste, in 10.5 hours and 9.3 hours, respectively (*43*).

Table 2. List of LCC Variants with Higher Thermostability and Improved Activity (*43*)

Nomenclature	*Mutations*
ICC	Phe243Ile/Asp238Cys/Ser283Cys
WCC	Phe243Trp/Asp238Cys/Ser283Cys
ICCG	Phe243Ile/Asp238Cys/Ser283Cys/Tyr127Gly
ICCM	Phe243Ile/Asp238Cys/Ser283Cys/Asn246Met
WCCG	Phe243Trp/Aps238Cys/Ser283Cys/Tyr127Gly
WCCM	Phe243Trp/Asp238Cys/Ser283Cys/Asn246Met

PETase and MHETase

PETase and MHETase, two of the enzymes isolated from *I. sakaiensis*, have recently been identified for their PET/MHET hydrolyzing activity (7). While PETase digests PET to MHET with trace amounts of TPA and bis(2-hydroxyethyl) terephthalate (BHET), MHETase converts MHET units to TPA and EG (Figure 3a) (7, *11*). Similar to cutinases, PETase belongs to the α/β hydrolase superfamily with PETase sharing 52% close identity to the cutinase derived from *T. fusca*. The core of the PETase consists of eight β-strands and six α-helices (*11*). Unlike cutinase that possess small patches of acidic and basic residues on the surface, PETase has a highly basic surface (*44*). The active site in PETase is three-fold wider than the cutinase; Ser238 of PETase (corresponding to Phe209 in cutinase) is responsible for broadening the active site cleft (Figure 3b and 3c) (*11*). The catalytic triad of PETase comprising of Ser160, Asp206, and His237 is more solvent accessible than thermophilic cutinases (Figure 3d and 3e). Residues Trp159 and Trp185 that are adjacent to the conserved nucleophile Ser160, confer substrate selectivity to PETases with Trp159, present exclusively in PETase, responsible for its enhanced PET-hydrolyzing activity (*11*).

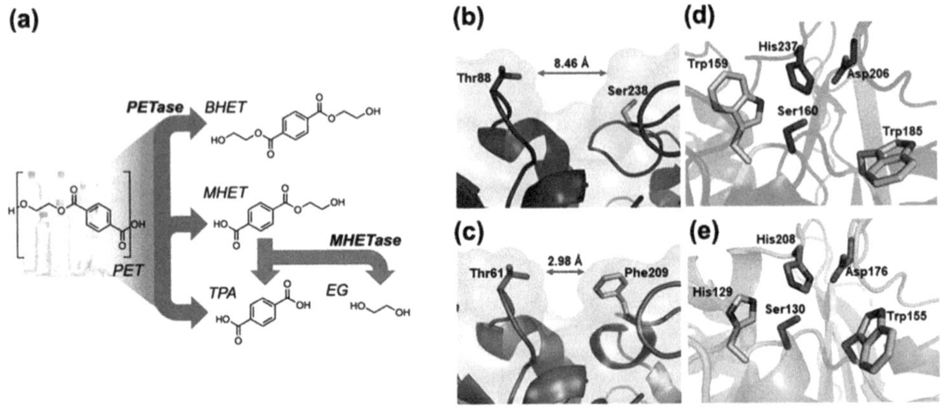

Figure 3. (a) PETase catalyzes the hydrolysis of PET to BHET, MHET, and TPA. MHET is further hydrolyzed by MHETase to its monomers, TPA and EG. Comparison of active site cleft (b) and (c) and catalytic triad (d) and (e) of PETase and cutinase, respectively. Trp159 is present exclusively in PETase. Adapted with permission from reference (11). Copyright 2018 National Academy of Sciences.

At 30 °C and pH 7, the activity of PETase against low-crystalline (1.9%) PET films was found to be 120, 5.5 and 88 times higher compared to purified TfH, LCC and FsC, respectively (7). To further increase the efficiency of PETase against crystalline PET and a biobased PET replacement, poly(ethylene furanoate) (PEF), Austin et al. developed a double mutant, Ser238Phe/Trp159His that closely resembles the active site of cutinase from *T. fusca* (*11*). When incubated with PET and PEF films (Figure 4), the double mutant exhibited higher surface erosion than wild-type PETase, with larger pits obtained for PEF films., Although having excellent biodegradability, aliphatic polyesters such as PLA and polybutylene succinate (PBS), were not digested by PETase, suggesting that the enzyme is more effective against aromatic or semi-aromatic polyesters (*11*).

Compared to PETase, MHETase contains an additional lid domain that is responsible for its substrate specificity (*12*). The catalytic triad of MHETase comprises of Ser225, His528, and Asp492 (Figure 1) with residues Gly132 and Glu226 forming the oxyanion hole. The residues Phe415, Leu254 and Trp397 of the lid domain and residues Phe495, Gly132 and Ala494 of the hydrolase domain are responsible for tight binding of MHETase with its nonhydrolyzable substrate, mono-

(2-hydroxyethyl) terephthalamide (MHETA) (*12*). MHETase or its variant (Arg411Lys/Ser416Ala/Phe424Ile) is inactive when incubated with amorphous PET films (Figure 5a). Pre-treatment of films with PETase generates carboxyl groups indicated by green fluorescence. With MHETase and its variant capable of hydrolyzing PET films with termini, a decrease in carboxyl groups is noted. The effect of MHETase variant was found to be better than wild-type MHETase (*45*). Recently, Sagong et al. have identified that MHETase is also capable of hydrolyzing BHET and a synthetic PET polymer, referred to as PET_5 that is comprised of five repeats of MHET with EG at both termini (Figure 5b) (*45*). Although the activity of MHETase is lower for BHET and PET (5) than MHET, it can still function as BHETase and exo-PETase (*45*). Given the inhibitory effect of intermediate hydrolysis products MHET and BHET on cutinase activity (*46*), MHETase can play a critical role in realizing complete enzymatic degradation of PET, resulting in the efficient plastic recycling process.

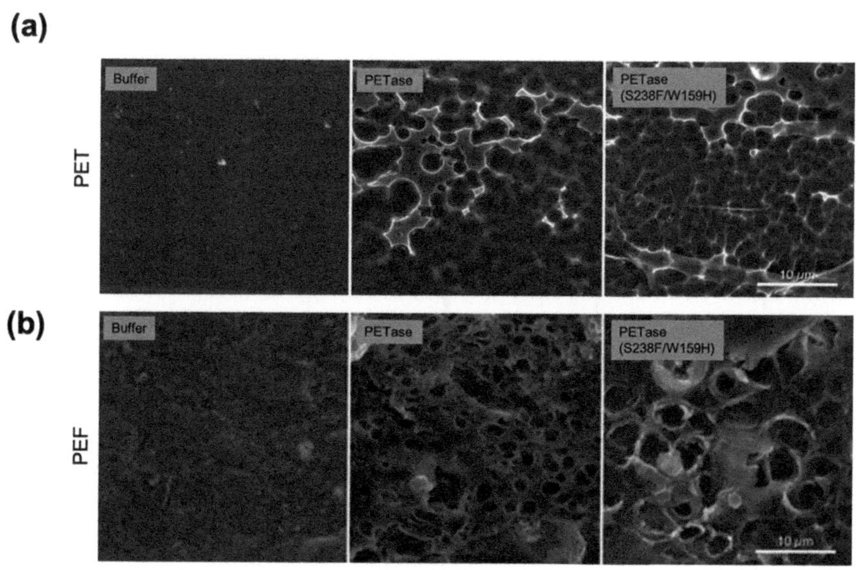

Figure 4. SEM micrographs of (a) PET and (b) PEF films in the presence of buffer, wild-type PETase and PETase double mutant, S238F/W159H. The films were imaged after 96 hours of incubation with either the enzyme or phosphate buffer. Adapted with permission from reference (11). Copyright 2018 National Academy of Sciences.

Increasing the Surface Adsorption of Polymer Degrading Enzymes

Redesigning the active site and broadening the substrate scope of enzymes are some of the common strategies that result in new and improved biocatalysts with better catalytic efficiencies. In some instances, the enzyme binding to hydrophobic polymeric films is a rate-limiting step in achieving complete hydrolysis. Engineering such enzymes with additional domains can tune their adsorption properties, resulting in effective hydrolysis (*2*). Some of the strategies that can increase the surface adsorption of polymer degrading enzymes are discussed below.

Many naturally occurring enzymes such as cellulases, chitinases, and even some of the polyester hydrolases have an additional substrate-binding domain that can enhance the hydrolysis of insoluble substrates (*47, 48*). Typically, cutinases lack the substrate-binding domains; however, Zhang et al. have designed a recombinant protein comprising TfH fused to a carbohydrate-binding module (CBM) of cellulase CenA from *Cellulomonas fimi*, facilitating the binding of the enzyme to the PET fibers (*49*). The CBM domain of CenA that naturally binds to cellulose is modified via site-directed mutagenesis, which enhances the binding and catalytic efficiency of cutinase–CBM_{CenA}. Of the different mutants, the catalytic efficiencies of variants Trp68Leu and Trp68Tyr demonstrate improvement by 1.4 and 1.5 fold respectively relative to the native enzyme against PET fibers hydrolysis (*49*).

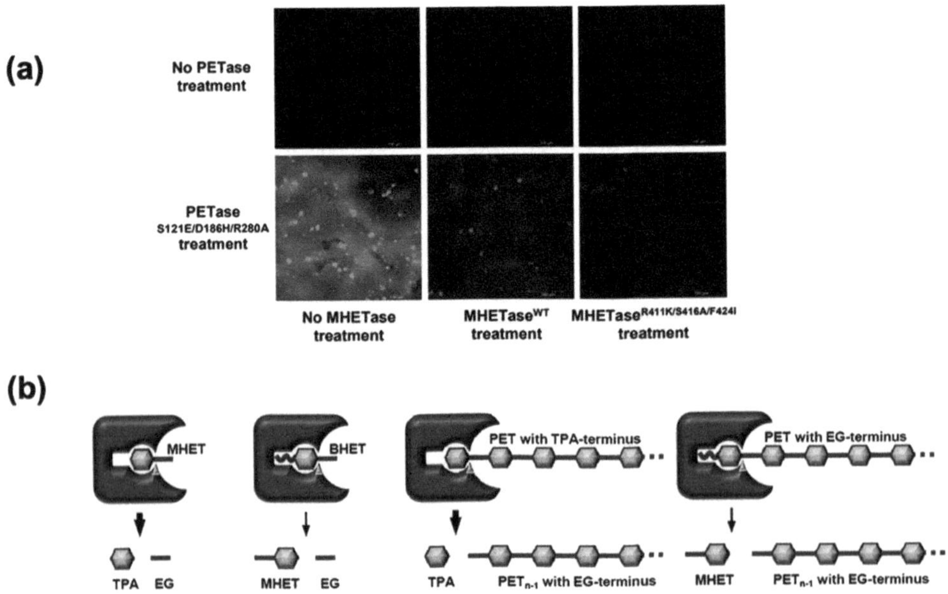

Figure 5. (a) Confocal images demonstrating the exo-PETase function of MHETase. (b) Schematic of MHETase hydrolysis in the presence of MHET, BHET or termini modified PET. Adapted with permission from reference (45). Copyright 2020 American Chemical Society.

A similar approach has been reported by Ribitsch et al., who investigated the effect of two different binding domains on the hydrolysis of PET films (*50*). Analogous to cutinase–CBM_{CenA}, Thc_Cut1 has been fused to the CBM of 1,4-beta-cellobiohydrolase I from *Hypocrea jecorin*. A second fusion protein of Thc_Cut1 is also designed by combining it with the substrate-binding domain (referred to as PBM) of the polyhydroxybutyrate depolymerase from *Alcaligenes faecalis* (PhaZAfa) (*50*). The amount of Thc_Cut1+CBM and Thc_Cut1+PBM adsorbed onto PET films are found to be 15.16 $\mu mol/cm^2$ and 18.95 $\mu mol/cm^2$, respectively. Owing to increased adsorption, Thc_Cut1+PBM can release 3.75-fold more hydrolytic products per mol of enzyme compared to wild-type Thc_Cut1 that showed an adsorption value of 14.27 $\mu mol/cm^2$. The activity of Thc_Cut1+CBM is intermediate between PBM bearing variant and the native enzyme (Figure 6a). Interestingly, the fusion proteins were effective only against PET films. For all the other substrates, in particular, for the soluble substrates (p-nitrophenyl butyrate (PNPB), MHET, and hydroxyethyl benzoate (HEB)), both Thc_Cut1+PBM and Thc_Cut1+CBM exhibits lower activity profiles (*50*).

Thc_Cut1+PBM has also been utilized by Perz et al. to investigate the hydrolysis of films composed of aliphatic polyesters (*16*). Compared to Thc_Cut1, the fusion protein of Thc_Cut1 with PBM domain shows faster hydrolysis of poly(1,4-butylene adipate) (PBA) films (Figure 6b) (*16*). Experiments using a Quartz crystal microbalance confirm that PBM domain increases the affinity of cutinase enzyme to the PBA films (*16*). Similar to cutinases, PBM when fused to a polyamidase from

Nocardia farcinica (PA) is less active for its soluble substrate but demonstrates a 4-fold improvement in hydrolyzing PU pellets, indicating an important role of enzyme adsorption in achieving efficient hydrolysis (*19*).

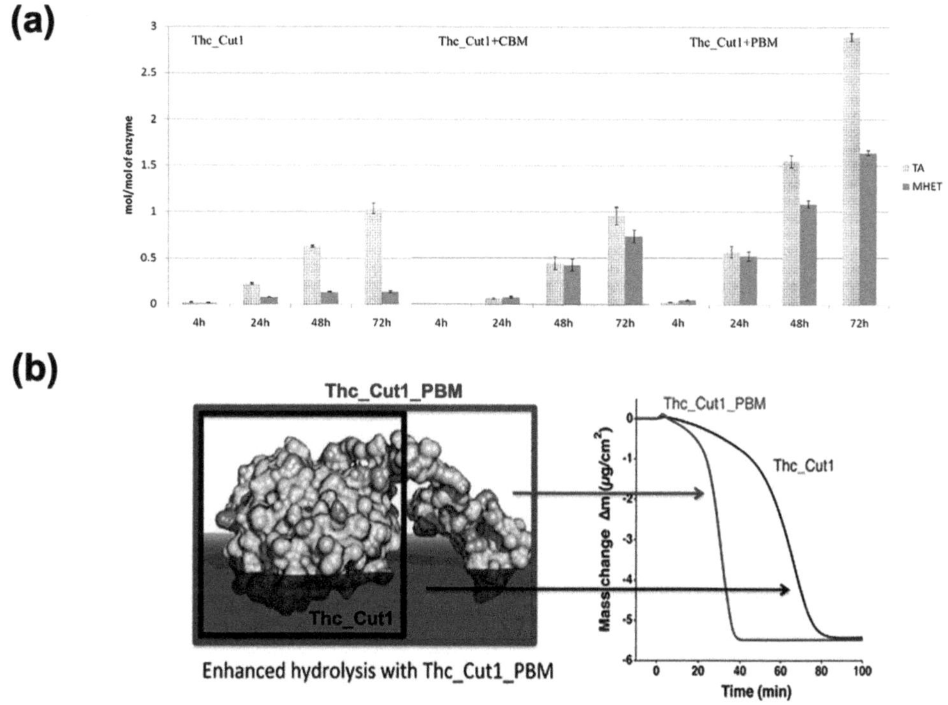

Figure 6. (a) Activity profile of PET films in the presence of Thc_Cut1, Thc_Cut1+CBM and Thc_Cut1+PBM. Reproduced with permission from reference (50). Copyright 2013 American Chemical Society. (b) Model of Thc_Cut1+PBM (left). ThC_Cut1 with PBM domain shows faster hydrolysis of PBA films compared to wt ThC_Cut1. Adapted with permission from reference (16). Copyright 2015 American Chemical Society.

In a separate study, Ribitsch et al. have designed three different fusion proteins of Thc_Cut1 with hydrophobins derived from *Trichoderma* species (*47*). Hydrophobins are small cysteine-rich proteins secreted by fungi that allows them to attach to hydrophobic surfaces (*51*). Presumably, hydrophobins adsorb to hydrophobic surfaces via their hydrophobic patch, exposing their hydrophilic side and thus altering the surface properties of polymers, making them more amenable to hydrolysis. These small surface-active proteins have been shown to stimulate the hydrolytic activity of cutinases against PET (*51*). Of the three fusion proteins, Thc_Cut1 fused to class II hydrophobins, HFB4 or HFB7, exhibits a >16-fold increase in PET hydrolysis relative to Thc_Cut1 alone or as a mixture with hydrophobins. Incubation of PET with a pseudo-class I hydrophobin HFB9, prior to treating with Thc_Cut1, shows better activity profile than a fusion protein of HFB9 with ThC_Cut1 (*47*).

Ulrich and colleagues have devoted considerable efforts towards directed evolution of anchor peptides (*52, 53*). These peptides, although tailored for functionalization of polymeric surfaces, have been fused to a cutinase Tcur1278 from *Thermomonospora curvata* (*54*). The peptide Tachystatin A2 (TA2), evolved from the directed evolution approach, binds to polyester-polyurethane (*54*). The fusion enzyme, Tcur1278-TA2, reveals a 6.6-fold faster degradation of diluted suspension of polyester-polyurethane nanoparticles (0.05% w/v of nanoparticles) relative to the wild-type enzyme. Degradation of the polyester-polyurethane films is faster in the presence of Tcur1278-TA2. By contrast, polymeric films in the presence of wild-type Tcur1278 does not show significant degradation, even after 40 hours of treatment (*54*) Antimicrobial peptides, such as adenoregulin, cecropin A and LCI derived from *Bacillus subtilis,* which have been previously evaluated to functionalize polypropylene (*53*), can also be utilized as anchor peptides to facilitate enzyme adsorption.

Expansins and swollenins are some of the other examples that have been suggested to increase enzyme accessibility to plastics (*55*). Expansins are small proteins isolated from plant tissues that can weaken the interactions between cell wall polysaccharides (such as cellulose, hemicellulose) (*56*). Swollenins, on the other hand, are expansin-like protein from Trichoderma reesei, which in addition to a C-terminal expansin-like domain, bear a carbohydrate-binding module family 1 domain (CBD) with cellulose-binding function (*57*). Although the properties of these proteins have not been assessed with any of the polymer degrading enzymes yet, the use of such domains can offer huge potential to improve the biocatalysts performance.

Increasing the Stability/Activity of Polymer Degrading Enzymes via Additives

Apart from protein engineering, a significant increase in catalytic activity has been observed in the presence of additives. The addition of surfactants can increase the local concentration of cutinase on a PET film; cationic surfactants (dodecyl trimethyl ammonium) can interact effectively with negatively charged TfCut2, via electrostatic interactions (*58*). Treatment of lcPET film with dodecyl trimethyl ammonium (C_{12}-$N(CH_3)_3^+$) prior to incubation with Tfcut2 at 40°C increases the reaction rate by 13-fold, while pre-treatment with an anionic surfactant, sodium dodecyl sulfate (C_{12}-OSO_3^-), increases the rate by 4.3 fold compared to TfCut2 in the absence of surfactants (Figure 7a) (*58*). At higher temperatures, smaller amounts of surfactants are required to accelerate the reaction rate. A specific mutant, Gly62Ala/Phe209Ala, designed by sequence comparison with PETase, exhibits an activity of 31 nmol $min^{-1}cm^{-2}$ in the presence of the surfactant, a 12.7-fold increase compared to wild-type TfCut2 (*58*). Complete degradation of lcPET film is observed within 30 hours of reaction conducted in the presence of Gly62Ala/Phe209Ala variant and cationic surfactant at 65°C (*58*).

PETase from *I. sakaiensis*, on the other hand, is a positively charged protein that shows an increase in activity when PET is pre-incubated with anionic surfactants bearing long alkyl chains (C_n-OSO_3^-, C_n-SO_3^- and C_n-COO^-) (*59*). With an increase in chain length, lower concentrations of surfactants are required as longer alkyl chains exhibit greater affinity towards lcPET surfaces. Amongst the different anionic headgroups, pre-treatment with C_n-OSO_3^- demonstrates the highest catalytic activity. Pre-incubation of lcPET films with dodecyl trimethyl ammonium (C_{12}-$N(CH_3)_3^+$), a cationic surfactant, adversely impacts the activity of PETase (*59*).

Eberl et al. have investigated the influence of a non-ionic detergent, Triton X-100, on the hydrolysis of a PET surrogate, bis(benzoyloxyethyl) terephthalate (3PET) (*60*). While Triton X-100 inhibits the hydrolytic activity of cutinases from *T. fusca* and *F. solani,* it enhances the activity of lipase from *Thermomyces lanuginosus* by seven-fold against 3PET. The increased activity of lipase is due to the conformational rearrangement of the lid domain in the presence of detergent. Since cutinase lacks the lid domain, Triton X-100 negatively affects its activity. In addition to Triton X-100, the effect of plasticizer is also studied (*60*). Both lipase and cutinases show increased hydrolysis of semi-crystalline PET films and fabrics in the presence of a plasticizer, N,N-diethylphenyl acetamide (DEPA) (1% v/v). The effect of DEPA is more pronounced on cutinases particularly in hydrolyzing PET fabrics (*60*).

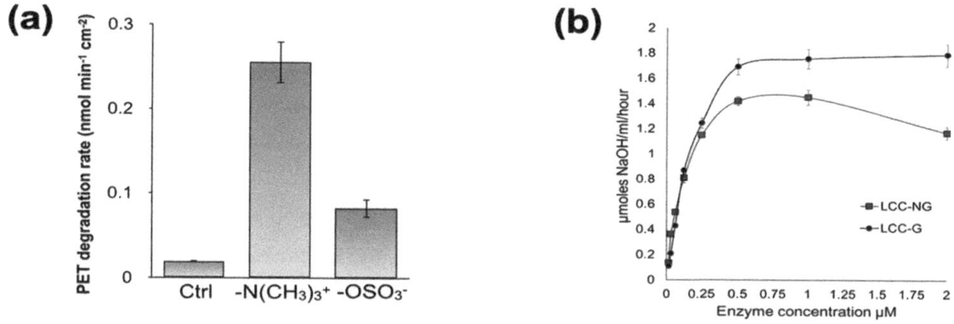

Figure 7. (a) TfCut2 mediated hydrolysis of lcPET films observed in the presence or absence of the surfactants C_{12}*-N(*CH_3*)*$_3^+$ *or* C_{12}*-*OSO_3^-*. Reproduced from reference (58). Copyright 2014 under Nature, Creative Commons Attribution 4.0. (b) Higher PET hydrolysis is observed in the presence of glycosylated LCC (LCC-G) than non-glycosylated enzyme (LCC-NG). Reproduced with permission from reference (61). Copyright 2018 American Chemical Society.*

Addition of divalent ions, such as calcium (Ca^{2+}), magnesium (Mg^{2+}), or manganese (Mn^{2+}), can enhance the thermostability of enzymes (*30*, *62*). Calcium-binding sites have been identified for esterase Est119 from *T. alba* (*63*) and LC-cutinase (*64*). Binding of calcium ions not only increases the T_m but also shifts the optimal temperature for enzymatic activity (*64*). Then et al. have studied the effect of calcium and magnesium ions on kinetic properties of different enzymes, TfH (also known as BTA1, identical to Tfu_0883), TfCut1 and TfCut2, Tfu_0882 and BTA2 (*62*). At 55 °C, all the enzymes except TfH showed similar activity profile in degrading PET films. As the temperature is increased to 60 °C, the cutinase enzymes in the presence of Mg^{2+} perform better than the enzymes with calcium chloride. Interestingly, at 65 °C native enzymes lose their activity while enzymes in the presence of Ca^{2+} exhibit better activity than with Mg^{2+} (*62*). Kawai et al. have also shown enhanced activity and thermostability of wild-type and mutant Cut190 in the presence of Ca^{2+} (*31*). Apart from calcium, Miyakawa et al. have demonstrated improved activity and thermostability of modified Cut190 (Ser226Pro) in the presence of Mn^{2+}, though Mg^{2+} was shown not to affect the enzymatic properties (*30*).

Another interesting approach utilized to increase the thermostability and associated kinetic stability has been reported by Shirke et al. (*61*) While bacterial cutinases lack glycosylation, cutinase from *F. solani* is glycosylated (*65*). Inspired by such glycosylated cutinases obtained naturally from fungal sources, Shirke et al. have expressed LCC, a bacterial cutinase in *Pichia pastoris* that allows for N-linked glycosylation (*61*). Glycosylated LCC (LCC-G) is more stable than non-glycosylated (LCC-NG), and exhibits 3-fold higher kinetic stability at 80 °C. Glycosylation inhibits thermal-induced aggregation and enables higher PET hydrolysis by LCC-G (Figure 7b) (*61*).

Conclusions and Future Outlook

Enzymatic degradation of polymers has shown great potential in plastic recycling and waste management. The recent work by Tournier et al. describes the LCC mediated degradation of PET to TPA, which is then used to recreate PET plastic bottles (*43*). This classic example of a circular economy has raised hopes and encourages the enzymatic reduction of plastic waste. In this chapter, we have discussed some of the enzymes that have been utilized for polymer degradation. We have also highlighted some of the strategies that can improve the catalytic efficiencies of these enzymes.

With enzymes being fragile molecules, there is a need to continuously improve and design new strategies, resulting in stable and efficient polymer active enzymes. Encapsulation or immobilization of enzymes is a promising approach that can be utilized to improve enzyme stability (*66*). A well-suited scaffold maintains the enzyme catalytic activity and provides a potential for enzyme recyclability (*67*). Recyclability is necessary to overcome the economic barrier associated with the industrialization of enzyme-mediated processes. Li et al. have demonstrated the potential of mesoporous metal-organic frameworks in immobilizing and stabilizing cutinase enzyme from *F. solani pisi* that can be recycled for up to five times (*67*). It is worth mentioning that not all immobilization strategies can preserve cutinase activity (*66*). Thus, it is essential to maintain a delicate balance between the thermal and kinetic stabilities of enzymes. Mixing two or more polymer degrading enzymes is another interesting approach in plastic waste management. Magnin et al. have recently reported the synergistic effect of mixing an amidase with an esterase in hydrolyzing polyurethanes (*68*). The approach of combining enzymes, having similar or different mechanisms of action, can possibly address the problems with mixed plastic waste and may prove useful for plastics that are recalcitrant to enzymatic degradation (such as polystyrene or polypropylene). High-throughput screening methods and computer-aided enzyme engineering also aid in rapid identification of enzymes that exhibit high specificity and activity (*43, 68*). Computer-aided enzyme engineering has witnessed tremendous growth with new tools not only being capable of predicting and designing novel enzymes but also expanding the substrate scope of existing enzymes.

With these great many tools at our disposal, it is now possible to promote plastic recycling and sustainability via enzyme-mediated processes. It is crucial to develop new and innovative technologies that can integrate these processes and transform the monomers into useful products including fuel, energy, hydrogen, or use them as raw materials for synthesizing new biodegradable plastics, ultimately closing the loop of the circular economy. Being a daunting challenge, this task will require concerted efforts across many different institutions to develop cost-effective and sustainable solutions.

Acknowledgments

This work was supported by the National Science Foundation award numbers IIP-1918981 and IIP-1644681.

References

1. Andrady, A. L.; Neal, M. A. Applications and societal benefits of plastics. *Philos. Trans. R. Soc. London B Biol. Sci.* **2009**, *364* (1526), 1977–1984.
2. Biundo, A.; Ribitsch, D.; Guebitz, G. M. Surface engineering of polyester-degrading enzymes to improve efficiency and tune specificity. *Appl. Microbiol. Biotechnol.* **2018**, *102* (8), 3551–3559.
3. Al-Sabagha, A. M.; Yehia, F. Z.; Eshaq, G.; Rabie, A. M.; ElMetwally, A. E. Greener routes for recycling of polyethylene terephthalate. *Egypt. J. Pet.* **2016**, *25* (1), 53–64.

4. Mahboob, M. The Plastic Problem - A Review of Current Methods of Plastic Waste Management and the Merits of Microbial Plastic Degradation. *iGEM Toronto*, https://2019.igem.org/wiki/images/e/ec/T--Toronto--iGEMLitReview.pdf 2019 (accessed on 2020-08-10).
5. Li, X. K.; Lu, H.; Guo, W. Z.; Cao, G. P.; Liu, H. L.; Shi, Y. H. Reaction kinetics and mechanism of catalyzed hydrolysis of waste PET using solid acid catalyst in supercritical CO2. *AIChE J.* **2014**, *61*, 200–214.
6. Danso, D.; Chow, J.; Streit, W. R. Plastics: Environmental and Biotechnological Perspectives on Microbial Degradation. *Appl. Environ. Microbiol.* **2019**, *85* (19), e01095-01019.
7. Yoshida, S.; Hiraga, K.; Takehana, T.; Taniguchi, I.; Yamaji, H.; Maeda, Y.; Toyohara, K.; Miyamoto, K.; Kimura, Y.; Oda, K. A bacterium that degrades and assimilates poly(ethylene terephthalate). *Science* **2016**, *351* (6278), 1196–1199.
8. Nimchua, T.; Eveleigh, D. E.; Sangwatanaroj, U.; Punnapayak, H. Screening of tropical fungi producing polyethylene terephthalate-hydrolyzing enzyme for fabric modification. *J. Ind. Microbiol. Biotechnol.* **2008**, *35* (8), 843–850.
9. Wallace, N. E.; Adams, M. C.; Chafin, A. C.; Jones, D. D.; Tsui, C. L.; Gruber, T. D. The highly crystalline PET found in plastic water bottles does not support the growth of the PETase-producing bacterium Ideonella sakaiensis. *Environ. Microbiol. Rep.* **2020**, Accepted Author Manuscript.
10. Kawai, F.; Kawabata, T.; Oda, M. Current knowledge on enzymatic PET degradation and its possible application to waste stream management and other fields. *Appl. Microbiol. Biotechnol.* **2019**, *103* (11), 4253–4268.
11. Austin, H. P.; Allen, M. D.; Donohoe, B. S.; Rorrer, N. A.; Kearns, F. L.; Silveira, R. L.; Pollard, B. C.; Dominick, G.; Duman, R.; El Omari, K.; Mykhaylyk, V.; Wagner, A.; Michener, W. E.; Amore, A.; Skaf, M. S.; Crowley, M. F.; Thorne, A. W.; Johnson, C. W.; Woodcock, H. L.; McGeehan, J. E.; Beckham, G. T. Characterization and engineering of a plastic-degrading aromatic polyesterase. *Proc. Natl. Acad. Sci. U. S. A.* **2018**, *115* (19), E4350–E4357.
12. Palm, G. J.; Reisky, L.; Bottcher, D.; Muller, H.; Michels, E. A. P.; Walczak, M. C.; Berndt, L.; Weiss, M. S.; Bornscheuer, U. T.; Weber, G. Structure of the plastic-degrading Ideonella sakaiensis MHETase bound to a substrate. *Nat. Commun.* **2019**, *10* (1), 1717.
13. Baker, P. J.; Poultney, C.; Liu, Z.; Gross, R.; Montclare, J. K. Identification and comparison of cutinases for synthetic polyester degradation. *Appl. Microbiol. Biotechnol.* **2012**, *93* (1), 229–240.
14. Ribitsch, D.; Acero, E. H.; Greimel, K.; Dellacher, A.; Zitzenbacher, S.; Marold, A.; Rodriguez, R. D.; Steinkellner, G.; Gruber, K.; Schwab, H.; Guebitz, G. M. A New Esterase from Thermobifida halotolerans Hydrolyses Polyethylene Terephthalate (PET) and Polylactic Acid (PLA). *Polymers* **2012**, *4*, 617–629.
15. Jaeger, K. E.; Steinbuchel, A.; Jendrossek, D. Substrate specificities of bacterial polyhydroxyalkanoate depolymerases and lipases: bacterial lipases hydrolyze poly(omega-hydroxyalkanoates). *Appl. Environ. Microbiol.* **1995**, *61* (8), 3113–3118.
16. Perz, V.; Zumstein, M. T.; Sander, M.; Zitzenbacher, S.; Ribitsch, D.; Guebitz, G. M. Biomimetic Approach to Enhance Enzymatic Hydrolysis of the Synthetic Polyester Poly(1,4-

butylene adipate): Fusing Binding Modules to Esterases. *Biomacromolecules* **2015**, *16* (12), 3889–3896.

17. Santo, M.; Weitsman, R.; Sivan, A. The role of the copper-binding enzyme – laccase – in the biodegradation of polyethylene by the actinomycete Rhodococcus ruber. *Int. Biodeterior. Biodegrad.* **2013**, *84*, 204–210.
18. Nakamiya, K.; Sakasita, G.; Ooi, T.; Kinoshita, S. Enzymatic degradation of polystyrene by hydroquinone peroxidase of Azotobacter beijerinckii HM121. *J. Fermentation Bioeng.* **1997**, *84* (5), 480–482.
19. Gamerith, C.; Acero, E. H.; Pellis, A.; Ortner, A.; Vielnascher, R.; Luschnig, D.; Zartl, B.; Haernvalla, K.; Zitzenbacher, S.; Strohmeier, G.; Hoff, O.; Steinkellner, G.; Gruber, K.; Ribitsch, D.; Guebitz, G. M. Improving enzymatic polyurethane hydrolysis by tuning enzyme sorption. *Polym. Degrad. Stab.* **2016**, *132*, 69–77.
20. Araujo, R.; Silva, C.; O'Neill, A.; Micaelo, N.; Guebitz, G.; Soares, C. M.; Casal, M.; Cavaco-Paulo, A. Tailoring cutinase activity towards polyethylene terephthalate and polyamide 6,6 fibers. *J. Biotechnol.* **2007**, *128* (4), 849–857.
21. Wei, R.; Zimmermann, W. Microbial enzymes for the recycling of recalcitrant petroleum-based plastics: how far are we? *Microb. Biotechnol.* **2017**, *10* (6), 1308–1322.
22. Kawai, F.; Kawabata, T.; Oda, M. Current State and Perspectives Related to the Polyethylene Terephthalate Hydrolases Available for Biorecycling. *ACS Sustainable Chem. Eng.* **2020**, *8* (24), 8894–8908.
23. Müller, R. J.; Schrader, H.; Profe, J.; Dresler, K.; Deckwer, W. D. Enzymatic Degradation of Poly(ethylene terephthalate): Rapid Hydrolyse using a Hydrolase from T. fusca. *Macromol. Rapid Commun.* **2005**, *26*, 1400–1405.
24. Chen, S.; Tong, X.; Woodard, R. W.; Du, G.; Wu, J.; Chen, J. Identification and characterization of bacterial cutinase. *J. Biol. Chem.* **2008**, *283* (38), 25854–25862.
25. Acero, E. H.; Ribitsch, D.; Steinkellner, G.; Gruber, K.; Greimel, K.; Eiteljoerg, I.; Trotscha, E.; Wei, R.; Zimmermann, W.; Zinn, M.; Cavaco-Paulo, A.; Freddi, G.; Schwab, H.; Guebitz, G. Enzymatic Surface Hydrolysis of PET: Effect of Structural Diversity on Kinetic Properties of Cutinases from Thermobifida. *Macromolecules* **2011**, *44* (12), 4632–4640.
26. Sulaiman, S.; Yamato, S.; Kanaya, E.; Kim, J. J.; Koga, Y.; Takano, K.; Kanaya, S. Isolation of a novel cutinase homolog with polyethylene terephthalate-degrading activity from leaf-branch compost by using a metagenomic approach. *Appl. Environ. Microbiol.* **2012**, *78* (5), 1556–1562.
27. Kitadokoro, K.; Thumarat, U.; Nakamura, R.; Nishimura, K.; Karatani, H.; Suzuki, H.; Kawai, F. Crystal structure of cutinase Est119 from Thermobifida alba AHK119 that can degrade modified polyethylene terephthalate at 1.76Å resolution. *Polym. Degrad. Stab.* **2012**, *97* (5), 771–775.
28. Hu, X.; Thumarat, U.; Zhang, X.; Tang, M.; Kawai, F. Diversity of polyester-degrading bacteria in compost and molecular analysis of a thermoactive esterase from Thermobifida alba AHK119. *Appl. Microbiol. Biotechnol.* **2010**, *87* (2), 771–779.
29. Wei, R.; Oeser, T.; Then, J.; Kuhn, N.; Barth, M.; Schmidt, J.; Zimmermann, W. Functional characterization and structural modeling of synthetic polyester-degrading hydrolases from Thermomonospora curvata. *AMB Express* **2014**, *4*, 44.

30. Miyakawa, T.; Mizushima, H.; Ohtsuka, J.; Oda, M.; Kawai, F.; Tanokura, M. Structural basis for the Ca(2+)-enhanced thermostability and activity of PET-degrading cutinase-like enzyme from Saccharomonospora viridis AHK190. *Appl. Microbiol. Biotechnol.* **2015**, *99* (10), 4297–4307.
31. Kawai, F.; Oda, M.; Tamashiro, T.; Waku, T.; Tanaka, N.; Yamamoto, M.; Mizushima, H.; Miyakawa, T.; Tanokura, M. A novel Ca2+-activated, thermostabilized polyesterase capable of hydrolyzing polyethylene terephthalate from Saccharomonospora viridis AHK190. *Appl. Microbiol. Biotechnol.* **2014**, *98*, 10053–10064.
32. Guo, Y.; Chen, S.; Su, L.; Chen, J. Cloning, expression, and characterization of polyamidase from Nocardia farcinica and its application to polyamide modification. *Biotechnol. Bioprocess Eng.* **2013**, *18*, 1067–1075.
33. Carniel, A.; Valoni, E.; Nicomedes Junior, J.; da Conceição Gomes, A.; deCastro, A. M. Lipase from Candida antarctica (CALB) and cutinase from Humicola insolens act synergistically for PET hydrolysis to terephthalic acid. *Process Biochem.* **2016**, *59*, 84–90.
34. Egmond, M. R.; de Vlieg, J. Fusarium solani pisi cutinase. *Biochimie* **2000**, *82* (11), 1015–1021.
35. Liu, Z.; Gosser, Y.; Baker, P. J.; Ravee, Y.; Lu, Z.; Alemu, G.; Li, H.; Butterfoss, G. L.; Kong, X. P.; Gross, R.; Montclare, J. K. Structural and functional studies of Aspergillus oryzae cutinase: enhanced thermostability and hydrolytic activity of synthetic ester and polyester degradation. *J. Am. Chem. Soc.* **2009**, *131* (43), 15711–15716.
36. Ronkvist, A. M.; Xie, W.; Lu, W.; Gross, R. A. Cutinase-Catalyzed Hydrolysis of Poly(ethylene terephthalate). *Macromolecules* **2009**, *42* (14), 5128–5138.
37. Liebminger, S.; Eberl, A.; Sousa, F.; Heumann, S.; Fischer-Colbrie, G.; Cavaco-Paulo, A.; Guebitz, G. M. Hydrolysis of PET and bis-(benzoyloxyethyl) terephthalate with a new polyesterase from Penicillium citrinum. *Biocatal. Biotransform.* **2007**, *25* (2–4), 171–177.
38. Longhi, S.; Czjzek, M.; Lamzin, V.; Nicolas, A.; Cambillau, C. Atomic resolution (1.0 A) crystal structure of Fusarium solani cutinase: stereochemical analysis. *J. Mol. Biol.* **1997**, *268* (4), 779–799.
39. Martinez, C.; De Geus, P.; Lauwereys, M.; Matthyssens, G.; Cambillau, C. Fusarium solani cutinase is a lipolytic enzyme with a catalytic serine accessible to solvent. *Nature* **1992**, *356* (6370), 615–618.
40. Fett, W. F.; Wijey, C.; Moreau, R. A.; Osman, S. F. Production of cutinase by Thermomonospora fusca ATCC 27730. *J. Appl. Microbiol.* **1999**, *86*, 561–568.
41. Dresler, K.; van den Heuvel, J.; Muller, R. J.; Deckwer, W. D. Production of a recombinant polyester-cleaving hydrolase from Thermobifida fusca in Escherichia coli. *Bioprocess Biosyst. Eng.* **2006**, *29* (3), 169–183.
42. Kleeberg, I.; Hetz, C.; Kroppenstedt, R. M.; Muller, R. J.; Deckwer, W. D. Biodegradation of aliphatic-aromatic copolyesters by Thermomonospora fusca and other thermophilic compost isolates. *Appl. Environ. Microbiol.* **1998**, *64* (5), 1731–1735.
43. Tournier, V.; Topham, C. M.; Gilles, A.; David, B.; Folgoas, C.; Moya-Leclair, E.; Kamionka, E.; Desrousseaux, M. L.; Texier, H.; Gavalda, S.; Cot, M.; Guemard, E.; Dalibey, M.; Nomme, J.; Cioci, G.; Barbe, S.; Chateau, M.; Andre, I.; Duquesne, S.; Marty, A. An engineered PET depolymerase to break down and recycle plastic bottles. *Nature* **2020**, *580* (7802), 216–219.

44. Liu, B.; He, L.; Wang, L.; Li, T.; Li, C.; Liu, H.; Luo, Y.; Bao, R. Protein Crystallography and Site-Direct Mutagenesis Analysis of the Poly(ethylene terephthalate) Hydrolase PETase from Ideonella sakaiensis. *ChemBioChem* **2018**, *19* (14), 1471–1475.
45. Sagong, H. Y.; Seo, H.; Kim, T.; Son, H. F.; Joo, S.; Lee, S. H.; Kim, S.; Woo, J. S.; Hwang, S. Y.; Kim, K. J. Decomposition of the PET Film by MHETase Using Exo-PETase Function. *ACS Catal.* **2020**, *10* (8), 4805–4812.
46. Gamerith, C.; Zartla, B.; Pellis, A.; Guillamot, F.; Marty, A.; Acero, E. H.; Guebitz, G. M. Enzymatic recovery of polyester building blocks from polymer blends. *Process Biochem.* **2017**, *59*, 58–64.
47. Ribitsch, D.; Acero, E. H.; Przylucka, A.; Zitzenbacher, S.; Marold, A.; Gamerith, C.; Tscheliessnig, R.; Jungbauer, A.; Rennhofer, H.; Lichtenegger, H.; Amenitsch, H.; Bonazza, K.; Kubicek, C. P.; Druzhinina, I. S.; Guebitz, G. M. Enhanced cutinase-catalyzed hydrolysis of polyethylene terephthalate by covalent fusion to hydrophobins. *Appl. Environ. Microbiol.* **2015**, *81* (11), 3586–3592.
48. Hiraishi, T.; Ohura, T.; Ito, S.; Kasuya, K.; Doi, Y. Function of the catalytic domain of poly(3-hydroxybutyrate) depolymerase from Pseudomonas stutzeri. *Biomacromolecules* **2000**, *1* (3), 320–324.
49. Zhang, Y.; Wang, L.; Chen, J.; Wu, J. Enhanced activity toward PET by site-directed mutagenesis of Thermobifida fusca cutinase-CBM fusion protein. *Carbohydr. Polym.* **2013**, *97* (1), 124–129.
50. Ribitsch, D.; Yebra, A. O.; Zitzenbacher, S.; Wu, J.; Nowitsch, S.; Steinkellner, G.; Greimel, K.; Doliska, A.; Oberdorfer, G.; Gruber, C. C.; Gruber, K.; Schwab, H.; Stana-Kleinschek, K.; Acero, E. H.; Guebitz, G. M. Fusion of binding domains to Thermobifida cellulosilytica cutinase to tune sorption characteristics and enhancing PET hydrolysis. *Biomacromolecules* **2013**, *14* (6), 1769–1776.
51. Espino-Rammer, L.; Ribitsch, D.; Przylucka, A.; Marold, A.; Greimel, K. J.; Herrero Acero, E.; Guebitz, G. M.; Kubicek, C. P.; Druzhinina, I. S. Two novel class II hydrophobins from Trichoderma spp. stimulate enzymatic hydrolysis of poly(ethylene terephthalate) when expressed as fusion proteins. *Appl. Environ. Microbiol.* **2013**, *79* (14), 4230–4238.
52. Rübsam, K.; Weber, L.; Jakob, F.; Schwaneberg, U. Directed evolution of polypropylene and polystyrene binding peptides. *Biotechnol. Bioeng.* **2018**, *115* (2), 321–330.
53. Rübsam, K.; Stomps, B.; Böker, A.; Jakob, F.; Schwaneberg, U. Anchor peptides: A green and versatile method for polypropylene functionalization. *Polymer* **2017**, *116*, 124–132.
54. Islam, S.; Apitius, L.; Jakob, F.; Schwaneberg, U. Targeting microplastic particles in the void of diluted suspensions. *Environ. Int.* **2019**, *123*, 428–435.
55. Maille, E.; Carbios, Process of recycling mixed PET plastic articles. 2015, US Patent #10,385,183.
56. Cosgrove, D. J. Loosening of plant cell walls by expansins. *Nature* **2000**, *407* (6802), 321–326.
57. Brotman, Y.; Briff, E.; Viterbo, A.; Chet, I. Role of swollenin, an expansin-like protein from Trichoderma, in plant root colonization. *Plant Physiol.* **2008**, *147* (2), 779–789.
58. Furukawa, M.; Kawakami, N.; Tomizawa, A.; Miyamoto, K. Efficient Degradation of Poly(ethylene terephthalate) with Thermobifida fusca Cutinase Exhibiting Improved Catalytic

Activity Generated using Mutagenesis and Additive-based Approaches. *Sci. Rep.* **2019**, *9* (1), 16038.

59. Furukawa, M.; Kawakami, N.; Oda, K.; Miyamoto, K. Acceleration of Enzymatic Degradation of Poly(ethylene terephthalate) by Surface Coating with Anionic Surfactants. *ChemSusChem* **2018**, *11* (23), 4018–4025.
60. Eberl, A.; Heumann, S.; Bruckner, T.; Araujo, R.; Cavaco-Paulo, A.; Kaufmann, F.; Kroutil, W.; Guebitz, G. M. Enzymatic surface hydrolysis of poly(ethylene terephthalate) and bis(benzoyloxyethyl) terephthalate by lipase and cutinase in the presence of surface active molecules. *J. Biotechnol.* **2009**, *143* (3), 207–212.
61. Shirke, A. N.; White, C.; Englaender, J. A.; Zwarycz, A.; Butterfoss, G. L.; Linhardt, R. J.; Gross, R. A. Stabilizing Leaf and Branch Compost Cutinase (LCC) with Glycosylation: Mechanism and Effect on PET Hydrolysis. *Biochemistry* **2018**, *57* (7), 1190–1200.
62. Then, J.; Wei, R.; Oeser, T.; Barth, M.; Belisario-Ferrari, M. R.; Schmidt, J.; Zimmermann, W. Ca2+ and Mg2+ binding site engineering increases the degradation of polyethylene terephthalate films by polyester hydrolases from Thermobifida fusca. *Biotechnol. J.* **2015**, *10* (4), 592–598.
63. Kawai, F.; Thumarat, U.; Kitadokoro, K.; Waku, T.; Tada, T.; Tanaka, N.; Kawabata, T. Comparison of Polyester-Degrading Cutinases from Genus Thermobifida. In *Green Polymer Chemistry: Biocatalysis and Materials II 2013*; Cheng, H. N., Gross, R. A., Smith, P. B., Ed.; American Chemical Society: Washington, DC, pp 111−120.
64. Sulaiman, S.; You, D. J.; Kanaya, E.; Koga, Y.; Kanaya, S. Crystal structure and thermodynamic and kinetic stability of metagenome-derived LC-cutinase. *Biochemistry* **2014**, *53* (11), 1858–1869.
65. Lin, T. S.; Kolattukudy, P. E. Structural studies on cutinase, a glycoprotein containing novel amino acids and glucuronic acid amide at the N terminus. *Eur. J. Biochem.* **1980**, *106* (2), 341–351.
66. Carvalho, C. M. L.; Aires-Barros, M. R.; Cabral, J. M. S. Cutinase structure, function and biocatalytic applications. *EJB J. Biotechnol.* **1998**, *1* (3), 28–29.
67. Li, P.; Modica, J. A.; Howarth, A. J.; Vargas, L. E.; Moghadam, P. Z.; Snurr, R. Q.; Mrksich, M.; Hupp, J. T.; Farha, O. K. Toward Design Rules for Enzyme Immobilization in Hierarchical Mesoporous Metal-Organic Frameworks. *Chem* **2016**, *1* (1), 154–169.
68. Magnin, A.; Pollet, E.; Perrin, R.; Ullmann, C.; Persillon, C.; Phalip, V.; Averous, L. Enzymatic recycling of thermoplastic polyurethanes: Synergistic effect of an esterase and an amidase and recovery of building blocks. *Waste Manag.* **2019**, *85*, 141–150.

Chapter 4

Crystal Structures of Polyethylene Terephthalate-Degrading Enzyme Cut190 in Substrate-Bound States Reveal the Enzymatic Reaction Cycle Accelerated by Calcium Ion

Nobutaka Numoto,[*,1] Narutoshi Kamiya,[2,3] Fusako Kawai,[4] and Masayuki Oda[5]

[1]Medical Research Institute, Tokyo Medical and Dental University, Tokyo 113-8510, Japan

[2]Institute of Protein Research, Osaka University, Osaka 565-0871, Japan

[3]Graduate School of Simulation Studies, University of Hyogo, Hyogo 650-0047, Japan

[4]Okayama University, Okayama 700-8530, Japan

[5]Graduate School of Life and Environmental Sciences, Kyoto Prefectural University, Kyoto 606-8522, Japan

[*]Email: numoto.str@mri.tmd.ac.jp

Cutinases from bacteria have an ability to hydrolyze polyethylene terephthalate (PET), which is known as a persistent plastic. Enzymatic treatment is expected to be one of the possible solutions for PET pollution, with a low environmental load. We reported that the cutinase from *Saccharomonospora viridis* AHK190 (Cut190) has unique characteristics of calcium (Ca^{2+})-dependent thermostability and activity that degrades the inner block of PET. We have determined the crystal structures of an improved mutant of Cut190, in complex with two small substrates. Both substrates are well accommodated in the active site and anchored at the oxyanion hole by the hydrogen bonds. One complex can be interpreted as being in a pre-reaction state, in which large conformational changes are observed at three loops in the vicinity of the active site. We designated this structure as a novel engaged form. The other complex can be interpreted as being in a post-reaction state, in which no remarkable structural changes are observed from the previously reported open form. We identified three Ca^{2+} binding sites, one of which is the same as that previously reported as being responsible for the structural change as an allosteric effector; the other two are novel binding sites associated with thermostability and/or activity. Although the engaged form binds the substrate, the allosteric Ca^{2+} is dissociated as in the case of the closed form. We propose a Ca^{2+}-dependent reaction cycle characteristic for Cut190. We have developed further engineered Cut190s with enhanced thermostability and activity based on the structural information.

Introduction

The enzymes known as cutinases are present in fungi and bacteria and belong to the lipase family (*1–4*). A cutinase facilitates the invasion of phytopathogenic microorganisms into plants by degrading the cutin layer (*5*). The structures of various types of cutinase have been reported, including the fungal cutinases from *Fusarium solani pici* (*6*), *Aspergillus oryzae* (*7*), and *Cryptococcus* sp. (*8*) as well as the bacterial cutinases from *Thermobifida alba* (*9*), a leaf compost (*10*), *T. fusca* (*11*), *T. cellulosilytica* (*12*), and *Ideonella sakaiensis* (*13*). Some of these cutinases have been reported to be able to hydrolyze plastic polyesters, such as polyethylene terephthalate (PET, Figure 1) (*10, 11, 13*). The active site of a cutinase is composed of the well-known catalytic triad with an active serine residue, and the active site is open to the solvent, whereas lipases have a lid-covered active site. We speculated that the large polymer substrates can easily gain access to the open active site of a cutinase. The efficient recycling of waste plastic is required as the environmental problem of plastic pollution continues to be a serious problem worldwide, and the enzymatic treatment of plastic is an attractive solution.

We have cloned a cutinase (Cut190) from a thermophilic isolate, *Saccharomonospora viridis* AHK190, and we demonstrated that Cut190 can degrade the inner block of PET (*14, 15*) at a temperature higher than the glass transition temperature (T_g) of PET at approx. 60°–65°C in aqueous solution. The remarkable nature of Cut190 is that its thermostability and enzymatic activity are regulated by calcium (Ca^{2+}) binding (*15, 16*). A crystallographic study of a Cut190 S226P mutant (*17*) elucidated that the Ca^{2+} binding site apart from the active site is responsible for the open/close transition of the active site as an allosteric effector. When a Ca^{2+} ion is bound at the Ca^{2+} binding loop, the active site is open and the oxyanion hole is formed to stabilize the reaction intermediate. In contrast, a release of Ca^{2+} causes large conformational changes at the Ca^{2+} binding loop and the adjacent loops, resulting in the disruption of the oxyanion hole and the sealing of the active site from solvents.

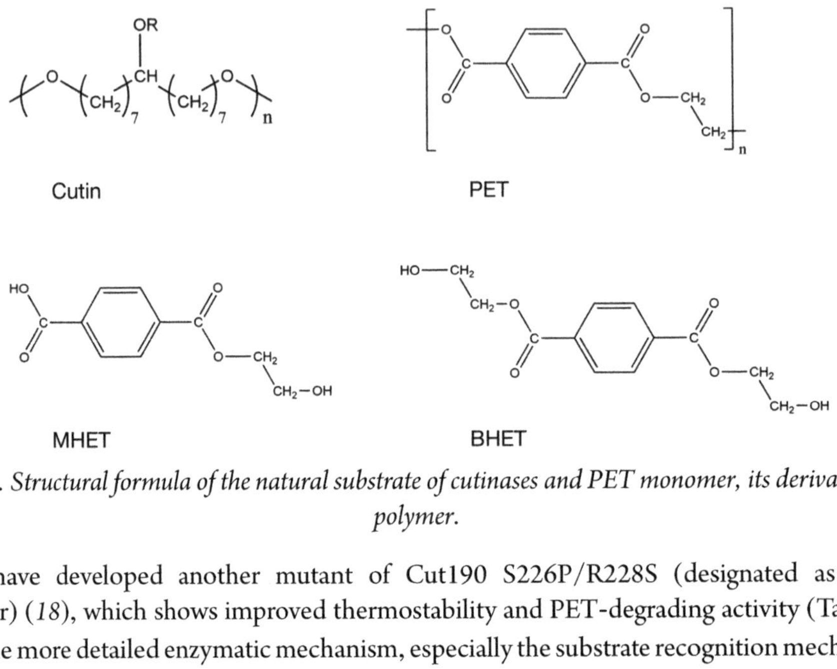

Figure 1. Structural formula of the natural substrate of cutinases and PET monomer, its derivative, and polymer.

We have developed another mutant of Cut190 S226P/R228S (designated as Cut190* hereinafter) (*18*), which shows improved thermostability and PET-degrading activity (Table 1). To address the more detailed enzymatic mechanism, especially the substrate recognition mechanism, we herein solved the crystal structures of two substrate-bound forms of Cut190* S176A mutant (*19*) in which the active Ser was mutated to Ala to inhibit the enzymatic reaction and entrap the substrate binding state. Although the substrates used in this study were small, straight-chain monoesters lacking an aromatic ring similar to PET, the obtained substrate complex structures show good agreement with the substrate recognition mechanisms and reaction schemes of the well-established serine proteases and esterases. In addition, the high-resolution structures obtained in this study allowed us to identify novel Ca^{2+} binding sites and examine the contributions to the thermostability and activity of Cut190*. We also developed a further improved mutant of Cut190 (*20*) based on the structural information, with no remarkable structural changes in the overall structure of Cut190*.

Table 1. Thermostabilities and Enzymatic Activities (*15, 16*) of the Wildtype and Mutants of Cut190

Cut190	T_m (°C)	*Activity (mkat/mg)*
Wildtype	70.6	2.8 ± 0.1
S226P	76.5	4.0 ± 0.1
S226P/R228S	76.8	5.9 ± 0.2
S226P/R228S/S176A	71.8 (250 mM Ca^{2+})	–

The values were measured in the presence of 300 mM Ca^{2+} otherwise specified. T_m, melting temperature.

Substrate-Binding Forms of Cut190*

Previous structural analyses of Cut190 S226P (*17*) demonstrated that Ca^{2+} is bound at the β1-β2 loop, and the Ca^{2+} binding induces the large conformational change of the β1-β2 loop concomitantly with the conformational changes at the β4-α3 and β3-α2 loops (Figure 2). The β3-α2 loop contains the key residue of Phe106. In the Ca^{2+} free form, the side chain of Phe106 seals the active Ser to prevent substrate interaction, and the oxyanion hole is disrupted. When a Ca^{2+} ion is bound at the β1-β2 loop, the conformational changes of the neighboring β4-α3 and β3-α2 loops induce the formation of an oxyanion hole with the main-chain amides of Phe106 and Met177. The side chain of Phe106 is relocated away from the active site, and the active site is open to the solvent, allowing access to large substrates. We therefore call the Ca^{2+} free and bound forms the closed and open forms, respectively.

We performed further structural analyses to obtain the substrate complexed structure in order to elucidate the detailed substrate recognition mechanism of Cut190. We used Cut190* S176A mutant to inhibit the enzymatic reaction and entrap the substrate binding state. We have confirmed in other experiments that mutating the active Ser residue to Ala has little effect on the overall structure (*21*). Here, we used a variety of substrate molecules including the "PET monomer" 2-hydroxyethyl terephthalic acid (MHET) and bishydroxyethyl terephthalate (BHET) to obtain the substrate-bound crystals by co-crystallization and/or soaking methods, but we succeeded in obtaining only two small substrates (Figure 3a) of monoethyl succinate (Et-succinate) and monoethyl adipate (Et-adipate) by a soaking method using the open-form crystals (*19*). Both compounds are straight-chain monoesters with no aromatic ring; however, these complexed crystal structures provide useful insights into the mechanisms underlying the recognition of the ester moiety, which is the cleaved site of PET, and the enzymatic reaction cycle of Cut190.

In the Et-succinate soaking crystal, the ligand molecule is accommodated at the active site and the ester group of the ligand is located adjacent to Ala176, which is the substituted residue from the active Ser forming the catalytic triad with His254 and Asp222 (Figure 3b). The carbonyl group of the ester is placed at the oxyanion hole formed by Phe106 and Met177. Surprisingly, we observed large conformational changes around the three loops of β3-α2, β4-α3, and β1-β2. The conformations of the β3-α2 and β4-α3 loops do not match those of the reported open or closed forms. The bound ligand is tightly held by relocated Thr107, forming a mildly closed conformation. We thus designated this structure as the engaged form, in which the bound ligand is accommodated closely to the active site but the cleavage has not yet occurred. That is, the engaged form indicates the pre-reaction state.

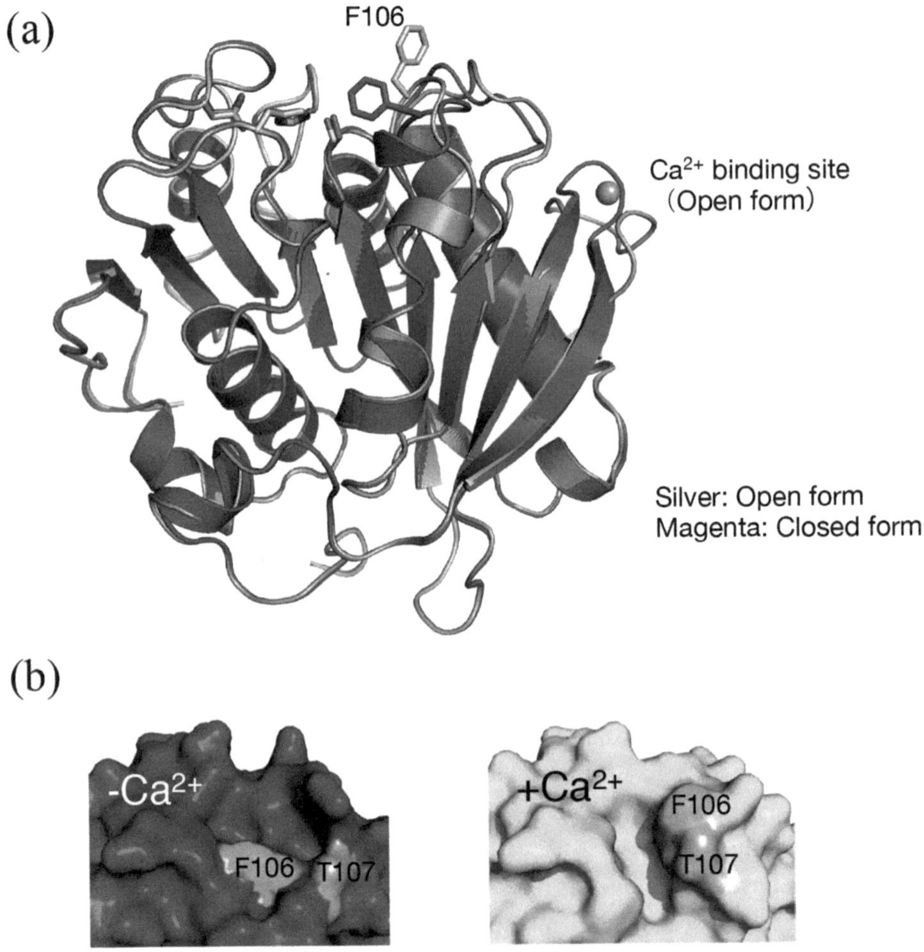

Figure 2. Crystal structures of Cut190 S226P mutant. (a) Superposition of the Ca^{2+}-free (magenta) and -bound (silver) forms. The sealing residue of Phe106 and the catalytic triad residues are shown as a stick model. Bound Ca^{2+} is drawn as a sphere model. (b) Surface representation around the active site of the closed (left) and open (right) forms. The conformational changes at the β3-α2 loop including Phe106 and Thr107 are highlighted as green and silver, respectively. Figures are prepared from the coordinates of PDB ID 4WFI and 4WFJ.

In the Et-adipate soaking crystal, the ligand is accommodated with no significant structural changes from the open form (Figure 3c). Unexpectedly, we observed that the carboxyl group is located proximally to the active site, and an oxygen atom of the carboxyl group is placed at the oxyanion hole. Alternatively, the ester group of the ligand is located at the distal side from the active site. Because the carboxyl group stabilized by the hydrogen bond to the oxyanion hole is in good agreement with the last step of the generally accepted reaction mechanism of well-studied proteases or esterases (*22, 23*), we speculate that the obtained structure demonstrates the post-reaction state in the enzymatic reaction cycle.

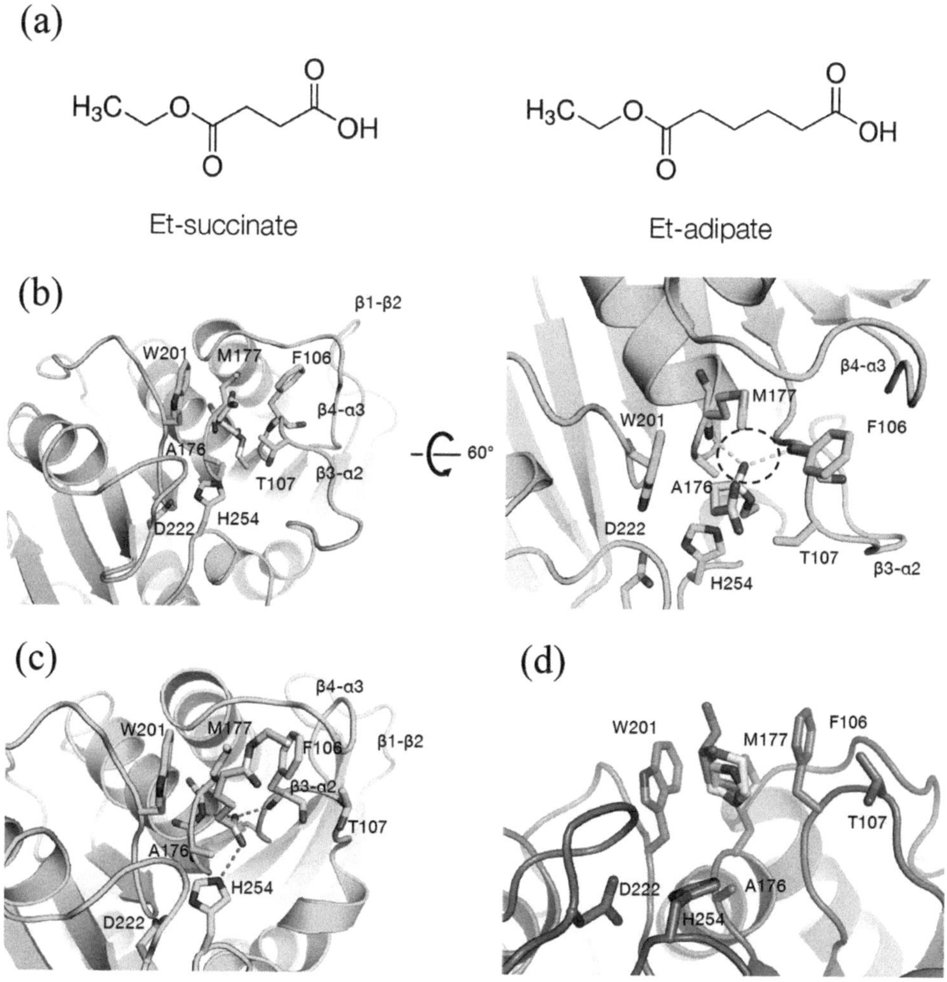

Figure 3. Crystal structures of Cut190 complexed with two substrates. (a) The substrates used in this study. (b) The Et-succinate bound structure. Residues responsible for the substrate recognition or active residues are displayed as a stick model. The three loops with large conformational changes via substrate binding are labeled. In the right panel, the oxyanion hole is indicated as a dashed circle, and the hydrogen bonds between the carbonyl group of the ester of the substrate and the oxyanion hole are shown as yellow dashed lines. (c) The Et-adipate bound structure. The same residues and loops as in (b) are labeled. The hydrogen bonds at the carboxyl group are shown as orange dashed lines. (d) The glycerol bound structure of Cut190* (orange). Bound glycerol and residues around the active site are represented as a stick model. Imidazole and dioxane bound in the structure of a leaf-branch compost cutinase (PDB ID 6THT and 6THS, respectively) are superimposed, and only the ligands are represented as a white stick model.*

We have tested various conditions to determine substrate complex structures, and we occasionally obtained the artifact complex in which a small molecule such as glycerol as a cryoprotectant is bound in the vicinity of the active site with the open form (Figure 3d). This is compatible with the idea that the open form is required to bind a large substrate polymer and is advantageous for the efficient exchange of substrates and products. Interestingly, a very recent report of the structures of a leaf-branch compost cutinase (*10*) in the open form revealed the bound solvent molecule of imidazole or dioxane from the crystallization condition in the vicinity of the active site, which is in almost the same position as our glycerol binding site. These findings suggest that the open form will provide a substrate waiting state as well as post-reaction state, as we proposed. In addition, the cyclic compound bound near the active site may reflect the location of the aromatic ring of the PET substrate.

Ca^{2+}-Mediated Thermostability and Activity

We identified three Ca^{2+} binding sites (Figure 4a) in the crystal structure of Cut190* S176A, designated as Sites 1–3 (*20*). Site 1 is the same as the reported Ca^{2+} binding site at the β1-β2 loop of Cut190 S226P (*17*), i.e., the allosteric Ca^{2+} binding site responsible for regulating the open/close conformational change at the active site. Sites 2 is composed of the side chains of Glu220, Asp250, and Glu296. Site 3 is composed from the side chains of Asp204 and Thr206 and the main-chain carbonyl of Thr206. These sites are also distal to the active site. We have demonstrated that the Ca^{2+} ions at Sites 1 and 2 contribute to the enzymatic activity and the thermostability, respectively, and the Ca^{2+} ions at Site 3 is related to both of these characteristics (*20*).

Site 1, which has the largest effect on activity, is associated solely with large structural changes of the enzyme. Because the substrate can gain access to the active site only in the Ca^{2+}-binding open form (*17*), the Ca^{2+}-dependent activity of Cut190 is easily understood. However, the Et-succinate bound form of Cut190* S176A clearly reveals that the electron density for Ca^{2+} disappears and the conformation of the β1-β2 loop is changed to the same as that of Ca^{2+}-free closed form (Figure 4b) (*19*). Since we used the crystals of the Ca^{2+}-binding open form for Et-succinate soaking, this conformational change and the Ca^{2+} release are triggered by the substrate binding, but are not due to the crystal packing. To investigate the Ca^{2+} binding on Cut190* (especially the Ca^{2+} binding on Site 1), we performed computational analyses using 3D reference interaction site model (RISM) calculations and molecular dynamics (MD) simulations (*19*). The results of the RISM confirmed the Ca^{2+} binding at Site 1 but showed that it ranked relatively low compared to other sites. The MD simulations also revealed a short average dissociation time at Site 1. Another confirmation of the relatively low affinity of Site 1 is our finding that the closed, Ca^{2+} unbound crystal structure at Site 1 was obtained under the condition of 100 mM $CaCl_2$ in the crystallization reagent (data not shown). Since Sites 2 and 3 are occupied by Ca^{2+} in this structure, the relative low affinity and easy-dissociation nature of bound Ca^{2+} at Site 1 were further validated. One possible cause of the formation of Ca^{2+} release under a high Ca^{2+} concentration is the formation of crystal packing forces of the closed form at Site 1 and the disruption of the coordinate site that is contributed by several main-chain carbonyls.

Combining the above findings, we propose a reaction cycle based in the obtained coarse snapshots of the crystal structures (Figure 4c). The closed, inactive form of Cut190 is induced to the open, substrate waiting form (I) accelerated by Ca^{2+} binding at Site 1. A substrate can approach the active site, but the open form would not be able to stably accommodate the substrate. A conformational change to the engaged form coupled with the Ca^{2+} release at Site 1 (II) is required to tightly hold the substrate and initiate the hydrolysis at the ester group. Once the hydrolysis reaction is complete, the carboxyl group is still tightly bound by the oxyanion hole. For the efficient exchange of the product to a new substrate, it is necessary to return to the open form again with Ca^{2+} re-binding at Site 1 (III).

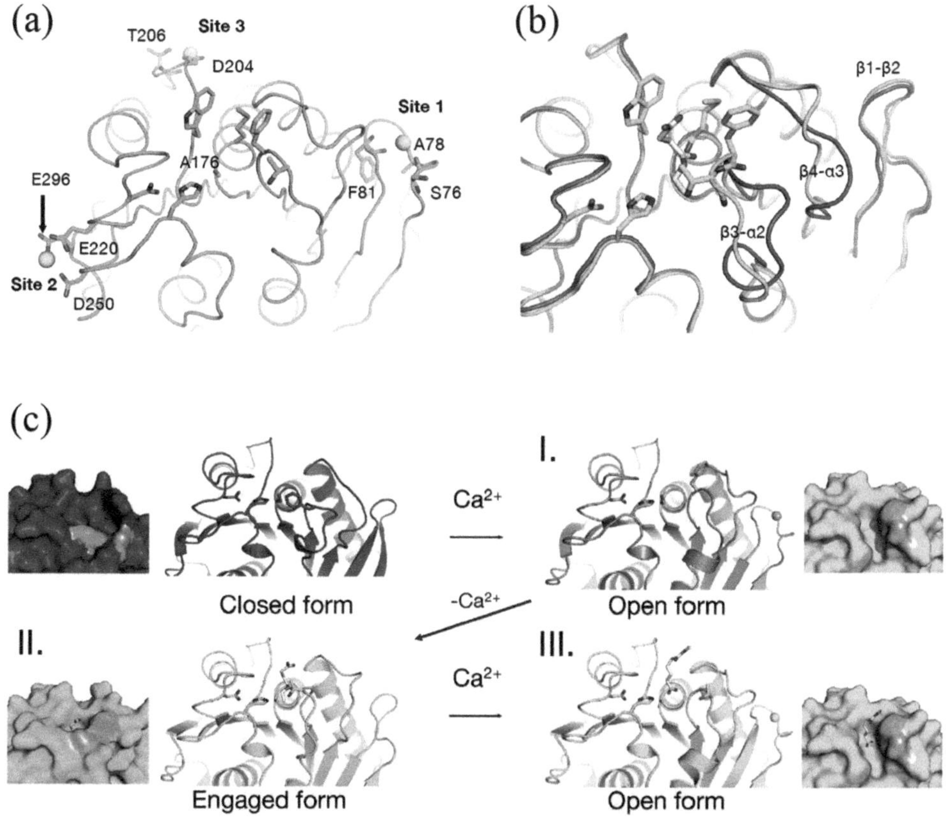

Figure 4. The Ca^{2+}-dependent regulation of Cut190. (a) Ca^{2+} binding sites identified in the structure of Cut190. Bound Ca^{2+} ions are shown as yellow spheres. Ca^{2+} coordinated residues are shown as a stick model and labeled. Residues around the active site are also represented as a stick model, but only Ala176 (substituted from Ser) is labeled. (b) Superposition of the Et-succinate bound structure (cyan) and that of the closed form (magenta). Although the $\beta3$-$\alpha2$ and $\beta4$-$\alpha3$ loops form different conformations in the open and closed forms, the $\beta1$-$\beta2$ loop reveals the same conformation as that of the closed form, associated with Ca^{2+} release. (c) Schematic diagram of the proposed reaction cycle of Cut190.*

Structure-Based Engineering of Cut190

For the efficient enzymatic hydrolysis of PET, a reaction at a temperature higher than the glass transition temperature (T_g) of approx. 60°–65°C in aqueous solution (*15*) is indispensable. To achieve such a high thermostability of Cut190 and also higher activity, we have attempted to engineer Cut190 by introducing mutations. Based on the structural information of Cut190*, the replacement of the Ca^{2+} bridge between Asp250 and Glu296 at Site 2 by a disulfide bond (i.e., D250C and E296C) results in a remarkable increase of the melting temperatures (T_m) of the mutant enzymes by >20°–30°C (*20*).

Substitution of the surface asparagine and glutamine is assumed to help improve thermostability because the deamidation of these residues is known to cause the instability of proteins (*24*, *25*). We tested various mutants and integrated the results of the disulfide bond introduction discussed above, and we eventually identified the mutants of Q138A/D250C-E296C/Q123H/ N202H, which showed reasonable enzymatic activity and the highest T_m values (*20*).

We recently determined the crystal structure of the improved mutant of Cut190* Q138A/ D250C-E296C/Q123H/ N202H (Figure 5) (*26*). The structure was solved as the Ca^{2+}-free closed form, and it is well matched with the previously reported closed form (*17*). Thus, our engineered design of Cut190 in which the surface residues are substituted including the introduction of a disulfide bond successfully achieved the improvement of the enzymatic profiles with respect to both the thermostability and activity, without significant structural changes.

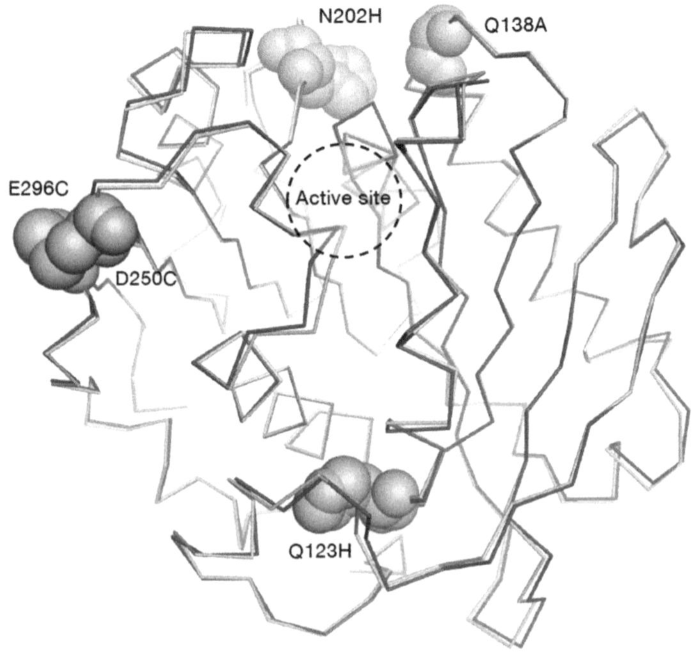

Figure 5. Structure of the engineered Cut190 (gold). Residues of the additional five substitutions from Cut190* are represented as a sphere model. The overall structure is superposed onto that of the closed form. The location of the active site is shown as a dashed circle.*

Conclusion

We have demonstrated in this chapter the potential advantages of the application of enzymatic treatment by cutinases to the efficient recycling of waste plastic. We have clarified the molecular mechanism of the Ca^{2+}-dependent enzymatic cycle of Cut190 based on the crystal structures complexed with two ligands and on the results of computational analyses. The overall structure of Cut190 is stable, but the large conformational changes at the three loops near the active site are enhanced by the Ca^{2+} binding and release at Site 1 during the enzymatic reaction cycle. These conformational changes are responsible for the open/close mechanism at the active site not only to permit the substrate access to the active residue but also to enhance the product dissociation and exchange with a new substrate to the subsequent reaction cycle.

Our analyses of the various mutants revealed that Ca^{2+} plays an important role in thermostability of Cut190, which is essential for the efficient degradation of PET. These structural bases enabled us to design an improved enzyme in its thermostability and activity. The introduction of a disulfide bond at Site 2 is likely to result in the disruption of Ca^{2+} binding, yet this substitution enhanced the thermostability of the enzyme. These findings suggest the possibility of achieving high thermostability in a calcium-independent manner. From the viewpoint of cost reduction, reducing the Ca^{2+} dependency will be an important milestone for future industrial applications of the enzymatic treatment of waste plastic.

Acknowledgments

We sincerely thank Ms. Yuri Yamagami and Ms. Akane Senga for performing the hard work of protein purification and crystallization when they were graduate students at Kyoto Prefectural University.

References

1. Carvalho, C. M. L.; Aires-Barros, M. R.; Cabral, J. M. S. *Biotechnol. Bioeng.* **1999**, *66*, 17–34.
2. Chin, I. S.; Murad, A. M. A.; Mahadi, N. M.; Nathan, S.; Abu Bakar, F. D. *Protein Eng. Des. Sel.* **2013**, *26*, 369–375.
3. Jaeger, K. E.; Dijkstra, B. W.; Reetz, M. T. *Annu. Rev. Microbiol.* **1999**, *53*, 315–351.
4. Pio, T. F.; Macedo, G. A. *Adv. Appl. Microbiol.* **2009**, *66*, 77–95.
5. Purdy, R. E.; Kolattukudy, P. E. *Biochemistry* **1975**, *14*, 2832–2840.
6. Martinez, C.; De Geus, P.; Lauwereys, M.; Matthyssens, G.; Cambillau, C. *Nature* **1992**, *356*, 615–618.
7. Liu, Z.; Gosser, Y.; Baker, P. J.; Ravee, Y.; Lu, Z.; Alemu, G.; Li, H.; Butterfoss, G. L.; Kong, X. P.; Gross, R.; Montclare, J. K. *J. Am. Chem. Soc.* **2009**, *131*, 15711–15716.
8. Kodama, Y.; Masaki, K.; Kondo, H.; Suzuki, M.; Tsuda, S.; Nagura, T.; Shimba, N.; Suzuki, E.; Iefuji, H. *Proteins* **2009**, *77*, 710–717.
9. Kitadokoro, K.; Thumarat, U.; Nakamura, R.; Nishimura, K.; Karatani, H.; Suzuki, H.; Kawai, F. *Polym. Degrad. Stab.* **2012**, *97*, 771–775.
10. Tournier, V.; Topham, C. M.; Gilles, A.; David, B.; Folgoas, C.; Moya-Leclair, E.; Kamionka, E.; Desrousseaux, M. L.; Texier, H.; Gavalda, S.; Cot, M.; Guemard, E.; Dalibey, M.; Nomme, J.; Cioci, G.; Barbe, S.; Chateau, M.; Andre, I.; Duquesne, S.; Marty, A. *Nature* **2020**, *580*, 216–219.
11. Roth, C.; Wei, R.; Oeser, T.; Then, J.; Follner, C.; Zimmermann, W.; Strater, N. *Appl. Microbiol. Biotechnol.* **2014**, *98*, 7815–7823.
12. Ribitsch, D.; Hromic, A.; Zitzenbacher, S.; Zartl, B.; Gamerith, C.; Pellis, A.; Jungbauer, A.; Lyskowski, A.; Steinkellner, G.; Gruber, K.; Tscheliessnig, R.; Herrero Acero, E.; Guebitz, G. M. *Biotechnol. Bioeng.* **2017**, *114*, 2481–2488.
13. Palm, G. J.; Reisky, L.; Bottcher, D.; Muller, H.; Michels, E. A. P.; Walczak, M. C.; Berndt, L.; Weiss, M. S.; Bornscheuer, U. T.; Weber, G. *Nat. Commun.* **2019**, *10*, 1717.
14. Hu, X. P.; Thumarat, U.; Zhang, X.; Tang, M.; Kawai, F. *Appl. Microbiol. Biotechnol.* **2010**, *87*, 771–779.

15. Kawai, F.; Oda, M.; Tamashiro, T.; Waku, T.; Tanaka, N.; Yamamoto, M.; Mizushima, H.; Miyakawa, T.; Tanokura, M. *Appl. Microbiol. Biotechnol.* **2014**, *98*, 10053–10064.
16. Inaba, S.; Kamiya, N.; Bekker, G. J.; Kawai, F.; Oda, M. *J. Therm. Anal. Calorim.* **2019**, *135*, 2655–2663.
17. Miyakawa, T.; Mizushima, H.; Ohtsuka, J.; Oda, M.; Kawai, F.; Tanokura, M. *Appl. Microbiol. Biotechnol.* **2015**, *99*, 4297–307.
18. Kawabata, T.; Oda, M.; Kawai, F. *J. Biosci. Bioeng.* **2017**, *124*, 28–35.
19. Numoto, N.; Kamiya, N.; Bekker, G. J.; Yamagami, Y.; Inaba, S.; Ishii, K.; Uchiyama, S.; Kawai, F.; Ito, N.; Oda, M. *Biochemistry* **2018**, *57*, 5289–5300.
20. Oda, M.; Yamagami, Y.; Inaba, S.; Oida, T.; Yamamoto, M.; Kitajima, S.; Kawai, F. *Appl. Microbiol. Biotechnol.* **2018**, *102*, 10067–10077.
21. Senga, A.; Numoto, N.; Yamashita, M.; Iida, A.; Ito, N.; Kawai, F.; Oda, M. *J. Biochem. Epub ahead of print https://doi.org/10.1093/jb/mvaa102.*
22. Hedstrom, L. *Chem. Rev.* **2002**, *102*, 4501–4523.
23. Rauwerdink, A.; Kazlauskas, R. J. *ACS Catal.* **2015**, *5*, 6153–6176.
24. Kaneko, H.; Minagawa, H.; Shimada, J. *Biotechnol. Lett.* **2005**, *27*, 1777–1784.
25. Wright, H. T. *Crit. Rev. Biochem. Mol. Biol.* **1991**, *26*, 1–52.
26. Emori, M.; Numoto, N.; Senga, A.; Bekker, G. J.; Kamiya, N.; Ito, N.; Kawai, F.; Oda, M. *Proteins: Struct., Funct., Bioinf.*, submitted.

Green Sustainable Building Blocks

Chapter 5

Cellulose-Derived Levoglucosenone, a Great Versatile Chemical Platform for the Production of Renewable Monomers and Polymers

Sami Fadlallah,[1] Louis M. M. Mouterde,[1] Gil Garnier,[1,2] Kei Saito,[3] and Florent Allais*,[1,2]

[1]URD Agro-Biotechnologies Industrielles (ABI), CEBB, AgroParisTech, 51100 Pomacle, France

[2]BioPRIA, Department of Chemical Engineering, Monash University, Clayton 3800, VIC, Australia

[3]School of Chemistry, BioPRIA, Monash University, Clayton 3800, VIC, Australia

***Email: florent.allais@agroparistech.fr**

Upon the raising of environmental awareness, the transition from unsustainable polymers based on fossil resources to sustainable bio-based materials has become a hot topic. Since then, researchers are developing new monomers derived from natural resources to meet the global need in commodity polymers and to replace those based on fossil feedstock. Levoglucosenone (LGO) is a readily accessible chiral molecule from the acid-catalyzed flash pyrolysis of cellulose that can be used as a versatile platform for a wide range of high value-added compounds. This chapter focuses on the new, yet rapidly evolving, field of LGO-derived monomers and their utility in the synthesis of polymers *via* different polymerization methods. Special considerations are given to the evaluation of the monomer syntheses and polymerization processes under the scope of green chemistry. The discussion is supported by the measurement of both atom economy and *E* factor, in order to assess the sustainability, as well as to provide a framework that allows the comparison between different approaches.

Introduction

Plastics are widespread valuable polymers with unique characteristics including attractive cost/performance ratio, low weight and corrosion resistance which make them used in various applications such as containers, automobiles and food packaging (*1*). Due to the massive worldwide demand, the global production of plastics (*2*), as a ubiquitous part of daily life, increased intensively to reach 359 million tons in 2018. This calls into question the environmental concerns as the vast majority of the synthetic polymers are produced from fossil fuels. It is thus required to look for new alternatives and bio-based resources for the polymer production (*3*).

Biomass is a natural feedstock that has been largely accepted as an alternative resource to fossil fuels to produce polymers (*4*). The reasons behind using biomass as an eco-friendly redundant starting material rely on two main inescapable facts: i) the urgent need to protect Earth from the severe environmental problems such as global warming and pollution generated due to the consumption of fossil fuels, ii) the inevitable progressive depletion of fossil resources within several hundred years. As a result, efforts were triggered to substitute the environmentally abusive resources by renewable ones for the production of sustainable polymers (*5*).

One of the main routes to produce biomass-based polymers is the use of bio-based monomers as building blocks with tailored structures (*6*). In this context, levoglucosenone (LGO) arises as a promising bio-candidate (*7–9*) and key-intermediate to produce bio-based monomers and their polymers. LGO is a sugar α,β-unsaturated ketone obtained from cellulose-based feedstock, such as bagasse and sawdust, through catalytic flash pyrolysis (*10–12*). The high potential of LGO relies on its ability to be chemically modified due to the presence of readily reactive functional moieties (*13*). For instance, the hydrogenation of its conjugated double bond was achieved to produce Cyrene (Scheme 1) (*14, 15*) by the company named Circa Group as a Registration, Evaluation, Authorization and Restriction of Chemicals (REACH)-compliant green aprotic polar solvent for various synthetic reactions (*16, 17*). This has encouraged Circa Group to develop the Furacell™ technology to produce LGO and Cyrene on a large scale (*18*). Moreover, the Baeyer-Villiger oxidation of LGO was developed to afford (*S*)-γ-hydroxymethyl-α,β-butenolide (HBO) (*19, 20*), a chiral butanolide product of high interest (Scheme 1) (*21*). Besides, (*S*)-γ-hydroxymethyl-γ-butyrolactone (2H-HBO) was prepared successfully either by the direct Baeyer-Villiger oxidation of Cyrene or the catalytic hydrogenation of HBO (Scheme 1) (*19, 20*).

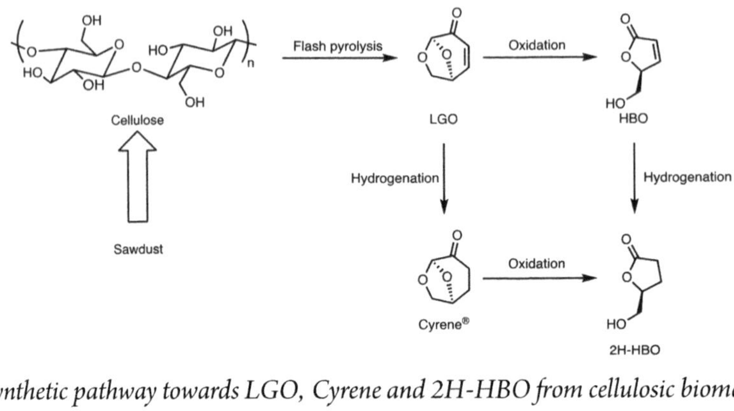

Scheme 1. Synthetic pathway towards LGO, Cyrene and 2H-HBO from cellulosic biomass (sawdust).

Zamzow and Höcker (*22*), used 2H-HBO to synthesize (*S*)-γ-hydroxymethyl-γ-butyrolactone methacrylate (m-2H-HBO) and reported its subsequent free-radical polymerization using azobisisobutyronitrile (AIBN) as an initiator (Scheme 2). Although 2H-HBO was not derived from LGO, the polymers produced were interesting and could undergo post-polymerization modification of the reactive lactone moiety to prepare polymers with pendent spiro-orthoesters groups. However, this work suffers from some drawbacks including the availability and greenness of the raw materials to access 2H-HBO as well as the use of methacryloyl chloride to form m-2H-HBO which results in a considerable amount of HCl released and the need of adding a base such as pyridine to trap the formed acid, thus making the synthesis less environmentally friendly.

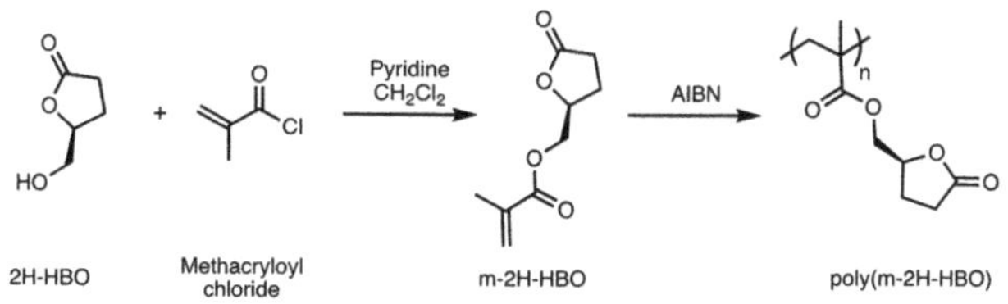

Scheme 2. Synthesis of poly(m-2H-HBO) from m-2H-HBO by free-radical polymerization.

Green Metrics: E Factor and Atom Economy

Identifying criteria to evaluate the greenness of a synthetic process is trivial when dealing with monomers and polymers issued from biomass. A direct way to do so, is to measure the green metrics, such as the environmental factor (*E* factor) (*23*) – usually expressed as kg of waster/kg of product – to evaluate the amount of waste generated during a chemical reaction. Ideally, the *E* factor of a sustainable process is equal to zero, which means no waste is generated. In contrast, a higher *E* factor implies a greater negative environmental impact (*23*). Another valuable metric, commonly used for rapid evaluation, is the atom economy (AE) proposed by Trost (*24*) to calculate how many atoms of the raw materials and reagents end up in the desired product. The *E* factor and AE can be calculated as follows:

$$E\ factor = \frac{\sum m(raw\ materials) + \sum m(reagents) + \sum m(solvents) - m(desired\ product)}{m(desired\ product)} \quad (1)$$

$$AE = \frac{MW(desired\ product)\ x\ 100}{\sum MW(raw\ materials) + \sum MW(reagents)} \quad (2)$$

To avoid any confusion, two derivatives of *E* factor will be employed in this chapter, the traditional *E* factor and the simple *E* factor (sEF) (*23*). The rationale behind our choice is that the sEF excludes solvents while calculating the waste generated and is useful for the primary evaluation of a process, whereas the traditional *E* factor takes into consideration the waste analysis of all reagents including solvents used in a reaction mixture (*23*).

$$sEF = \frac{\sum m(raw\ materials) + \sum m(reagents) - m(desired\ product)}{m(desired\ product)} \quad (3)$$

Water will be counted in the calculations whenever used as a solvent and a contribution percentage to the value will be given. To allow comparison between the different reports, only solvents and water used in the synthesis procedure are included, *i.e.* the work-up chemicals are excluded as the corresponding experimental details are not always provided.

Levoglucosenol

The development of readily accessible bio-derived monomers is more critical than ever in polymer chemistry. Previous reports have shown the preparation and polymerization of functional monomers from renewable resources, however, in most cases, the syntheses were accomplished using complex multistep processes which often required the use of toxic reagents (*25*). Schlaad and coworkers (*26, 27*) demonstrated recently that cellulose-derived LGO could be chemically modified and used for the synthesis of thermoplastic polyacetal using ring-opening metathesis polymerization (ROMP) (*28, 29*) or cationic ring-opening polymerization (CROP) (*30*). The same research group also investigated the production of polymers from LGO using radical or anionic polymerization, however, only oligomers with low molecular weight (M_w) were obtained.

Ring-Opening Metathesis Polymerization of HO-LGO

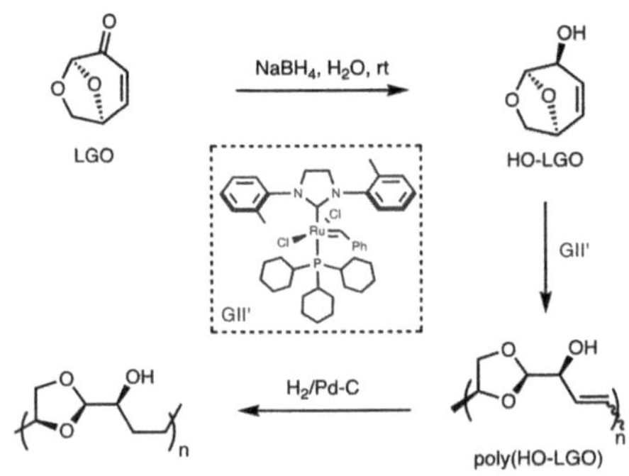

Scheme 3. Synthesis of polyacetals by ROMP of readily accessible LGO-derived monomers.

One of the interesting structural features of LGO is the enclosed strained double bond that can be used in metathesis polymerization. The attempts of Schlaad and coworkers (*26*) to conduct the ROMP of LGO proved unsuccessful, most probably because of the existence of polar carbonyl moiety that hampers the reactivity of the catalyst with the olefin (*31*). Such behavior was typically established for the early transition metal catalysts and showed that the tolerance of the catalysts to polar functional groups is inversely proportional to its reactivity (*31*). Thus, to overcome this issue, the reduction of the LGO keto moiety was realized using $NaBH_4$ in water to yield quantitatively HO-LGO (Scheme 3A). Using the data provided by Schlaad and coworkers (*26*), we calculated the AE that is found to be 78% (Table A1, entry 1), knowing that the calculation of AE assumes stoichiometric quantities of the starting materials, thus it does not count the excess of $NaBH_4$ used (1.6 equiv. with respect to LGO) and the reaction yield. Also, we estimated the *E* factor of the process as 9.5 (Table A1, entry 1) with water contribution of 84% to the value, nonetheless, a low value of *E* factor is registered if water is excluded (sEF = 0.6, Table A1, entry 1). Although water is non-toxic at the outset and is preferably used over the common organic solvents, the mentioned high contribution (84%), evidently shows its impact on the waste generated from a chemical reaction. This validates the necessity of using the least possible amount of water as a solvent for a synthesis.

Five different Ru-based catalysts were initially screened for the ROMP of HO-LGO, however, none of them was proved active. However, notably, the polymerization was successfully initiated with a derivative of Grubbs 2nd generation catalyst (GII) bearing mono-*ortho*-substituted *N*-heterocarbene (NHC) as a ligand (GII', Scheme 3). Although no clear explanation of the difference in reactivity between GII and GII' was given, the latter could be due to the lower steric hindrance of the NHC mono-*ortho* ligand around the Ru center, especially since GII' was reported as a highly efficient catalyst for the formation of bulky tetrasubstituted olefin by ring-closing metathesis.

The first attempts of HO-LGO polymerization using GII' were performed at room temperature in 1,4-dioxane. Although 1,4-dioxane is a toxic solvent, the authors tried to use concentrated solutions with up to 4 M of $[HO\text{-}LGO]_0$ which in turn decreased the amount of solvent waste generated and thus diminished the *E* factor of the process from 17.6 (for 2 M) to 4.3 (for 4 M)

(Table A2, entries 1 and 2a). Interestingly, at room temperature, and for monomer-to-catalyst ratio ([M]/[C]) of 100, the best yield (59%) was obtained when $[HO\text{-}LGO]_0 = 4$ M (compared to 28% at 2 M). Moreover, although a range of temperature (0-90 °C) was studied and showed that the best results were obtained at 0-5 °C, the first attempts of ROMP performed at room temperature showed a common but good activity of the metathesis catalyst even when [M]/[C] was increased to 1000 (50% yield in 24 h). This can help in saving energy as no heating or cooling is applied. The unreacted HO-LGO monomer was recovered at the end of the polymerization in quantitative yield and high purity upon purification by sublimation. Considering this efficient recovery, the *E* factor of the polymerization was reduced. For instance, we chose a typical experiment with 59% yield and the calculations showed that a decrease of *E* factor from 4.3 to 2.1 and of sEF from 0.8 to 0.06 is detected when the residual monomer is recovered, (Table A2, entries 2a,b). This further demonstrates the importance of recovering the residual monomers to minimize the amount of waste formed.

Concerning the molecular structure of the polyacetal (*32*) formed, different isomers of poly(HO-LGO) were observed by NMR analysis. This is due to the chirality of HO-LGO which can add in different orientations to the growing polymer chain, thus leading to the formation of sequence isomers such as head-to-head, head-to-tail and tail-to-tail. Despite the occurrence of stereocenters and olefin bonds along the backbone chain, an amorphous character and high glass transition temperature (T_g = 92-100 °C) were registered.

The importance of these new LGO-derived polymers relies also on the presence of functional groups in the polymer chain such as hydroxy and olefin groups. For instance, the hydrogenation of the poly(HO-LGO) double bonds was carried out in 98% yield to the alkane analog using H_2/Pd-C (Scheme 3). Furthermore, these polymers contain degradable acetal units that degrade completely within 40 days at room temperature in a 1,4-dioxane solution in the presence of *p*-toluenesulfonic acid.

Cationic Ring-Opening Polymerization of MeO-LGO

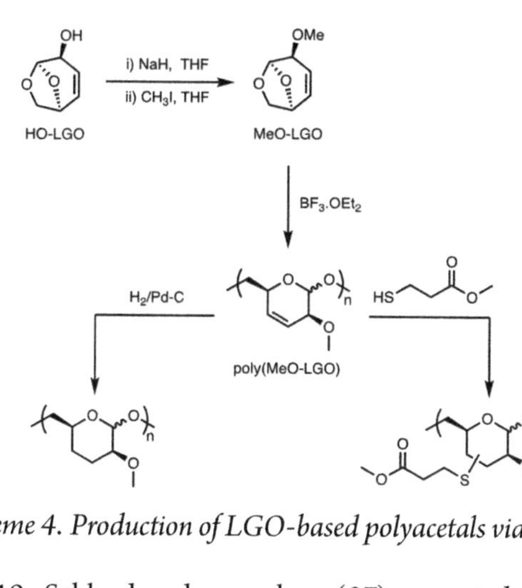

Scheme 4. Production of LGO-based polyacetals via CROP.

Later in the year 2019, Schlaad and coworkers (*27*), reported the reactivity of LGO in the cationic ring-opening polymerization (CROP). Their attempts to polymerize HO-LGO *via* a Lewis acid-catalyzed CROP process failed which led them to protect the hydroxy function and isolate a new monomer, MeO-LGO, in high yield (81%) (Scheme 4). The methylated analog of HO-LGO was readily obtained through Williamson-etherification in presence of sodium hydride and methyl iodide

(Scheme 4). As for the HO-LGO monomer, we measured the already described green metrics to evaluate the process. The evaluation of AE shows that 48% of the starting materials ended up in the MeO-LGO product (Table A1, entry 2). In addition, the necessity of adding an excess of CH_3I for the methylation to occur (1.75 equiv. with respect to HO-LGO) gives a relatively higher sEF of 2.0 and *E* factor of 9.0 (Table A1, entry 2) when the solvent (THF) is considered. It is important to note that while evaluating the greenness of a process by *E* factors, the toxicity of the reagents and solvents, *e.g.* THF *vs.* a green solvent such as 2-MeTHF (*33*), are not examined, which makes *E* factors less representative regarding this issue.

In contrast to HO-LGO, MeO-LGO monomer was found convenient for the CROP using triflic acid (TfOH) and boron trifluoride etherate ($BF_3.OEt_2$) in dichloromethane under various reaction conditions. TfOH appeared as a more efficient initiator for the CROP than $BF_3.OEt_2$ both at room temperature and 0 °C, resulting in high experimental M_w polymers in yields within the range of 84 to 92%. In the case of $BF_3.OEt_2$, much higher catalyst loading was required (20 and 30 mol% with respect to monomer), otherwise low conversion of monomer (as low as 2%) was obtained. This was rationalized by the necessity of having traces amount of water to provide the proton that acts as the initiation species for the polymerization (*34*). A comprehensive explanation is presented in the original publication including a mechanism for the proposed pathway of the proton-initiated CROP (*27*). As TfOH is a very strong acid, the authors chose to work with $BF_3.OEt_2$ to increase the scope of polymerization experiments at various conditions. However, generally, high catalyst loading is accompanied with some drawbacks including larger amounts of chemicals used which render the process less environmentally friendly. For instance, we calculated the *E* factors of two typical ($[MeO\text{-}LGO]_0 = 4$ M and assuming 100% yield) polymerization experiments with 0.5 mol% of TfOH as an initiator *vs.* 30 mol% of $BF_3.OEt_2$ and we obtained the following *E* factors, 2.2 (Table A2, entry 3a) and 2.5 (Table A2, entry 3b), respectively. Although the difference is not significant, the mass of acids used should be taken into account; 1.5 mg TfOH *vs.* 85 mg $BF_3.OEt_2$ (Table A2, entry 3a,b), especially when the process is performed on a large scale. As evidence, we calculated the sEF (by excluding the mass of DCM) of the two aforementioned polymerizations and they were found to be 0.3 in the case of 30 mol% $BF_3.OEt_2$ (Table A2, entry 3b) and 0.005 (almost zero waste generated) with 0.5 mol% triflic acid (Table A2, entry 3a) .

Contrary to poly(HO-LGO), the polyacetals obtained from the CROP (annotated as poly(MeO-LGO) in Scheme 4) contained only one sequence isomer. Moreover, poly(MeO-LGO) displayed the characteristics of semi-crystalline thermoplastic polymers with $T_g \sim 35$ °C and melting transition (T_m) at 40-120 °C. Such behavior is different from what was observed for poly(HO-LGO) (amorphous polyacetal with $T_g = 92\text{-}100$ °C). This shows the high impact of the polymerization method used to access the polymers, *e.g.* ROMP *vs.* CROP, which leads to distinct architectures and properties, and thus different potential applications of the synthesized polymers.

Post-polymerization modifications of poly(MeO-LGO) were also performed and afforded straightforward routes to new polymers through hydrogenation (Scheme 4) and thiol-ene chemistry (Scheme 4). Furthermore, poly(MeO-LGO) polymers were found to degrade within several hours in dichloromethane containing $BF_3.OEt_2$ and methanol.

Dihydrolevoglucosenol

As stated in the introduction, the reduction of the alkene moiety of LGO gives access to dihydrolevoglucosenone (aka Cyrene) (*35*) which is the precursor of an interesting monomer, m-Cyrene. Two steps are necessary to synthesize the latter (*36*). The first one consists in the reduction of the ketone moiety of Cyrene using $LiAlH_4$ in diethyl ether to access to levoglucosanol (aka HO-Cyrene) in 95% yield (Scheme 5). It has to be noted that this method produces a mixture of two enantiomers with an enantiomeric excess of 70% in favor of the (*S*) enantiomer. The levoglucosanol can then be converted into m-Cyrene™ in presence of methacrylic anhydride (MAN) and 4-dimethylaminopyridine (DMAP) in ethyl acetate in 85% isolated yield (81% overall yield over the two steps). Using the data provided by Saito, Simon and coworkers (*36*), we calculated the AE for the production of m-Cyrene to be 45% (Table A1, entry 3). Again, this calculation assumes stoichiometric quantities of the starting materials and does not consider the reaction yields. Also, we estimated the *E* factor of the process being 9.7 (Table A1, entry 3) with a solvent contribution of 73% to the value, nonetheless, a low value of *E* factor is obtained if the solvents are excluded (sEF = 1.9, Table A1, entry 3).

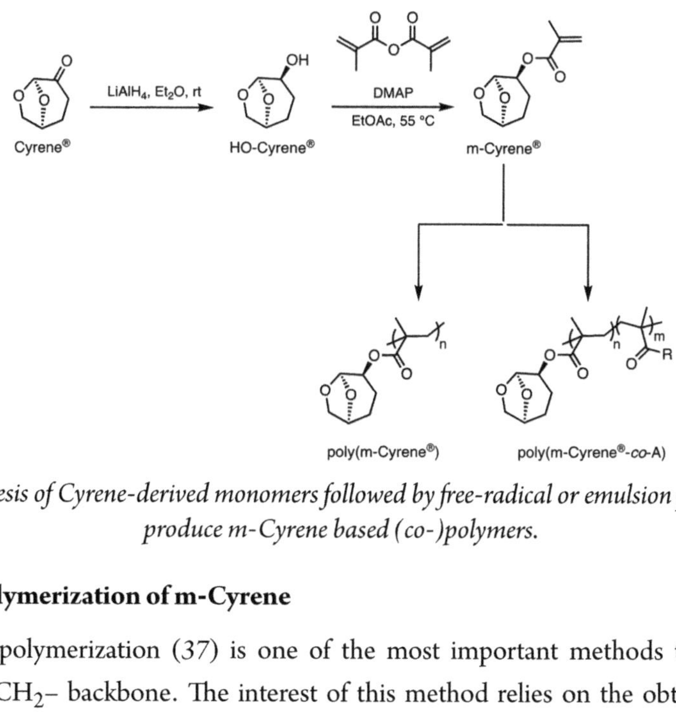

Scheme 5. Synthesis of Cyrene-derived monomers followed by free-radical or emulsion polymerization to produce m-Cyrene based (co-)polymers.

Free-Radical Polymerization of m-Cyrene

Free-radical polymerization (*37*) is one of the most important methods to make high M_w polymers with $-CH_2-$ backbone. The interest of this method relies on the obtaining of uniform polymer while minimizing the concentration of residual monomer. More than 50% of all plastics are industrially produced using this method. The ability of m-Cyrene to polymerize through AIBN-initiated free-radical polymerization was introduced by Saito, Simon and coworkers (Scheme 5) (*36*). The polymerization of m-Cyrene occurred successfully and led to highly valuable polymers (poly(m-Cyrene)) while conserving the cyclic rings.

The authors applied different reaction conditions for the polymerization experiments. Firstly, the bulk polymerization of pure m-Cyrene monomer was carried out by adding a soluble initiator such as AIBN at 72 °C. It is known that the best solvent is no solvent, thus this method can be considered as the greenest when compared to the solution polymerization of the same type, despite the low yield obtained (45%). Apart from some well-controlled polymerization techniques (*38*), a yield of < 50% is common and is known as one of the drawbacks when dealing with bulk polymerization. We looked at the experimental data provided in the original publication to measure the *E* factor that is equal to 1.2 (Table A2, entry 4a). It is important to note that the relatively higher value of *E* factor is due to the low conversion of the monomer, for instance, if we assume 100% conversion, only 0.01 kg waste/kg of polymer is recorded (Table A2, entry 4b). Solution polymerization of m-Cyrene was also performed with different solvents (*39*), *n*-butanol, dimethyl sulfoxide, ethylene glycol, methyl isobutyl ketone, γ-valerolactone and Cyrene (*36*). Better yields along with high M_w polymers were obtained especially in Cyrene (85% and 38 kg.mol^{-1}). Looking at the *E* factor of 4.9 (Table A2, entry 5), one can conclude that the use of a solvent increased the *E* factor significantly compared to bulk polymerization (Table A2, entries 4a,b). Surely, the toxicity of the reaction solvent has to be taken into account where Cyrene does not have the same environmental impact as dimethyl sulfoxide for example. It is important to note that the copolymerization ability of m-Cyrene with common methacrylic, such as isobornyl methacrylate, was also studied in Cyrene, and showed that a feed ratio of 1:9 (m-Cyrene: comonomer) is sufficient for the comonomer to insert in high proportion (up to 85%) into the m-Cyrene polymer chain (Scheme 5D).

In most cases, poly(m-Cyrene) with high thermal stability was obtained (20% degradation at 290 °C) with high T_g up to ~ 210 °C which makes them one of the best-reported high T_g vinyl polymers up-to-date, especially for their commercial utilization, both in industrial and personal care applications.

Emulsion Polymerization of m-Cyrene™

Saito, Simon and coworkers (*36*), then chose to study the behavior of m-Cyrene in emulsion polymerization (*40*) which is known as one of the most important methods to produce several commercial polymers. One of the biggest advantages of this technique is the use of water as a solvent which is non-toxic and abundant at the same time. Sodium dodecyl sulfate (SDS) was used as the initiator and afforded poly(m-Cyrene) in the highest yield (92%) compared to bulk (45%) and solution (85%) free-radical polymerization, along with high M_w (39 kg.mol^{-1}). The increased overall reactivity of the reaction is due to the combined solution and bulk effects in the emulsion-based system. Indeed, water acts as a good heat dispersant, even better when compared to other organic solvents, and each formed micelle behaves like a mini-bulk reactor. The *E* factor of the emulsion polymerization process is 4.5 (Table A2, entry 6), which is quite similar to that of the Cyrene-based polymerizations (*E* factor = 4.9, Table A2, entry 5), however, with water contribution of 79% to the waste. Notably, the *E* factor is 0.2 if water is not considered (Table A2, entry 6).

(S)-γ-Hydroxymethyl-γ-butyrolactone (HBO)

Although reduction reactions involving LGO yielded valuable monomers, oxidation reaction can also be used to produce novel bio-based monomers. Allais and coworkers showed that LGO could be efficiently converted to HBO through a Baeyer-Villiger oxidation using 4.9 equiv. of a 30% aqueous solution of H_2O_2 in 71% yield (Scheme 6) (*20*). The alkene moiety of the latter can then be reduced by dihydrogen (H_2) in the presence of a catalytic amount (10 % w/w) of 10 % palladium on carbon (Pd/C) in 87% yield (*19*). The resulting 2H-HBO is the precursor of two monomers, m-2H-HBO and M-THP-2H-HBO. m-2H-HBO is readily obtained by reacting 2H-HBO with methacrylic anhydride in ethyl acetate in presence of DMAP and triethyl amine (TEA) (79% yield) (*41*). This three-step synthesis from LGO has an overall yield of 49%. Using the data provided by Allais and coworkers (*19, 20*), and Saito, Simon and coworkers (*41*), we calculated the AE for the production of m-2H-HBO to be 29% (Table A1, entry 4). Also, we estimated the *E* factor of the process being 29.4 (Table A1, entry 4) with a solvent contribution of 55.5% to the value. A low value of *E* factor is obtained if the solvents are excluded (sEF = 12.5, Table A1, entry 4).

The second monomer of interest, M-THP-2H-HBO, can be obtained from 2H-HBO in a two-step process. The first step consists in the protection of the alcohol moiety of 2H-HBO using pyridinium *p*-toluenesulfonate (PPTS) and 3,4-dihydro-2H-pyran (DHP) in dichloromethane in quantitative yield. The α-methylenation of the latter then provides M-THP-2H-HBO in 62% yield. This four-step synthesis from LGO has an overall yield of 38%. Using the data provided by Allais and coworkers (*19, 20, 42*), we calculated the AE and the *E* factor for the production of M-THP-2H-HBO to be 27% and 35.6, respectively (Table A1, entry 5). A low value of *E* factor is obtained if solvents are excluded (sEF = 9.4, Table A1, entry 5).

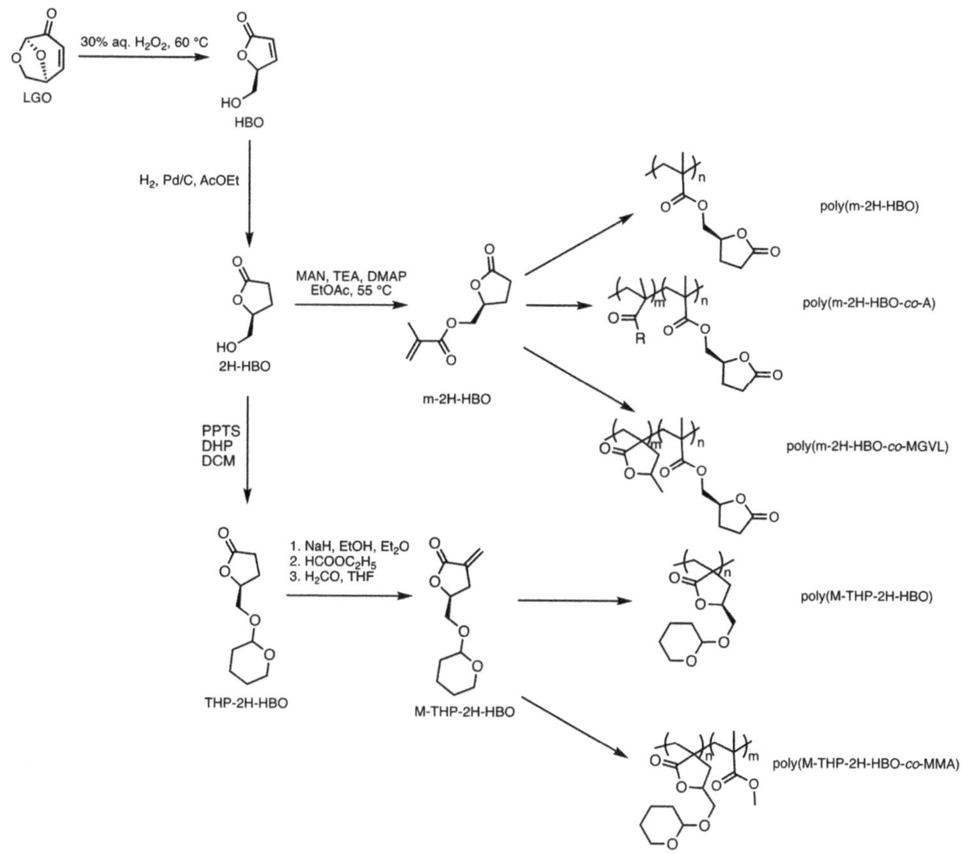

Scheme 6. Synthesis and polymerizations of HBO-derived monomers.

Free-Radical Polymerization of m-HBO, m-2H-HBO and M-THP-2H-HBO

Saito, Simon and coworkers (*41*) studied the ability of m-2H-HBO to polymerize in the presence of AIBN initiator in bulk conditions while ensuring the miscibility of both components before starting the polymerization at 65 °C. As in the case of m-Cyrene (Scheme 5), the solventless free-radical polymerization converted only 46% of the m-2H-HBO monomers to poly(m-2H-HBO) (Scheme 6D). Also, high dispersities were obtained (up to 4.2) which shows the uncontrolled behavior of the polymer growth. The *E* factor of the process is 1.2 (Table A2, entry 7) where 53% of the value corresponds to the unconverted monomers. For instance, if the residual monomers were recovered, this could diminish the *E* factor of the process to almost 0.01. Thus, it is of high importance to recover the residual monomers especially in the case of low conversions, this will in turn decrease the negative impact on the environment and, at the same time, decrease the cost of the synthesis.

Solution free-radical polymerizations were also attempted by the same workers (*41*) in different solvents such as xylene, toluene, ethyl acetate, *n*-butanol, ethylene glycol and dimethyl sulfoxide. Regardless of the difference in toxicity of the different solvents, the *E* factor is 7.2 if 61% of the monomer was converted and sEF is 0.7 (Table A2, entry 8). Later in the same year, Miller, Allais and coworkers (*42*) reported the free-radical polymerization of m-2H-HBO (synthesized by the chemo-enzymatic route) with AIBN as initiator. Although they followed almost the same polymerization conditions as Saito, Simon and coworkers (*41*), the results obtained were better with regards to the waste produced (*42*). Notably, we recorded a decrease of *E* factor from 7.2 to 5.2 and sEF from 0.7 to 0.1 (Table A2, entries 8 and 9 respectively) due to the enhancement of the monomer conversion described as quantitative in the original publication (*42*). The same recent work reported the use of other radical initiators such as V-65 and V-70; however, low yields were obtained. The produced polymers showed comparable T_g values ~ 96 °C and thermal stability up to ~ 200 °C with 20% degradation at 300 °C. Such behavior is very close to that of poly(methyl methacrylate) and hence 2H-HBO can be considered as a suitable candidate to replace the methyl methacrylate (MMA) (*43*). Interestingly, the copolymerization of m-2H-HBO with methacrylamide (MAA) (Scheme 6, R = NH_2) was conducted in DMF (for solubility issues) and led to copolymers with higher T_g values, up to 187 °C when 89% of the MAA was incorporated in the backbone chain. Furthermore, m-2H-HBO was successfully copolymerized with the bio-based monomer methylene-γ-valerolactone (MGVL) yielding also valuable copolymers with high comparable T_g. On the other hand, the attempts to use M-THP-2H-HBO as a monomer were accomplished and yielded homopolymers and copolymers (with MMA) having the same T_g ~ 148 °C.

Emulsion Polymerization of m-2H-HBO

With consideration to the environmental impact of the solvent-based polymerizations, which can be detrimental especially when dealing with very toxic solvents such as DMF, emulsion polymerization in water appears as a greener method and would be preferred if leading to polymers with similar properties to those done in solution. Interestingly, Saito, Simon and coworkers (*41*) showed that the polymerization of m-2H-HBO performed in emulsion afforded polymers of higher M_w, yield and T_g when compared to bulk and solution polymerization reported in the same work. Additionally, the copolymerization capability in water with other commercially available monomers was reported and showed how the comonomer polarity and steric hindrance can affect its insertion into the polymer chain (*41*). Polar monomers such as MMA were found to copolymerize well with m-2H-HBO. Differently, lower copolymer composition ratios were obtained in the case of non-polar monomers such as styrene. To achieve more control along the polymerization process, reversible addition-fragmentation chain transfer (RAFT) polymerization (*44–46*) was investigated in the presence of chain transfer agent (CTA) such as 4-Cyano-4-(((dodecylthio)-

carbonothioyl)thio)pentanoic acid (BM1432) (*41*). CTA is often added to control the chain length during the polymerization to achieve certain properties. RAFT polymerization produced poly(m-2H-HBO) with defined and controlled M_w fitting well with the experimental ones. Moreover, authors showed that this method – under identical conditions to the previously mentioned emulsion polymerizations – gave low dispersity as low as 1.06, which is another indication of improved control of the polymerization in the presence of the CTA (BM1432).

LGO-Cyrene-Based Bicyclic 2H-HBO-HBO

As shown above, LGO can be transformed into new bio-based monomers and polymers. Nevertheless, LGO can also undergo dimerization *via* a Michael addition catalyzed by a primary alcohol such methanol or ethanol. This dimerization reaction was first described by Shafizadeh and coworkers in 1982 (*47*) and was then optimized by Allais and coworkers in 2020 (*48*). Indeed by dissolving LGO in methanol in the presence of a catalytic amount of safe K_2CO_3, the homocoupling of LGO yielded LGO-Cyrene dimer in quantitative yield (Scheme 7). The latter can then undergo Bayer-Villiger oxidation in the same reaction conditions than for the oxidation of LGO to HBO (30% aq. H_2O_2) and provides the 2H-HBO-HBO dimer in 60% yield (Scheme 7B). 2H-HBO-HBO was subsequently used to access new bio-based polyesters. Using the data provided by Allais and coworkers (*48*), we calculated the AE and the *E* factor for the production of 2H-HBO-HBO to be 76% and 3.8, respectively (Table A1, entry 6). Considering a solvent contribution of 24%, the *E* factor calculated if solvents are excluded is sEF = 2.6 (Table A1, entry 6).

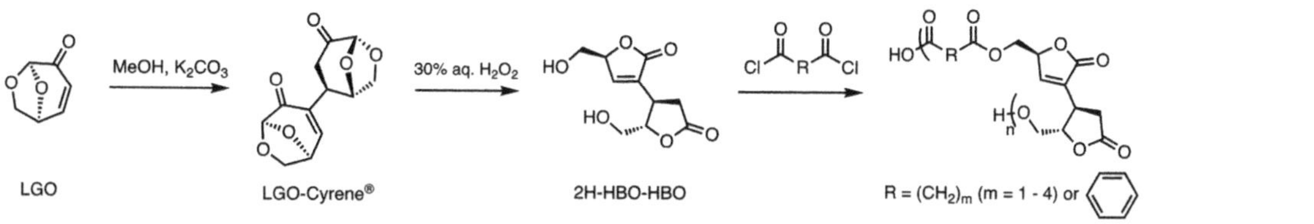

Scheme 7. Synthesis of 2H-HBO-HBO by the Baeyer-Villiger oxidation of the LGO-Cyrene dimer followed by polycondensation to produce HBO-based polyester.

The bio-based bicyclic diol monomer, 2H-HBO-HBO, was reacted with diacyl chlorides through polycondensation polymerization to yield a family of 2H-HBO-HBO based polyesters (*48*). Three general criteria led to the selection of the diacid monomers in that approach: they are i) commercially available, ii) difunctional to allow the formation of linear polymers rather than crosslinked or branched ones, iii) with different carbon chain lengths to tune the thermal properties of the resulting polymers. Different diacyl chlorides were employed for this purpose, ranging from malonyl chloride (m = 1) to adipoyl chloride (m = 4), in addition to the aromatic monomer terephthaloyl chloride.

In this study, Allais and coworkers (*48*) considered the preparation of the polymers in bulk conditions due to the poor solubility of 2H-HBO-HBO in common organic solvents. We calculated the *E* factor of the polyester obtained from 2H-HBO-HBO and malonyl chloride and found it to be 1.5 (Table A2, entry 10) which is relatively low when compared to the *E* factors of the solution polymerizations (Table A2, entries 5, 6, 8 and 9). However, especially in the case of polycondensation, bulk polymerization is usually accompanied by drawbacks including the low control of the process over the M_w distribution, the occurrence of side reactions such as cyclization, as well as, the production of oligomers which decrease the homogeneity of the material produced.

Besides, upon each condensation reaction and while creating new bonds to form the polymer chain, the monomers lose small molecules as by-products such as water, alcohol or even acids, which generally increase the *E* factor and decrease the AE, especially when dealing with industrial-scale processes (*49*).

As expected, the thermal properties of the 2H-HBO-HBO-based polymers showed that the T_g is greatly affected by the length of the aliphatic chain of the diacyl chloride. The T_g attained 71 °C when succinyl chloride was engaged and 81 °C in the case of terephthaloyl chloride, thus competing with commodity plastics such as poly(ethylene terephthalate) with high stability up to 300 °C where 50% of the polymer degradation occurred (*48*).

Norbornene-(iso-)levoglucosenone

LGO contains a reactive conjugated double bond which can be reacted with cyclopentadiene (Cp) by Diels-Alder reaction to produce norbornene-containing LGO molecules. The synthesis of this molecules was described in the 1980s by Shafizadeh et al. (*50*) and Horton et al. (*51*) and showed that LGO can act as dienophile to afford two isomers, *endo*-N-LGO and *exo*-N-LGO, as major and minor products respectively. In recent times, Banwell and coworkers (*52*) reported the synthesis of a library of monomers derived from *endo*-N-LGO (named as N-LGO in this chapter) and N-isoLGO (isoLGO is the pseudo-enantiomer of N-LGO) (*53*) (Scheme 8). The synthesis of these precursors (N-LGO and N-isoLGO) was carried out in DCM with a 70% yield. Using the data provided by Banwell and coworker, we calculated the AE and *E* factor for the production of N-LGO and N-isoLGO to be 100% and 6.83, respectively (Table A1, entry 7), considering a solvent contribution of 57%. The *E* factor calculated if solvents are excluded is sEF = 2.36 (Table A1, entry 7). It has to be noted that 3 supplementary steps are necessary to access isoLGO from LGO which greatly affect its AE, *E* factor and sEF. Banwell and coworker also synthesized a series of N-LGO derivatives in order to produce new monomers. Indeed, N-LGO can be reduced to the corresponding alcohol using $NaBH_4$ in an aqueous ethanol solution with a 92% yield. The AE and the *E* factor for the production of N-LGO alcohol were calculated to be 84.5% and 85.3, respectively and with a sEF = 0.375 (Table A1, entry 8). It has to be noted that the high *E* factor value is due to the large quantity of ethanol involved in the reaction. However, the sEF value is very comparable to the one calculated from Schlaad and coworkers' studies on the reduction of LGO to OH-LGO (Scheme 3). The two epimeric alcohols can be isolated and used to perform nucleophilic substitution (S_N1) in presence of ethyl α-bromoacetate in dry THF to produce the corresponding esters. The latter can be reduced to the corresponding carboxylic acids using LiOH in methanol or to the corresponding alcohols using $LiBH_4$ in THF. Finally, these alcohols intermediates can overgo fluorination using Diethylaminosulfur trifluoride (DAST) in DCM. The yields of these different transformations in functions of the precursor (N-LGO or N-isoLGO) are summarized in Table 1.

Banwell and coworkers showed that N-LGO, N-isoLGO and their derivatives can be used to produce a diverse functional and stereochemical polymers *via* ROMP (Scheme 8). The ability of the strained norbornene-adduct to be opened by ROMP was tested using the following metathesis catalysts, GII, third-generation Grubbs' (GIII), second-generation Hoveyda-Grubbs' (HGII) and the Stewart-Grubbs' (HG) catalysts. The four catalysts were able to fully convert N-isoLGO into the corresponding polymer, poly(N-isoLGO), where GIII showed the best control over the polymerization process. Thus, GIII was selected to go further in the ROMP of N-LGO, N-isoLGO and their derivatives (Scheme 8). The authors performed a series of experiments in diluted conditions in DCM with substrate concentration ranging from 0.05 to 1 M. A minimum concentration of 0.1 M was needed to allow the full conversion of the substrate, in addition, more catalyst loading was required in more diluted solutions. To highlight the effect of solvent, especially toxic solvents such as DCM, on the *E* factor, we calculated the *E* factor (excluding the work-up solvents) of three selected polymerization experiments with 100% conversion of the monomers, performed at 0.1, 0.5 and 1 M in DCM using GIII and 250 equiv. of N-isoLGO, as expected, the amount of waste generated from the polymerization processes dramatically increased as the concentration was reduced. At 0.1 M, *E* factor was found to be 69.2 kg of waster/kg of polymer with 99% of solvent contribution to waste (Table A2, entry 11). A remarkable decrease in *E* factor was registered in more concentrated medium to reach 13.9 and 6.9 at 0.5 and 1 M respectively (Table A2, entries 12 and 13, respectively).

Table 1. Yields in Function of the Precursor ((*S*) or (*R*) Enantiomer)

	(S)-N-LGO alcohol	*(R)-N-LGO alcohol*
S_N1	65%	69%
LiOH reduction	95%	95%
$LiBH_4$ reduction	66%	68%
Fluorination	54%	42%

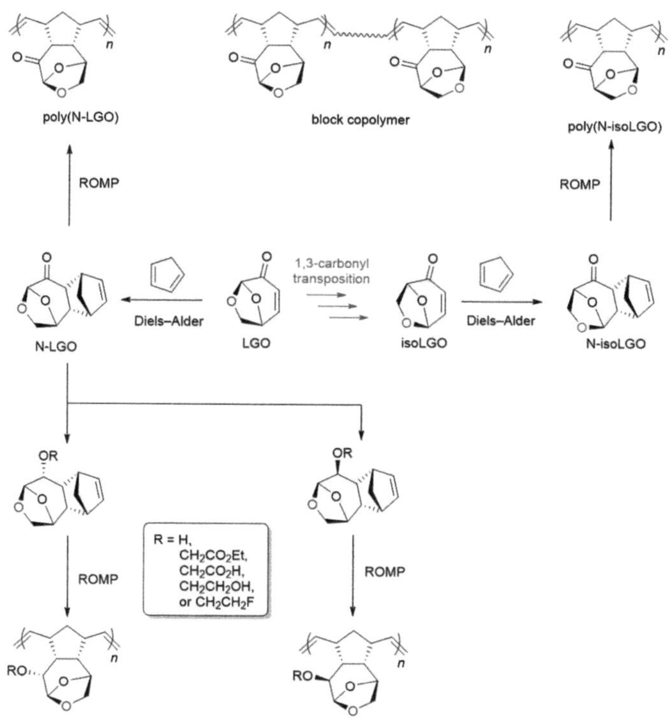

Scheme 8. Synthesis of a library of monomers and polymers from N-LGO and N-isoLGO.

The TGA analysis of the prepared polymers showed that they are thermostable up to 140 °C and 168 °C in the case of N-LGO and N-isoLGO-based polymers respectively. Notably, the same authors showed the capacity of the pseudo-enantiomeric monomers to form highly thermostable copolymers ($T_{d5\%}$ up to 278 °C) based on N-LGO and N-isoLGO.

Conclusion

LGO is as a promising bio-candidate and key-intermediate to produce bio-based monomers and their corresponding polymers. LGO can provide Cyrene *via* hydrogenation, and the latter can be used as a monomer for further polymerization or can be transformed into another interesting monomer, m-Cyrene. Oxidation of LGO can produce HBO, and the latter can be converted further to m-2H-HBO or M-THP-2H-HBO monomers. LGO can also undergo dimerization and subsequent Bayer-Villiger oxidation to provide 2H-HBO-HBO monomer. Moreover, the double bond of the LGO was employed to synthesize a library of norbornene-containing LGO-based monomers. All these bio-derived monomers from LGO provide corresponding polymers *via* different polymerization techniques such as ring-opening, free-radical, RAFT, emulsion, and condensation polymerizations.

The sustainability of all these monomer and polymer syntheses was evaluated using both AE and *E* factor. For the monomers syntheses, AE decreased as LGO is modified. The synthesis of N-LGO and N-isoLGO had the best AE (100%) and the synthesis of m-2H-HBO and M-THP-2H-HBO monomer had the lowest AE due to the further modification. For polymerization, the *E* factor increased as solvent was used. Bulk polymerization showed the lowest *E* factor, however, the conversion of the monomer was depressed. These results indicate that the selection of a sustainable solvent is the essential factor for the green polymerization of renewable monomers.

Appendix

Table A1. Calculation of the AE, sEF and *E* Factor for the Monomer Syntheses

Entry	*Input*			*Output*	*Yield*	*AE*	*sEF*	*E factor* [b]	*Ref*
	material	*m (g)*	*n (mol)*	*m (g)*	*(%)* [a]	*(%)*			
1	LGO	5 g	0.04	4.5	90	78	0.6	9.5	(26)
	$NaBH_4$	2.4 g	0.06						
	water	40 g	2.22						
2	HO-LGO	10.24	0.08	10.2	90	48	2.0	9.0	(25)
	THF	71.2	0.98						
	NaH	0.45	0.02						
	CH_3I	20	0.14						
3	Cyrene	10	0.78	8.6	81	45	1.9	9.7	(36)
	$LiAlH_4$	3.26	0.86						
	ET_2O	35	0.48						
	DMAP	0.09	0.0008						
	MAN	11.85	0.077						
	EtOAc	31.6	0.36						

Table A1. (Continued). Calculation of the AE, sEF and *E* Factor for the Monomer Syntheses

Entry	*Input*			*Output m (g)*	*Yield (%)* [a]	*AE (%)*	*sEF*	*E factor* [b]	*Ref*
	material	*m (g)*	*n (mol)*						
4	LGO	14.5	0.115	7.9	49	29	12.5	29.4	(*19, 21, 41*)
	30% aq. H_2O_2	63.8	0.563						
	Pd/C	1.45	0.014						
	DMAP	0.1	0.00082						
	TEA	13.1	0.129						
	MAN	13.9	0.09						
	EtOAc	133.4	1.51						
5	LGO	5	0.040	3.94	38	27	9.4	35.6	(*19, 21, 42*)
	30% aq. H_2O_2	22	0.195						
	Pd/C	0.5	0.005						
	EtOAc	7.75	0.088						
	PPTS	0.785	0.0031						
	DHP	4.69	0.0558						
	DCM	45.75	0.538						
	60% NaH in mineral oil	1.20	0.030						
	EtOH	0.142	0.0030						
	$HCOOC_2H_5$	2.20	0.030						
	Et_2O	32	0.43						
	H_2CO	4.5	0.15						
	THF	17.75	0.246						
6	LGO	2.59	0.0205	1.536	60	76	2.6	3.8	(*48*)
	K_2CO_3	0.45	0.0003						
	MeOH	1.74	0.054						
	30% aq. H_2O_2	2.52	0.0318						
7	LGO or isoLGO	1.4	0.0111	1.49	70	100	2.36	6.83	(53)
	Cyclopentadiene	3.615	0.0547						
	DCM	6.65	0.0783						

Table A1. (Continued). Calculation of the AE, sEF and *E* Factor for the Monomer Syntheses

Entry	*Input*			*Output*	*Yield*	*AE*	*sEF*	*E factor* [b]	*Ref*
	material	*m (g)*	*n (mol)*	*m (g)*	*(%)* [a]	*(%)*			
8	N-LGO	3.2	0.0166	3.017	93	84.5	0.37	85.30	(53)
	$NaBH_4$	0.948	0.0251						
	Ethanol	14.8	0.8218						
	Water	241.43	5.2405						

[a] Yield recorded in the original publications, unless stated otherwise. [b] Including process water in the calculations.

Table A2. Calculations of the sEF and *E* Factor for the Polymerization Processes

Entry	*Input*			*Output m (g)*	*Yield (%)* [a]	*sEF*	*E factor* [b]	*Ref*
	material	*m (g)*	*n (mmol)*					
1	HO-LGO [c]	0.128	0.9	0.035	28	2.9	17.6	(26)
	Dioxane	0.515	5.8					
	GII'	0.008	1.10^{-5}					
2	HO-LGO [d]	0.128	0.9					(26)
	Dioxane	0.258	2.9					
	GII'	0.008	1.10-5					
a				0.075	59	0.8	4.3	
b				0.128	100 [e]	0.06	2.1	
3	MeO-LGO [f]	0.298	2					(27)
	DCM	0.665	7.8					
a	TfOH [g]	0.0015	0.01	238	100 [i]	0.005	2.2	
b	BF_3OEt_2 [h]	0.085	0.6	298	100 [i]	0.3	2.5	
4	m-Cyrene	1	5					(36)
	AIBN	0.01	0.06					
a				0.45	45	1.2	1.2	
b				1	100 [i]	0.01	0.01	
5	m-Cyrene	1	5	0.85	85	0.2	4.9	(36)
	AIBN	0.01	0.06					
	Cyrene	4	30					

Table A2. (Continued). Calculations of the sEF and *E* Factor for the Polymerization Processes

Entry	Input			Output m (g)	Yield (%) [a]	sEF	E factor [b]	Ref
	material	m (g)	n (mmol)					
6	m-Cyrene	1	5	0.92	92	0.2	4.5	(36)
	SDS	0.05	0.17					
	Water	4	222					
	APS	0.01	0.004					
7	m-2H-HBO	1	5.4	0.46	46	1.2	1.2	(41)
	AIBN	0.01	0.06					
8	m-2H-HBO	1	5.4	0.61	61	0.7	7.2	(41)
	AIBN	0.01	0.06					
	toluene	4	43.4					
9	m-2H-HBO	0.5	2.7	0.45	90 [j]	0.1	5.2	(42)
	AIBN	0.005	0.03					
	toluene	2.3	24.9					
10	2H-HBO-HBO	0.885	3.75	0.57	40	1.5	1.5	(48)
	Malonyl chloride	0.536	3.80					
11	N-isoLGO	0.048	0.25	0.048	100	0.02	69.2	(53)
	GIII	0.0009	0.001					
	DCM	3.3	39.2					
12	N-isoLGO	0.048	0.25	0.048	100	0.02	13.9	(53)
	GIII	0.0009	0.001					
	DCM	0.665	7.8					
13	N-isoLGO	0.048	0.25	0.048	100	0.02	6.9	(53)
	GIII	0.0009	0.001					
	DCM	0.332	3.9					

[a] Yields recorded in the original publications, unless stated otherwise. [b] Including process water in the calculations. [c] [HO-LGO] = 2 M. [d] [HO-LGO] = 4 M. [e] Assuming 100% yield as the residual monomers were recovered. [f] [MeO-LGO] = 4 M. [g] 0.5 mol% of TfOH. [h] 30 mol% of BF_3OEt_2. [i] Assuming 100% conversion of the monomer to allow the comparison. [j] The authors mentioned quantitative yield.

References

1. *Applications of Polymers*; Seymour, R. B., Mark, H. F., Eds.; Springer US: Boston, MA, 1988. https://doi.org/10.1007/978-1-4684-5448-2.

2. Lebreton, L. C. M.; van der Zwet, J.; Damsteeg, J.-W.; Slat, B.; Andrady, A.; Reisser, J. River Plastic Emissions to the World's Oceans. *Nat. Commun.* **2017**, *8* (1), 15611. https://doi.org/10.1038/ncomms15611.
3. Wang, W.; Ge, J.; Yu, X.; Li, H. Environmental Fate and Impacts of Microplastics in Soil Ecosystems: Progress and Perspective. *Sci. Total Environ.* **2020**, *708*, 134841. https://doi.org/10.1016/j.scitotenv.2019.134841.
4. Iwata, T. Biodegradable and Bio-Based Polymers: Future Prospects of Eco-Friendly Plastics. *Angew. Chem., Int. Ed.* **2015**, *54* (11), 3210–3215. https://doi.org/10.1002/anie.201410770.
5. Mecking, S. Nature or Petrochemistry?—Biologically Degradable Materials. *Angew. Chem., Int. Ed.* **2004**, *43* (9), 1078–1085. https://doi.org/10.1002/anie.200301655.
6. Wang, B.; Ma, S.; Li, Q.; Zhang, H.; Liu, J.; Wang, R.; Chen, Z.; Xu, X.; Wang, S.; Lu, N.; Liu, Y.; Yan, S.; Zhu, J. Facile Synthesis of "Digestible", Rigid-and-Flexible, Bio-Based Building Block for High-Performance Degradable Thermosetting Plastics. *Green Chem.* **2020**, *22* (4), 1275–1290. https://doi.org/10.1039/C9GC04020J.
7. Sarotti, A. M.; Zanardi, M. M.; Spanevello R. A. *Recent Applications of Levoglucosenone as Chiral Synthon* https://www.eurekaselect.com/100919/article (accessed 2020-07-16).
8. Comba, M. B.; Tsai, Y.; Sarotti, A. M.; Mangione, M. I.; Suárez, A. G.; Spanevello, R. A. Levoglucosenone and Its New Applications: Valorization of Cellulose Residues. *Eur. J. Org. Chem.* **2018**, *2018* (5), 590–604. https://doi.org/10.1002/ejoc.201701227.
9. Kühlborn, J.; Groß, J.; Opatz, T. Making Natural Products from Renewable Feedstocks: Back to the Roots? *Nat. Prod. Rep.* **2020**, *37* (3), 380–424. https://doi.org/10.1039/C9NP00040B.
10. Halpern, Y.; Riffer, R.; Broido, A. Levoglucosenone (1,6-Anhydro-3,4-Dideoxy-.DELTA.3-.Beta.-D-Pyranosen-2-One). Major Product of the Acid-Catalyzed Pyrolysis of Cellulose and Related Carbohydrates. *J. Org. Chem.* **1973**, *38* (2), 204–209. https://doi.org/10.1021/jo00942a005.
11. Sui, X.; Wang, Z.; Liao, B.; Zhang, Y.; Guo, Q. Preparation of Levoglucosenone through Sulfuric Acid Promoted Pyrolysis of Bagasse at Low Temperature. *Bioresour. Technol.* **2012**, *103* (1), 466–469. https://doi.org/10.1016/j.biortech.2011.10.010.
12. Oyola-Rivera, O.; He, J.; Huber, G. W.; Dumesic, J. A.; Cardona-Martínez, N. Catalytic Dehydration of Levoglucosan to Levoglucosenone Using Brønsted Solid Acid Catalysts in Tetrahydrofuran. *Green Chem.* **2019**, *21* (18), 4988–4999. https://doi.org/10.1039/C9GC01526D.
13. Sarotti, A. M.; Zanardi, M. M.; Spanevello, R. A. Recent Applications of Levoglucosenone as Chiral Synthon. *Curr. Org. Synth.* **2012**, *9* (4), 439–459. https://doi.org/10.2174/157017912802651401.
14. Krishna, S. H.; McClelland, D. J.; Rashke, Q. A.; Dumesic, J. A.; Huber, G. W. Hydrogenation of Levoglucosenone to Renewable Chemicals. *Green Chem.* **2017**, *19* (5), 1278–1285. https://doi.org/10.1039/C6GC03028A.
15. Mouterde, L. M. M.; Allais, F.; Stewart, J. D. Enzymatic Reduction of Levoglucosenone by an Alkene Reductase (OYE 2.6): A Sustainable Metal- and Dihydrogen-Free Access to the Bio-Based Solvent CyreneTM. *Green Chem.* **2018**, *20* (24), 5528–5532. https://doi.org/10.1039/C8GC03146K.

16. Sherwood, J.; De bruyn, M.; Constantinou, A.; Moity, L.; McElroy, C. R.; Farmer, T. J.; Duncan, T.; Raverty, W.; Hunt, A. J.; Clark, J. H. Dihydrolevoglucosenone (Cyrene) as a Bio-Based Alternative for Dipolar Aprotic Solvents. *Chem. Commun.* **2014**, *50* (68), 9650–9652. https://doi.org/10.1039/C4CC04133J.
17. Salavagione, H. J.; Sherwood, J.; Bruyn, M. D.; Budarin, V. L.; Ellis, G. J.; Clark, J. H.; Shuttleworth, P. S. Identification of High Performance Solvents for the Sustainable Processing of Graphene. *Green Chem.* **2017**, *19* (11), 2550–2560. https://doi.org/10.1039/C7GC00112F.
18. Court, G. R.; Lawrence, C. H.; Raverty, W. D.; Duncan, A. J. *Method for Converting Lignocellulosic Materials into Useful Chemicals*. US20120111714A1, May 10, 2012.
19. Flourat, A. L.; Peru, A. A. M.; Teixeira, A. R. S.; Brunissen, F.; Allais, F. Chemo-Enzymatic Synthesis of Key Intermediates (S)-γ-Hydroxymethyl-α,β-Butenolide and (S)-γ-Hydroxymethyl-γ-Butyrolactone via Lipase-Mediated Baeyer–Villiger Oxidation of Levoglucosenone. *Green Chem.* **2015**, *17* (1), 404–412. https://doi.org/10.1039/C4GC01231C.
20. Bonneau, G.; Peru, A. A. M.; Flourat, A. L.; Allais, F. Organic Solvent- and Catalyst-Free Baeyer–Villiger Oxidation of Levoglucosenone and Dihydrolevoglucosenone (CyreneTM): A Sustainable Route to (S)-γ-Hydroxymethyl-α,β-Butenolide and (S)-γ-Hydroxymethyl-γ-Butyrolactone. *Green Chem.* **2018**, *20* (11), 2455–2458. https://doi.org/10.1039/C8GC00553B.
21. Flourat, A. L.; Haudrechy, A.; Allais, F.; Renault, J.-H. (S)-γ-Hydroxymethyl-α,β-Butenolide, a Valuable Chiral Synthon: Syntheses, Reactivity, and Applications. *Org. Process Res. Dev.* **2019**. https://doi.org/10.1021/acs.oprd.9b00468.
22. Zamzow, M.; Höcker, H. Synthesis of Polymers with Pendant Spiro Orthoester Groups. *Macromol. Chem. Phys.* **1994**, *195* (7), 2381–2400. https://doi.org/10.1002/macp.1994.021950708.
23. Sheldon, R. A. The E Factor 25 Years on: The Rise of Green Chemistry and Sustainability. *Green Chem.* **2017**, *19* (1), 18–43. https://doi.org/10.1039/C6GC02157C.
24. Trost, B. The Atom Economy—A Search for Synthetic Efficiency. *Science* **1991**, *254* (5037), 1471–1477. https://doi.org/10.1126/science.1962206.
25. Williams, C. K. Synthesis of Functionalized Biodegradable Polyesters. *Chem. Soc. Rev.* **2007**, *36* (10), 1573. https://doi.org/10.1039/b614342n.
26. Debsharma, T.; Behrendt, F. N.; Laschewsky, A.; Schlaad, H. Ring-Opening Metathesis Polymerization of Biomass-Derived Levoglucosenol. *Angew. Chem.* **2019**, *131* (20), 6790–6793. https://doi.org/10.1002/ange.201814501.
27. Debsharma, T.; Yagci, Y.; Schlaad, H. Cellulose-Derived Functional Polyacetal by Cationic Ring-Opening Polymerization of Levoglucosenyl Methyl Ether. *Angew. Chem., Int. Ed.* **2019**, *58* (51), 18492–18495. https://doi.org/10.1002/anie.201908458.
28. Bielawski, C. W.; Grubbs, R. H. Living Ring-Opening Metathesis Polymerization. *Prog. Polym. Sci.* **2007**, *32* (1), 1–29. https://doi.org/10.1016/j.progpolymsci.2006.08.006.
29. Schrock, R. R. Synthesis of Stereoregular Polymers through Ring-Opening Metathesis Polymerization. *Acc. Chem. Res.* **2014**, *47* (8), 2457–2466. https://doi.org/10.1021/ar500139s.

30. Nuyken, O.; Pask, S. Ring-Opening Polymerization—An Introductory Review. *Polymers* **2013,** *5* (2), 361–403. https://doi.org/10.3390/polym5020361.
31. Hillmyer, M. A.; Laredo, W. R.; Grubbs, R. H. Ring-Opening Metathesis Polymerization of Functionalized Cyclooctenes by a Ruthenium-Based Metathesis Catalyst. *Macromolecules* **1995,** *28* (18), 6311–6316. https://doi.org/10.1021/ma00122a043.
32. Hufendiek, A.; Lingier, S.; Du Prez, F. E. Thermoplastic Polyacetals: Chemistry from the Past for a Sustainable Future? *Polym. Chem.* **2019,** *10* (1), 9–33. https://doi.org/10.1039/C8PY01219A.
33. Prat, D.; Wells, A.; Hayler, J.; Sneddon, H.; McElroy, C. R.; Abou-Shehada, S.; Dunn, P. J. CHEM21 Selection Guide of Classical- and Less Classical-Solvents. *Green Chem.* **2016,** *18* (1), 288–296. https://doi.org/10.1039/C5GC01008J.
34. Collins, G. L.; Greene, R. K.; Berardinelli, F. M.; Garruto, W. V. Observations on the Initiation of Trioxane with Boron Trifluoride Dibutyl Etherate. *J. Polym. Sci. Polym. Lett. Ed.* **1979,** *17* (10), 667–671. https://doi.org/10.1002/pol.1979.130171008.
35. Camp, J. E. Bio-Available Solvent Cyrene: Synthesis, Derivatization, and Applications. *ChemSusChem* **2018,** *11* (18), 3048–3055. https://doi.org/10.1002/cssc.201801420.
36. Ray, P.; Hughes, T.; Smith, C.; Hibbert, M.; Saito, K.; Simon, G. P. Development of Bio-Acrylic Polymers from Cyrene^{TM}: Transforming a Green Solvent to a Green Polymer. *Polym. Chem.* **2019,** *10* (24), 3334–3341. https://doi.org/10.1039/C9PY00353C.
37. Hutchinson, R. A.; Penlidis, A. Free-Radical Polymerization: Homogeneous Systems. In *Polymer Reaction Engineering*; Asua, J. M., Ed.; Blackwell Publishing Ltd: Oxford, UK, 2007; pp 118–178. https://doi.org/10.1002/9780470692134.ch3.
38. Li, H.; Shakaroun, R. M.; Guillaume, S. M.; Carpentier, J. Recent Advances in Metal-Mediated Stereoselective Ring-Opening Polymerization of Functional Cyclic Esters towards Well-Defined Poly(Hydroxy Acid)s: From Stereoselectivity to Sequence-Control. *Chem.—Eur. J.* **2020,** *26* (1), 128–138. https://doi.org/10.1002/chem.201904108.
39. Coote, M. L.; Davis, T. P. SOLVENT EFFECTS ON FREE RADICAL POLYMERIZATION. In *Handbook of Solvents*; Elsevier, 2014; pp 811–833. https://doi.org/10.1016/B978-1-895198-64-5.50021-0.
40. Oh, J. K. Recent Advances in Controlled/Living Radical Polymerization in Emulsion and Dispersion. *J. Polym. Sci. Part Polym. Chem.* **2008,** *46* (21), 6983–7001. https://doi.org/10.1002/pola.23011.
41. Ray, P.; Hughes, T.; Smith, C.; Simon, G. P.; Saito, K. Synthesis of Bioacrylic Polymers from Dihydro-5-Hydroxyl Furan-2-One (2H-HBO) by Free and Controlled Radical Polymerization. *ACS Omega* **2018,** 3 (2), 2040–2048. https://doi.org/10.1021/acsomega.7b01929.
42. Diot-Néant, F.; Rastoder, E.; Miller, S. A.; Allais, F. Chemo-Enzymatic Synthesis and Free Radical Polymerization of Renewable Acrylate Monomers from Cellulose-Based Lactones. *ACS Sustainable Chem. Eng.* **2018,** *6* (12), 17284–17293. https://doi.org/10.1021/acssuschemeng.8b04707.
43. Kuo, S.-W.; Kao, H.-C.; Chang, F.-C. Thermal Behavior and Specific Interaction in High Glass Transition Temperature PMMA Copolymer. *Polymer* **2003,** *44* (22), 6873–6882. https://doi.org/10.1016/j.polymer.2003.08.026.

44. Prescott, S. W.; Ballard, M. J.; Rizzardo, E.; Gilbert, R. G. RAFT in Emulsion Polymerization: What Makes It Different? *Aust. J. Chem.* **2002**, *55* (7), 415. https://doi.org/10.1071/CH02073.
45. Maiti, B.; De, P. RAFT Polymerization of Fatty Acid Containing Monomers: Controlled Synthesis of Polymers from Renewable Resources. *RSC Adv.* **2013**, *3* (47), 24983. https://doi.org/10.1039/c3ra45541f.
46. Perrier, S. *50th Anniversary Perspective*: RAFT Polymerization—A User Guide. *Macromolecules* **2017**, *50* (19), 7433–7447. https://doi.org/10.1021/acs.macromol.7b00767.
47. Shafizadeh, F.; Furneaux, R. H.; Pang, D.; Stevenson, T. T. Base-Catalyzed Oligomerization of Levoglucosenone. *Carbohydr. Res.* **1982**, *100* (1), 303–313. https://doi.org/10.1016/S0008-6215(00)81044-7.
48. Diot-Néant, F.; Mouterde, L.; Fadlallah, S.; Miller, S. A.; Allais, F. Sustainable Synthesis and Polycondensation of Levoglucosenone-Cyrene™-Based Bicyclic Diol Monomer: An Access to Renewable Polyesters. *ChemSusChem* **2020**. https://doi.org/10.1002/cssc.202000680.
49. Fukushima, K.; Fujiwara, T. New Routes to Tailor-Made Polyesters. In *Polymers for Biomedicine*; John Wiley & Sons, Ltd, 2017; pp 149–189. https://doi.org/10.1002/9781118967904.ch6.
50. Ward, D. D.; Shafizadeh, F. Cycloaddition "4 + 2" Reactions of Levoglucosenone. *Carbohydr. Res.* **1981**, *95* (2), 155–176. https://doi.org/10.1016/S0008-6215(00)85573-1.
51. Bhaté, P.; Horton, D. Stereoselective Synthesis of Functionalized Carbocycles by Cycloaddition to Levoglucosenone. *Carbohydr. Res.* **1983**, *122* (2), 189–199. https://doi.org/10.1016/0008-6215(83)88330-X.
52. Nikahd, M.; Mikusek, J.; Yu, L.-J.; Coote, M. L.; Banwell, M. G.; Ma, C.; Gardiner, M. G. Exploiting Chitin as a Source of Biologically Fixed Nitrogen: Formation and Full Characterization of Small-Molecule Hetero- and Carbocyclic Pyrolysis Products. *J. Org. Chem.* **2020**, *85* (7), 4583–4593. https://doi.org/10.1021/acs.joc.9b03438.
53. Banwell, M. G.; Liu, X.; Connal, L. A.; Gardiner, M. G. Synthesis of Functionally and Stereochemically Diverse Polymers via Ring-Opening Metathesis Polymerization of Derivatives of the Biomass-Derived Platform Molecule Levoglucosenone Produced at Industrial Scale. *Macromolecules* **2020**, *53* (13), 5308–5314. https://doi.org/10.1021/acs.macromol.0c01305.

Chapter 6

Bio-Based Aromatics: Aminobenzoic Acid Derivatives for High-Performance Bioplastics

Aniruddha Nag, Noriyoshi Matsumi, and Tatsuo Kaneko*

Graduate School of Advanced Science and Technology, Energy and Environment Area, Japan Advanced Institute of Science and Technology, 1-1 Asahidai, Nomi, Ishikawa 9231292, Japan

***Email: kaneko@jaist.ac.jp**

The conventional method for the preparation of aromatic precursor molecules involves the chemical synthesis of fossil-fuel-based starting materials, such as benzene and xylene. This process is considerably energy-intensive, and many byproducts are generated. To make this process environmentally benign, aromatic building blocks must be extracted from natural resources; however, most of the reported approaches are expensive and have low yield. Alternately, the microbial fermentation of renewable feedstock to produce aromatic molecules through the shikimate pathway is inexpensive and is an ideal alternative method for establishing a sustainable society. This chapter mainly focuses on the microbial production of aromatic aminobenzoic acid (3-amino-4-hydroxybenzoic acid) using non-edible feedstock, and its use in various polymer products, such as polyaniline, polyester, and polybenzazole. Various bio-based functional polymers are reviewed here, along with their structure–property relationships. Furthermore, the advancement of various aspects of society using such high-performance bio-based aromatic polymers is discussed.

Introduction

In the effort to achieve a sustainable society, the production of bio-based polymers using renewable materials is considered the right strategy; as such, this approach has gained significant research interest in the past few decades. White biotechnological methods enhance the production of bio-derived precursors and play a significant role in lowering the final product costs (*1*). Cellulosic feedstock, which is a ubiquitous, abundant, and renewable natural resource, is potentially the best sustainable alternative to existing fossil-based resources for the production of commercial plastics (2).

Most monomers are produced via biosynthesis, using edible sugars or starch as starting materials. Considering the current limited food resources, inedible renewable feedstock, obtained from genetically engineered microorganisms, is potentially the best alternative for the production of biomonomers. Waste products from sugarcane, corn stover, wheat stew, and kraft pulp are examples of lignocellulosic feedstocks, from which biomonomers have been extracted, such as lactic acid and succinic acid, for their respective polymer productions, to establish a biopolymer-based economy (*3–5*).

Polyethylene terephthalate (PET), important commercial polyester, can be used for the production of various bio-based plastics. Dupont produced partially renewable Sorona® EP and Hytrel® RS polymers using a precursor derived from corn sugar and other non-edible feedstocks (*6*). Carbohydrates and lignin are the most abundant sources of aromatic biomonomers, such as furandicarboxylic acid (FDCA), which serves as a substitute for terephthalic acid, isophthalic, vanillic, caffeic, and ferulic acid-based polyesters (*7–9*). The biomonomer, FDCA can be further converted to various other monomers and can form various thermoplastic polyesters (*10–13*). Bio-derived vanillin reacts with acetic anhydride to produce acetyldihydroferulic acid, which can be further polymerized to poly (dihydroferulic acid) with better thermal properties than the commodity plastic, PET (*14*). A family of polyesters with considerably high thermal stability was prepared using lignin-based precursors, with various methoxy groups such as; vaniline, syringaldehyde, 4-hydroxybenzaldehyde (*15*). Ferulic acid is another abundant naturally available phenolic compound obtained from various plant food sources, such as cereal, rice, peanut, and apple, as well as lignin sources (*16*). The chemical structure of ferulic acid resembles those of naturally occurring cinnamic acids such as; coumaric, caffeic acid (*17*). Several bio-based polyesters and polyamides have been recently reported, produced by the polycondensation of such skeletons, and they can be good alternatives to PET (*14*). Caffeic acid can be decarboxylated and protected to produce various vinyl catechols, which are feasible monomers for controlled radical polymerization and living anionic polymerization (*18*). The naturally occurring ferulic acid can be decarboxylated to obtain various vinylphenolic compounds, including 4-vinylguaiacol (4VG), which can be radically polymerized to poly(vinylguaiacol) and poly(vinylcatechol) and their copolymers (*19*). Pinocarvone, derived from the plant-based compound, α-pinene, can be subjected to radical polymerization under visible light irradiation to yield thermally stable, optically active materials (*20*).

Poly[(*R*)-3-hydroxybutyrate] (P-3HB) is another type of polyester synthesized from bio-derived monomers with a high melting temperature close to that of polypropylene (*21*). Copolymers, including 3HB, and various stereoisomeric monomers (lactates), having a wide range of mechanical properties, i.e., from crystalline to elastic, depending on the backbone structure, were prepared. Additionally, Phenyl lactic acid (PhLA) is another naturally abundant bio-based polymer precursor that can be polymerized to poly-PhLA, which exhibits high ultraviolet absorption and better thermo-mechanical stability than PLA (*22*). PhLA can be biosynthesized as an intermediate product from glucose during phenylalanine metabolism through the shikimate pathway (*23*). PhLA bioproduction from cellulosic feedstock by simultaneous saccharification and fermentation (SSF) using an *E. coli* strain is effective for achieving enhanced yield due to the reduced accumulation of fermentation inhibitors (*24, 25*). The bioavailability of *p*-coumaric acid (4HCA) is high; in the enzymatic conversion of lignin into cinnamate, 4HCA is produced over phenylalanine ammonia-lyase and Cyt P450 (*24*), and it also exists as a protein component in photosynthetic bacteria, such as *Rhodobacter capsulatus and Rhodobacter sphaeroids* (*25, 26*). Homopolymers and copolymers of 4HCA with other bio-derived phytomonomers have been synthesized, and they show high thermo-mechanical stability among degradable bioplastics, along with liquid crystalline behavior (*27–29*).

Several bio-derived monomers synthesized through microbial fermentation can be used as building blocks to establish bio-based polymer society as an alternative to the current petroleum-based society. To achieve the similar performance and application requirements of petroleum-based plastics in terms of thermo-mechanical stability, it is important to synthesize aromatic ring building blocks similar to those used in commercial super-engineering polymers, with low production cost. The preparation process of ferulic acid-based biopolymers (*30*), which are components of gramineous plant, and thermoplastic copolymer (*31*) production using the phenolic phytomonomers, *p*-coumaric acid and caffeic acid, are both energy-intensive, and the yields are low. In contrast, microbial fermentation methods are fairly efficient for the preparation of bio-based building blocks, in terms of energy consumption and yield. Aromatic biopolyamides, showing ultrahigh strength comparable to those of glass and steel, were prepared from 4ACA, a biomonomer derived from glucose using recombinant *E. coli* strains from 4-aminophenylalanine (*32*). Monoamine 4ACA can be further dimerized into diamine to obtain high-performance bio-based polyimides while reacting with various dianhydrides (*32*). Furthermore, a new monomer structure was prepared through the microbial hydrogenation of 4ACA to obtain 4-aminohydrocinnamic acid (4AHCA), which can be used as a novel building block for biopolyamide synthesis (*33*).

In another category, several semi-bio-based plastics were developed, where a combination of bio- and petroleum-based monomers was used. Poly(butylenesuccinate) (PBS) (*34*) can be produced from bio-derived succinic acid and petroleum-based butanediol. Wang et al. (2014) reported the synthesis of butanediol using bio-based succinic acid, which makes PBS a completely bio-based plastic (*35*). Similarly, 1,3-propanediol, a bio-derived monomer, in combination with petroleum-based terephthalic acid produces semi-bio-based poly(trimethylene terephthalate) (PTT), which is potentially the best alternative to the commercially available PET. Nylon 6, 10, which can be produced from bio-based sebacic acid and commercially available hexamethylenediamine, is a partial bio-based plastic with mechanical strength comparable to those of commercial nylon-6 and nylon-6, 6 (*36*, *37*).

Bioproduction of 3-Amino-4-hydroxybenzoic Acid

In 2006, Suzuki et al. reported 3-amino-4-hydroxybenzoic acid (3,4-AHBA) as an intermediate product, obtained through the biosynthesis of grixazone using the shikimate pathway (*38*). The biosynthesis of 3,4-AHBA from *Streptomyces griseus* via two steps involves two primary metabolites, C3 and C4, and two enzymes, GriI and GriH. Dihydroxyacetone phosphate (DHAP) and aspartate semialdehyde (ASA) are used as the starting materials for the biosynthesis of 3,4-AHBA. In the first step, a C7 compound, 3-acetylamino-4-hydroxybenzoic acid (3,4-AcAHBA), is formed. In the second step, deacetylation occurs to yield the final product, 3,4-AHBA; these steps involve two enzymes, GriI and GriH, respectively.

Suzuki et al. (2007) reported that the genetic engineering method for removing the N-acetyltransferase gene (natA) can inhibit the formation of 3,4-AcAHBA to increase the stoichiometric amount of 3,4-AHBA (*39*).

Shozui et al. (2017) reported an inexpensive method for producing 3,4-AHBA from *E. coli* by inhibiting 3,4-AcAHBA production while reducing the activity of N-hydroxyarylamine O-acetyltransferase gene (nhoA) (*40*).

The starting materials used for these bio-based building block extractions are mainly edible feedstocks, which establish competition for the currently limited food resources. Thus, the development of various bioplastics using non-edible feedstock is important for achieving a sustainable society (*41*). The utilization of lignocellulosic feedstock in the fermentation process was limited because of the extremely low yield and the numerous byproducts formed other than the target aromatic biomonomers even with following the complex shikimate pathway (*42*). Several bioprocessing approaches have been developed by various research groups to produce bio-based aliphatic chemicals from renewable polysaccharides and can be classified as separate hydrolysis and fermentation (SHF), simultaneous saccharification and fermentation (SSF) and consolidated bioprocessing (CBP). Being most economically feasible CBP approach, it can be well accepted over SHF and SSF, however, continuous technological innovation to increase cellulolytic activities and reaction productivities, and reduce fermentation inhibition is needed to overcome the limited productivity and feedstock availability for CBP-based approaches (*3*).

Polymerization of 3,4-AHBA

The aromatic π-electron-rich biomonomer, 3,4-AHBA, can be extracted from the renewable precursor, *S. griseus*. This biomonomer is a precursor to several secondary metabolites, and it has medicinal significance in antibiotics, immunosuppressants, etc. Polyaniline (PANI) is an example of intrinsically conductive polymers (ICPs), which are important for various applications, including in microelectronics, light-emitting diodes (LEDs), etc. (*43, 44*) Conductive PANIs with high thermal and chemical resistance can be combined with various other polymers, such as polyethylene oxide (PEO), polymethyl methacrylate (PMMA), and polyvinyl alcohol (PVA), to prepare composites to enhance conductivity. Conductive liquid crystalline polymers are receiving significant interest owing to their thermo-mechanical stability. However, due to their poor solubility and processability, their applications are limited. To overcome these issues, polymer chain structures with polar terminal groups should be utilized (*45*).

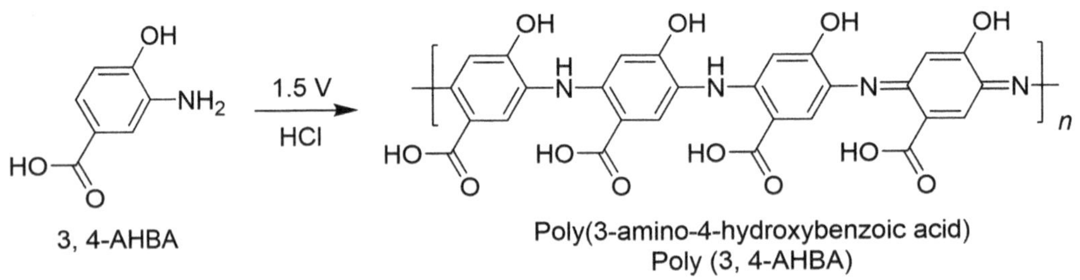

Figure 1. Electropolymerization scheme of 3,4-AHBA for polyaniline production.

Thus, the bio-derived 3,4-AHBA monomer was electropolymerized to produce PANI polymers (Figure 1), which are highly soluble in polar solvents, including water, and also show solvatochromism, because of their polar groups (–OH and –COOH) (*46*). This type of polymer shows a halochromic pattern different from those of usual PANI polymers; for example, the poly(3,4-AHBA) solution shows a brown color at pH 3–6, black at pH 6–7, and dark brown at pH 7–12 (Figure 2). Solvatochromism can be observed when the film is dissolved in an HCl/aq. solution, in which it shows brown but changes to green after a few minutes. When DMF and DMSO are added, it changes again to brown and turns pink with the addition of NMP and distilled water (Figure 2). The ultraviolet–visible (UV–vis) spectrum shows quinoid and polaron structures at 520 and 660 nm, respectively. In DMF, DMSO, and NMP solvents, the polaron band is weakened due to the self-doping effect of poly(3,4-AHBA), and the band disappears after the de-doping phenomenon. Contrarily, the quinoid bands exhibit improved strength in the same solvents, which indicates solvatochromism, with the interaction of the polar solvents with carboxylate groups inducing de-doping. Freestanding films cannot be prepared, and hydroxypropyl cellulose was used as a supporter to cast a green transparent membrane with high conductivity comparable to those of semiconductors.

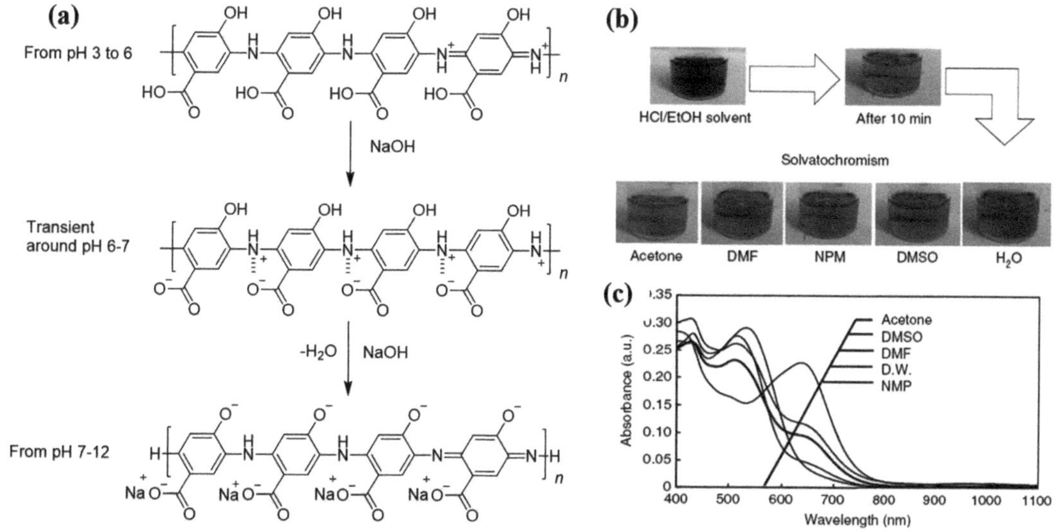

Figure 2. (a) Schematic structure change of poly(3,4-AHBA) under different pH, (b) Solvatochromic behavior of poly(3,4-AHBA) in various solvents, (c) UV-vis spectra of various solutions.

Preparation of Polyesters Using 3,4-AHBA

Thermotropic liquid crystalline polymers (LCPs) exhibit excellent superior thermo-mechanical properties along with low density; thus, they are suitable for application in targeted fields (*47–50*). The mesogenic units in LCPs with a mixture of ordered crystals and isotropic liquids create an anisotropic system. Aromatic polyester fibers, such as LCPs (Xyder® and Vectar®), can be highly oriented to enhance their mechanical strength to be comparable to those of Kevlar™ and Zylon™ (*51, 52*). Aromatic polyesters, a class of LCPs, exhibit high heat resistance, mechanical strength, and dimensional stability, making them suitable commodity plastics, e.g., PET; the production of PET may reach as high as 21.1 million tons by 2021. However, wholly aromatic rigid-rod LCPs, particularly in the case of homopolymers, are not conventionally employed due to their processability issues, caused by their high melting temperature and low chain flexibility. Aliphatic-aromatic LCPs with mesogenic groups can be ideal candidates, considering their relatively good processability. Various methods have been employed to lower the melting temperatures of rigid-chain aromatic LCPs, including using non-linear comonomers (e.g., *m*-hydroxybenzoic acid) or flexible side alkyl groups (*53–55*).

Conversely, polymers with photoluminescence properties are attractive for applications in photolithography and metal ion sensors. Polymers with such photosensitivity must contain photoactive chromophores, such as aromatic rings with π-electron density, in their main chain, which are cross-linked or cleaved under UV irradiation (*56–58*). Aromatic rigid-rod backbones are ideal for producing the π-conjugated structures of photosensitive polymers; however, the solubility defect and high softening temperature of such polymers inhibit their industrial applications.

The 3,4-AHBA monomer was firstly converted into 3-benzylidene amino-4-hydroxybenzoic acid (3,4-BAHBA) by the reaction with bio-based benzaldehyde followed by copolymerization with another bio-derived building block *trans*-coumaric acid (4-HCA) (Figure 3). The rigid chain homopolymer prepared from 4-HCA showed thermotropic LC properties and was capable of inducing LC behavior in the copolymer poly (3,4-BAHBA-*co*-4HCA) prepared using the 3,4-AHBA biomonomer containing chromophore (Figure 3). The glass transition temperature (T_g) of such thermotropic copolymers reaches up to 135 °C, while the 10% weight-loss temperature T_{d10} can reach up to 350 °C, which are higher than those of most conventional bio-based plastics (59). The 4-HCA unit copolymers result in a nematic LC phase formation due to their mesogenic effect; further, due to their flexibility, they increase the processability of the copolymers to overcome solubility defects. While soluble in NMP (0.1%), the copolymers exhibit photoluminescence behavior, while 4-HCA incorporation dominates over 75% and polarized light emission occurs under both parallel and crossed polarizers.

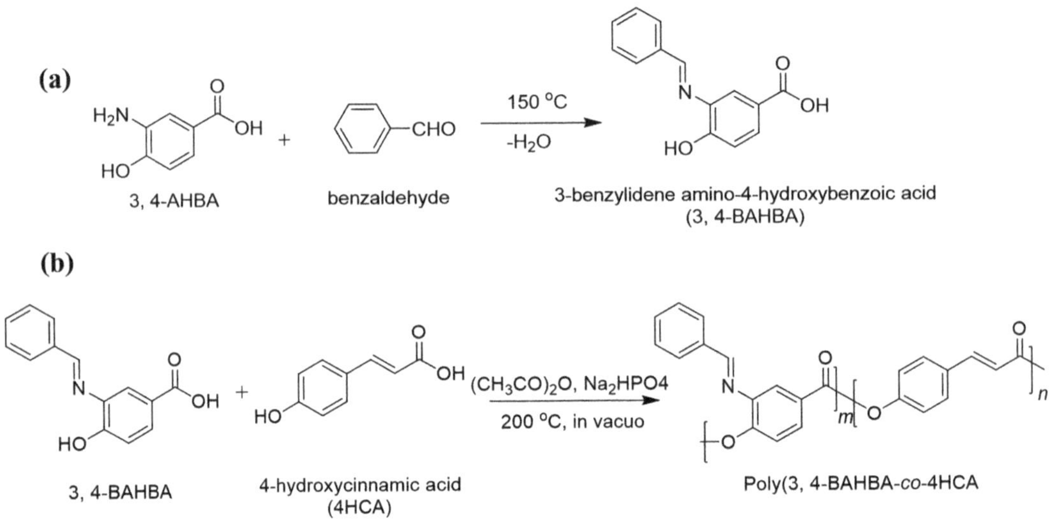

Figure 3. (a) Synthetic scheme of 3,4-BAHBA production, (b) preparation of copolymer polyester using biomonomers.

Bio-Based Polybenzoxazole

Polybenzoxazole (PBOs), since invention, have attracted the attention of researchers owing to their ultrahigh thermo-mechanical, chemical and environmental stability. Moreover, owing to their very low dielectric constant, attributed to their non-polarizable structure, PBOs have various applications, including as high heat-resistant insulating materials in high-performance motor coils; they are also useful in the electronics industry, in integrated circuits or multichip production. The most common method for synthesizing PBOs is polycondensation using dicarboxylic acid or dialdehyde derivatives and *bis*(*o*-aminophenol). The first step involves producing poly(o-hydroxyamide)s, followed by their cyclodehydration to yield PBOs (57, *60–62*). The homopolymerization of 3,4-AHBA and its isomer, 4-amino-3-hydroxybenzoic acid (4,3-AHBA), can produce the PBO with the simplest structure, poly(2,5-benzoxazole) (ABPBO), and poly(2,6-benzoxazole), respectively, through the formation of low molecular weight polyester amides, followed by thermal ring-closing to obtain high molecular weight PBOs (*63*).

The homopolymerization of the bio-derived 3,4-AHBA monomer to produce polybenzoxazole was first reported in a patent (*46*). This polymer has a rigid-rod backbone with an aromatic chain precursor, and it exhibits high thermo-mechanical stability and chemical resistance comparable to those of poly(p-phenylenebenzobisoxazole) (Zylon) (*64*); however, it faces processability and fabrication limitations.

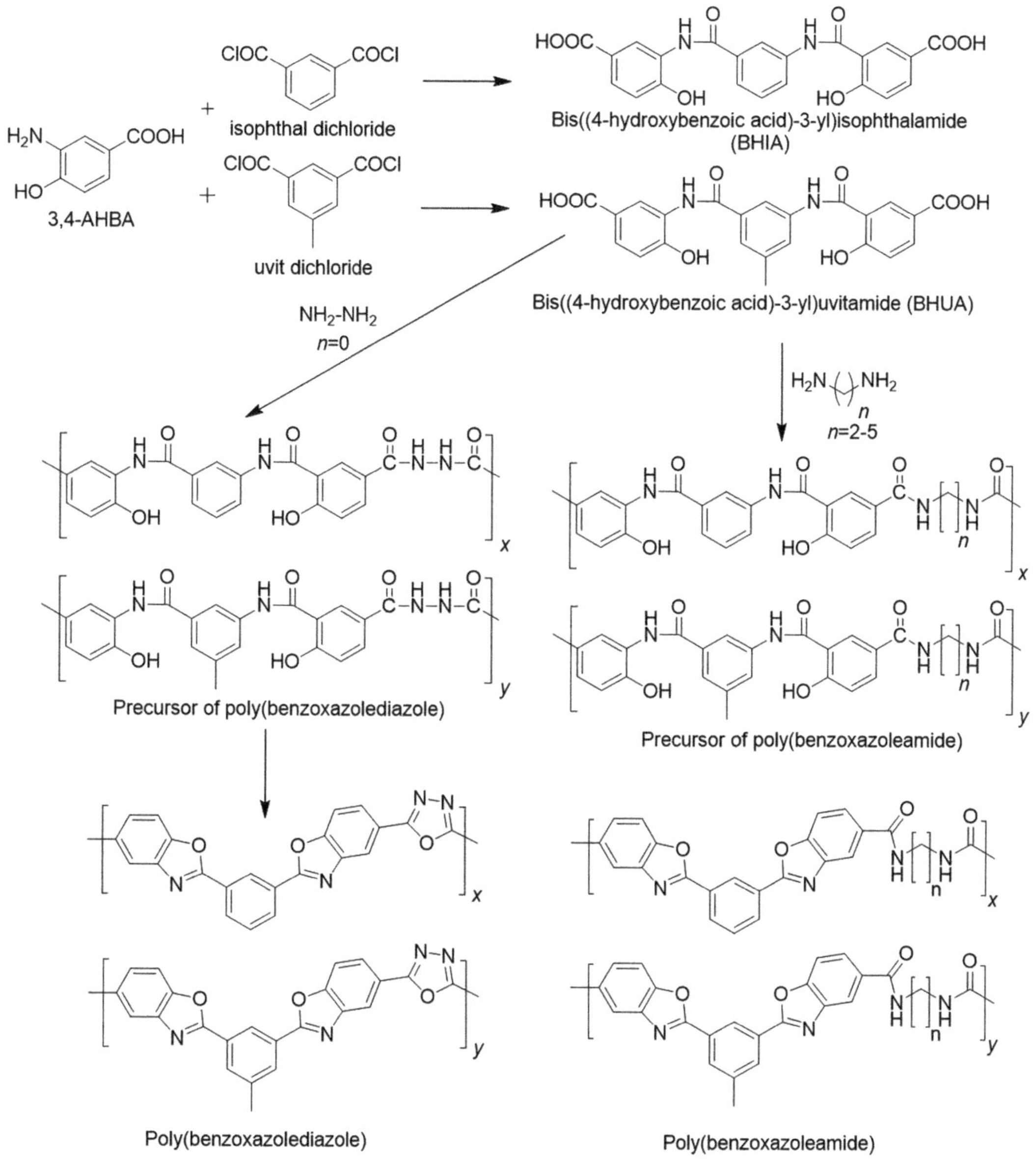

Figure 4. Synthetic routes for various thermotropic polybenzoxazoles derived using 3,4-AHBA.

To derive flexible-chain PBOs, diacid biomonomers (e.g., *ortho*-hydroxyamide monomeric units) can be prepared using isophthalic acid chloride and uvitic acid chloride with bio-derived 3,4-AHBA, followed by polycondensation with various diamines. Poly(benzoxazolediazole) can be prepared by the reaction between various *ortho-hydroxyamides* and hydrazine, whereas different poly(benzoxazoleamide)s can be prepared by the polymerization of various alkyl chain diamine moieties with different *ortho*-hydroxyamides (Figure 4) (*65*). These alkyl chain containing PBO structures show better solubility and processability than Zylon while maintaining moderately high thermal stability and mechanical strength. Almost all the prepared PBO samples show liquid crystallinity above their respective glass transition temperatures (T_g); this special bisbenzoxazole structure attached to the *meta*-position of the phenylene ring behaves as a mesogenic unit utilized for the preparation of LCPs. With these samples, light transmission was observed through a crossed-Nicol polarizer microscope, and for the relatively long-chain diamine, the PBO moiety clearly exhibited schlieren chain texture, which indicated the nematic phase LC behavior.

Applications of Aromatic PBO

Commercial PBOs are applied mainly as high-modulus strength fibers and photosensitive PBO systems; accordingly, organo-soluble PBOs have been prepared by incorporating flexible linkages and bulky groups (*66–68*). Until now, few studies have been conducted on PBO films and their application as flexible LC displays (LCDs) with high T_g, toughness, low water absorption, and high solubility (*69*).

Recently, aromatic PBOs are gaining continuous research interest due to their high optical transparency and low dielectric constant, which are important for their use as photoresistors and insulators. With reducing the molecular polarizability, the dielectric constant (k) can be lowered by introducing fluorinated moieties or bulky alicyclic groups into the PBO backbone (*70*). Polyethylene and polypropylene, which are low-k polymers, have limited applications owing to their low thermo-mechanical resistance, which shows the importance of high-performance low-k materials (*71*, *72*). Other different approaches to reducing the k value include preparing porous materials, which contain air with a very low dielectric constant, and incorporating SiO_2 into the organic polymer structure, which facilitates pore creation. However, the mechanical properties are compromised (*73*, *74*). Organic materials with low-k properties are always interesting because of their attractive thermo-mechanical resistance and easy manufacturing process with high processability, e.g., polyimides (PIs); these properties extend their application in substrate coating as insulating membranes (*75*, *76*).

PIs cannot meet the desirable properties for low-k or ultra-low k applications ($k<2$) due to their relatively high moisture absorbance capacity and high k values (*77*, *78*). Alternately, compared to PIs, PBOs are ideal low-k materials since they contain polar carbonyl groups and exhibit higher chemical resistance, better metal surface adhesion, and lower coefficient of thermal expansion (CTE). Zylon® is considered one of the best-performing polymers considering its high mechanical strength and very low CTE value; however, owing to its extremely poor processability, its applications are limited in the low-k field. Fukumaru et al. derived a *tert*-butoxycarbonyl (*t*-Boc)-substituted Zylon precursor, where *t*-Boc can be easily removed during heat treatment of *t*-Boc-*pre*-PBO to obtain a ring-cyclized PBO structure, thereby enhancing the porosity of the final structure (*79*). Moreover, the *t*-Boc-substituted-PBO can be used for photolithography using a photoacid generator, since *t*-Boc is an acid-sensitive protective group that is soluble in alkaline developers (*80*). In another approach, aromatic hyperbranched PBOs (HBPBO) were synthesized and inserted into linear fluorinated polymer chains, such as fluorinated poly(hydroxyamide) (FPHA), to prepare high-performance low-k polymers (*81*). Alicyclic PBOs with adamantyl units exhibit fewer conjugates, which lower the molecular density in polymer chains; as such, they are the ideal candidates for application in optoelectronics and interlayer dielectric materials owing to their high transparency and low k value (*82*).

Considering all the previous studies related to low-k PBOs with high-performance properties and good processability, a combination of highly conjugated non-polar ring structures can be a promising candidate. The bio-derived monomer, 3,4-AHBA, can be used as the precursor of

ABPBO; conversely, another monomer, 3,4-DABA, was prepared from 3,4-AHBA by simple chemical conversion and can be used as the precursor of poly(2,5-benzimidazole) ABPBI (*83*). The Smiles rearrangement, which is known as intramolecular nucleophilic aromatic substitution, can be very useful for this conversion. In this process, suitable functional groups (*e.g.*, sulfone, sulfide, and ether) that can dislodge from the arene group carrying a negative charge and a nucleophilic group (*e.g.*, alcohol, amine, and thiol) that can act as a strong nucleophile must be present. The hydroxyl (-OH) group of 3,4-AHBA can be modified to an easily substitutable group, containing a nucleophilic moiety that can initiate the Smiles rearrangement to yield diaminobenzoic acid (3,4-DABA).

Polybenzimidazoles are well-established proton exchange membranes in high-temperature proton exchange membrane fuel cells (PEMFCs). Poly[2,2'-(*m*-phenylene)-5,5'-bibenzimidazole] (*m*-PBI, Celazole®) and ABPBI have attracted research interest owing to their ability to undergo proton conduction at high temperatures, oxidative and chemical stability, mechanical stability, lower gas permeability and high acid uptake ability for proton conductivity (*84–88*). ABPBI has several advantages over commercial Celazole®. The film forming property of ABPBI enhances processability. Additionally, a single monomer unit, 3,4-DABA is required to prepare ABPBI, and higher stability is achieved with acidic solvent doping, compared to that of Celazole® (*89*). These polybenzazole structures are very effective for creating low-*k* materials while maintaining high performance; further, polarity control of the structures can be important for balancing between the thermo-mechanical performance and dielectric property (*90*). ABPBO homopolymer units with an ultra-low *k* value of 1.9 at 1 MHz, exhibiting high thermal stability and considerable mechanical strength, were prepared.

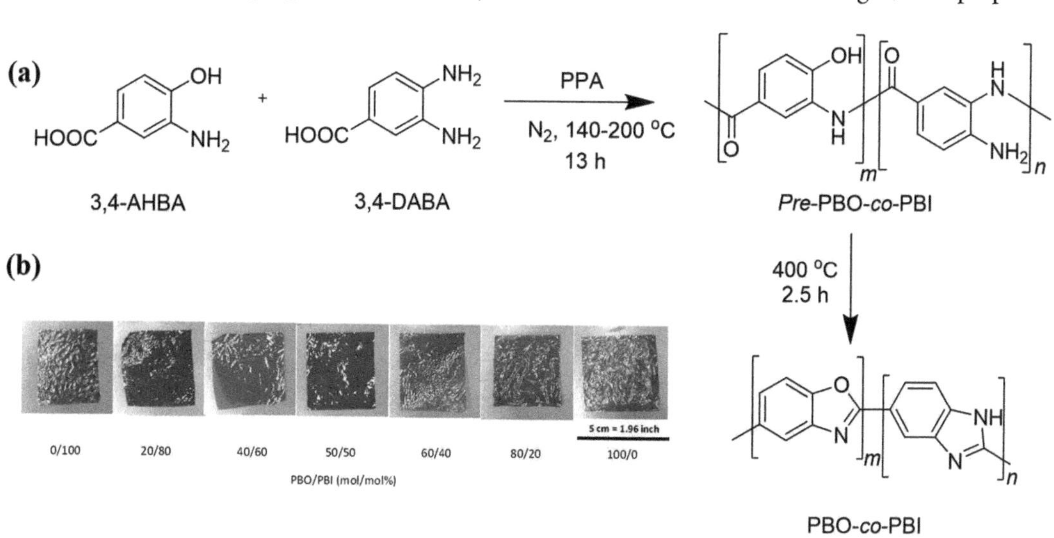

Figure 5. (a) Schematic route for copolymer synthesis using two bio-based monomers 3,4-AHBA and 3,4-DABA; (b) film pictures for various copolymer combinations.

When another monomer 3,4-DABA was copolymerized with 3,4-AHBA to prepare the copolymers (PBO-*co*-PBI), (Figure 5) which showed higher thermo-mechanical stability than ABPBO, with dielectric properties comparable to those of Kapton® (*91*) For ABPBO homopolymer units, including benzazole units, when the -O- group changes to -NH- group, the polarity increases; consequently, processability improves, and films can be cast for every copolymer composition. This approach to preparing low-*k* materials has several advantages over the preparation processes involving porous materials or fluorinated polymers, and thus far, it is considered the ideal method for low-k material preparation. Such ultralow-k high-performance polymer films can be used to interconnect technology for minimizing cross-talk and as interlayer dielectric polymers; thus, they are evidently better candidates than fluorinated or SiO_2 hybrid low-k polymers (*92, 93*).

Synthesis of Bio-Based Polyelectrolyte

Considering the present environmental scenario, the world is in desperate need of various alternative sustainable, efficient energy sources. Ion conducting polymers can be employed in electrochemical storage devices including batteries, photoelectrochemical cells and supercapacitors (*94*). *Grätzel* cells or dye-sensitized solar cells (DSSC) is effective in terms of power conversion efficiency, cost-efficiency and simplicity of fabrication technology. Rhuthenium-based photosensitizers are predominant in DSSCs because of their relatively high oxidation stability and good photoelectrochemical properties (*95–97*). To meet the growing demand for high energy density storage systems for electric vehicles and other electronics, lithium metal-based batteries (LIBs) are attracting interest (*98, 99*), considering that the anodes of these batteries can produce high theoretical capacities of up to 3860 mA hg^{-1}; in addition, they have low density and high negative potential of −3.04 V versus standard hydrogen electrode (*100*). LIBs have light-weight, long cycle life, and minimum memory effect, which make them ideal candidates for various applications in the current electronic industry. LiBs consist of three major components: cathode, anode, and electrolyte (Figure 6), which are undergoing constant innovation and are receiving extremely high research attention. Even after extensive research on electrolyte materials for next-generation LIBs, liquid electrolytes still dominate commercial usage. They are mostly prepared by dissolving available lithium salts in various organic solvents such as; propylene carbonate, diethyl carbonate (DEC) and ethylene carbonate (EC) (*101*). The safety issues associated with liquid electrolytes are unavoidable and thoroughly scrutinized to establish safe LIBs for use in our daily life. Various issues have been considered, including leakage and flammability during over-heating, toxic organic solvents, and, most importantly, dendrite growth in liquid electrolytes, which are inhibiting factors in the pathway toward preparing high-energy-density LIBs (*102*). To mitigate these barriers and prepare flexible, safe advanced energy-storage devices, solid-state polymer electrolytes have emerged as the best alternatives to conventional liquid electrolytes. Solid-state electrolytes have several advantages, including their wide electrochemical window and high thermal stability, as well as low risk of leakage during over-heating and corrosion of the electrodes. Dendrite growth can also be minimized to obtain stable LIBs (*103–108*).

Polymer electrolytes can be categorized into two segments: solid polymer electrolytes (SPEs) and gel polymer electrolytes (GPEs). SPEs are more advantageous than GPEs in terms of accommodating the electrode volume change and flexibility to be employed in all kinds of solid-state batteries. If the ionic movement in SPEs is thoroughly studied, it would be observed that lithium ions can travel through the motion of the polymer chain (*109*). However, the ionic conductivity of SPEs at ambient temperature (10^{-8}– 10^{-5} S cm^{-1}) is many orders of magnitude lower than that of liquid electrolytes (10^{-3} S cm^{-1}). On the contrary, GPEs have relatively high ionic conductivity at room temperature; however, due to the presence of solvents in GPEs, leakage problems are unavoidable, and corrosion around the electrode is also possible after prolonged usage (*110*). PEO and poly(ethylene glycol) (PEG)-based SPEs have already been reported; however, they have a low lithium transference number (t_{Li^+}) and ionic conductivity at ambient temperature because of the low mobility of lithium ions (*111*). Various approaches can be adopted to improve the t_{Li^+} of SPEs, such as covalent interaction of dissociable lithium salts (*112*) or introducing anion receptors for selective lithium-ion transport (*113, 114*).

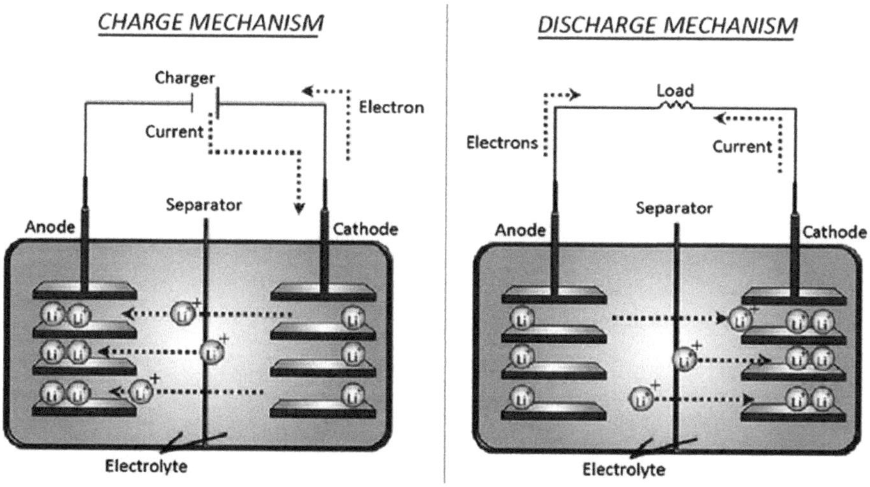

Figure 6. Charge-discharge mechanism of secondary lithium ion batteries (89).

A category of harmless, heat and chemical-resistant, electrochemically stable borate salts containing organoborate anions with aromatic and aliphatic diols or carboxylic acids are continuously being studied by Gores et al. (*115–118*) Boron chemistry, more precisely organoborane moieties, are promising candidates for preparing dissociable lithium salts or anion receptors (*119, 120*). Polymer electrolytes, where lithium cations are attached with delocalized polyanions, can attain high t_{Li^+} and deliver high ionic conductivity simultaneously (*121, 122*). The best candidate for such a category is a pseudo-solid electrolyte with a highly dissociable lithium-salt structure and plasticization with suitable ionic liquids (ILs) (*123–125*).

The thermo-mechanical stability and structural advantages of the polybenzimidazole series (PBI) of polymers are already well-known for various energy applications, and the side-chain modification of the imidazole moiety is also important for various applications in relevant fields (*126–129*). The biomonomer, 3,4-AHBA, can easily be transformed into the PBI precursor material, 3,4-DABA, just by converting the –OH group to the $–NH_2$ group following simple chemical steps. High-performance poly(2,5-benzimidazole) (ABPBI) can be obtained through the polycondensation of the bio-based monomer, 3,4-DABA, in the presence of polyphosphoric acid as a polycondensation agent [Figure 7 (a)] (*130*). The imidazole ring was modified to lithium boron imidazolate (B-PBI) by substituting the imidazole proton with lithium hydride (LiH), followed by the introduction of triethylborane (BEt_3) [Figure 7(b)] (*131*). The quaternary borate structure containing the lithium cation associated with the larger polyanion promotes the dissociation and stabilization of the negative charge over the polyanion, which enables facile lithium cation transport.

Plasticizers are very useful for increasing the amorphous phase in polymer electrolytes; thus, they can promote segmental motion, which further increases ion dissociation to increase ionic conductivity (*132, 133*). Various plasticizers are available, such as organic solvents (e. g., EC and DEC) or low molecular weight polymers (PEG, etc.); however, ILs are potentially the best choice for use as plasticizers owing to their high thermal stability, electrochemical stability, low volatility, etc. (*134–137*) Various ILs comprising cations based on pyridinium, imidazolium, and quaternary ammonium, and anions based on tetrafluoroborate $[BF_4]^-$, hexafluorophosphate $[PF_6]^-$, $[N(CF_3SO_2)_2]^-$, and $[CF_3SO_3]^-$ have been considered as liquid electrolytes. The identical structure of lithium salts and ILs can result in superior electrochemical performance. ILs comprising a combination of 1-butyl-3-methylimidazolium (BMIM) with different anions ($TFSI^-$, BF_4^-, $CF_3SO_3^-$) were studied to determine the influence of anions on the electrolyte, and BMImTFSI was found to exhibit the highest electrochemical stability and low interfacial resistance (*137, 138*). After careful investigation of BMImTFSI, a suitable IL with a structure similar to that of B-PBI was added to prepare various SPE combinations while maintaining the solid nature and tailor the ionic conductivity while maintaining the IL composition [Figures 8(a, b)].

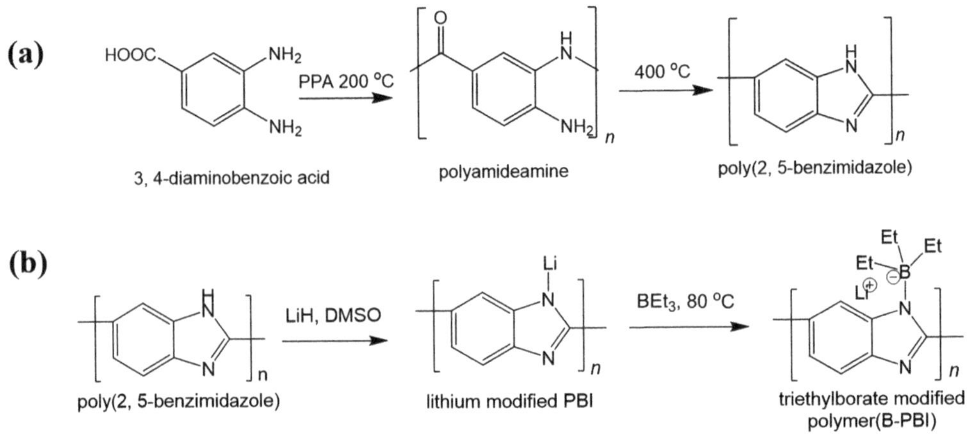

Figure 7. (a) Synthetic route of ABPBI using bio-based 3,4-DABA as monomer, (b) Structural modification of ABPBI to bornated polybenzimidazole as a novel SPE.

The thermal stability of the prepared SPEs was confirmed by thermogravimetric (TGA); the results revealed that they were stable up to 250 °C, and the T_{d10} was found to be 350 °C, which was the highest. A wide potential window with a maximum value of 5.45 V was determined by linear sweep voltammetry (LSV) measurements, which is high enough to be used in high-voltage battery applications. The ionic conductivity reached a maximum value of 8.8 mS/cm^{-1} while t_{Li^+} reached a maximum of 0.63 for a particular combination of SPE. Compared with other reported SPEs, such as $(PEO)_8LiClO_4$, PEG-borate ester, and Nafion-based SPE (*139–141*), the B-PBI/IL-based SPE exhibited higher ionic conductivity and t_{Li^+} (*131*). An anodic half-cell was fabricated using the SPE and silicon anode to check the charge–discharge profile and cycling performance. A maximum discharge capacity of 1300 mA h g^{-1} was observed at a charging rate of 0.1 C. Under a higher charging rate of 2 C, a discharge capacity of 800 mA h g^{-1} was reported. Moreover, the formation of a stable solid electrolyte interface (SEI) layer was confirmed with prolonged cycling time, and a real-time interfacial study was performed using a novel technique, dynamic electrochemical impedance spectroscopy (DEIS) (*142, 143*). This method is advantageous over the conventional electrochemical impedance spectroscopy (EIS) method, which records impedance data after the battery reaches a steady-state. Thus, in the EIS method, the DC potential becomes zero when the battery is in an equilibrium state. In contrast, in the DEIS method, real-time investigations can be performed, where the variable frequency response of an AC signal superimposed with the potentiodynamic DC voltage at regular intervals of potentials is recorded. Moreover, lithium-nickel manganese cobalt oxide (Li-MNC) cathodic half-cells were also prepared to investigate the high-voltage charge–discharge cycling property. The results evidenced the electrochemical stability of the electrolyte and its potential safe usage in high-performance batteries.

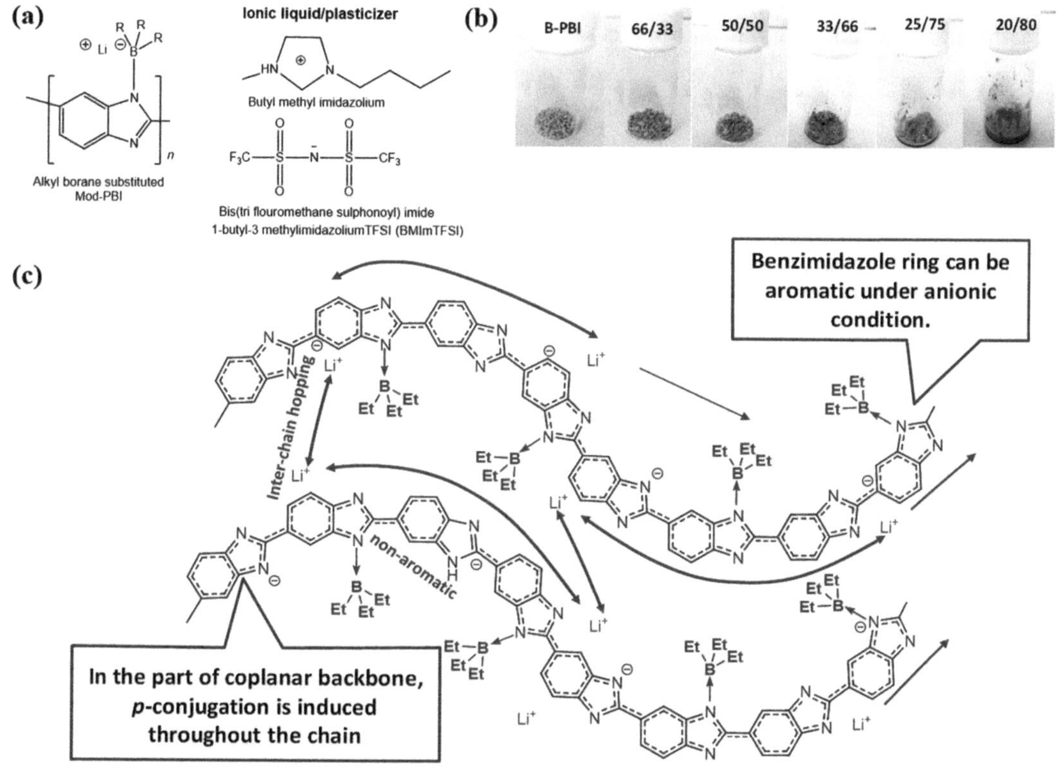

Figure 8. (a) Composite of SPE and ionic liquid BMImTFSI, (b) pictures of the SPEs with various combinations of ionic liquid, (c) plausible mechanism of lithium ion conduction.

If an in-depth analysis is performed on the novel structure of the B-PBI/IL system, it would be observed that ABPBI has a pseudo-aromatic structure. However, the imidazolate anion formed through the lithiation and boronation of the PBI imidazole moiety, has six π-electrons, which induce aromaticity. Therefore B-PBI with lithiated and boronated imidazole moiety can be regarded as a unique type of π-conjugated polymer electrolyte. The relatively larger polyanion delocalizes the negative charge over the polymer chain, which reduces the site-hopping energy of lithium ions [Figure 8(c)]. Thus the novel conceptual SPE exhibits a high ionic conductivity and lithium transference number.

Conclusion

Bio-based plastics produced from non-edible feedstock are ideal for a sustainable society as they exhibit comparable and even superior performance to commodity plastics. Aromatic building blocks derived from biomass and their microbial conversion to heterocyclic monomers have several advantages over aliphatic bioplastics, including their high thermo-mechanical stability and multifunctional structure–property relationships. The biomonomer, 3,4-AHBA, is an aromatic building block obtained through genetic engineering, and it can be employed to prepare homopolymers and copolymers with other bio-derived aromatic moieties, such as conductive PANI and photoresponsive polyesters, respectively. Polybenzazoles are being employed in heavy-duty engineering goods owing to their relatively high performance, chemical stability, and light-weight metal alternatives. However, the bio-based precursor materials for polybenzazole production are not easily available because they must involve aromatic building blocks that are difficult to derive from natural sources. Bio-based polybenzoxazoles with various structures can be applied extensively in the preparation of advanced materials, including ultra-low-*k* devices, flexible LCDs, and bio-based polybenzimidazole (ABPBI), which is already well-established as a proton-conducting membrane in PEMFCs and extremely high thermoresistant plastic, and it can also be used as a polyelectrolyte in LIBs.

Significance of Bio-Derived Aromatics

To truly effectively replace the fossil-based economy, apart from aliphatic-based bioplastics, aromatic building blocks must be derived from natural resources. Bio-based aromatic polymers with equivalent and even better performance compared to commodity plastics offer high safety and environmental characteristics, which are required for a truly sustainable society. Biomass related to food resources such as straw, grains, and starch creates competition for the currently limited food resources; therefore, wood, foliage, roots, pulps, and other waste-based resources would be more appropriate for establishing a sustainable society. Microbial fermentation is an effective strategy for producing genetically engineered biomonomers and for their further functionalization.

Acknowledgments

The authors acknowledge the financial support of Japan Science and Technology, under the project: Cross-ministerial Strategic Innovation Promotion Program (SIP), "Smart-bio" (Bio-oriented Technology Research Advancement Institution, NARO).

References

1. Dwivedi, S.; Kaneko, T. Aromatic Bioplastics with Heterocycles. In *Green Polymer Chemistry: New Products, Processes, and Applications*; ACS Symposium Series; American Chemical Society, 2018; Vol. 1310, pp 201−218.
2. Lynd, L. R.; Cushman, J. H.; Nichols, R. J.; Wyman, C. E. Fuel Ethanol from Cellulosic Biomass. *Science* **1991**, *251*, 1318–1323.
3. Kawaguchi, H.; Hasunuma, T.; Ogino, C.; Kondo, A. Bioprocessing of Bio-Based Chemicals Produced from Lignocellulosic Feedstocks. *Curr. Opin. Biotechnol.* **2016**, *42*, 30–39.
4. Hasunuma, T.; Kondo, A. Development of Yeast Cell Factories for Consolidated Bioprocessing of Lignocellulose to Bioethanol through Cell Surface Engineering. *Biotechnol. Adv.* **2012**, *30*, 1207–1218.
5. Hasunuma, T.; Kondo, A. Consolidated Bioprocessing and Simultaneous Saccharification and Fermentation of Lignocellulose to Ethanol with Thermotolerant Yeast Strains. *Process Biochem.* **2012**, *47*, 1287–1294.
6. Renewably Sourced Automotive Materials, https://www.dupont.com/knowledge/renewably-sourced-interiors.html (accessed 2020-06-10).
7. Holladay, J. E.; White, J. F.; Bozell, J. J.; Johnson, D. *Top Value Added Chemicals from Biomass - Volume II, Results of Screening for Potential Candidates from Biorefinery Lignin*; PNNL-16983; Pacific Northwest National Lab. (PNNL), Richland, WA (United States); National Renewable Energy Lab. (NREL), Golden, CO (United States), 2007.

8. Bozell, J. J.; Petersen, G. R. Technology Development for the Production of Biobased Products from Biorefinery Carbohydrates—the US Department of Energy's "Top 10" Revisited. *Green Chem.* **2010**, *12*, 539–554.
9. Marshall, A.-L.; Alaimo, P. J. Useful Products from Complex Starting Materials: Common Chemicals from Biomass Feedstocks. *Chem.—Eur. J.* **2010**, *16*, 4970–4980.
10. Gandini, A.; Belgacem, M. N. Chapter 1 - The State of the Art. In *Monomers, Polymers and Composites from Renewable Resources*; Belgacem, M. N., Gandini, A., Eds.; Elsevier: Amsterdam, 2008; 1–16.
11. Gandini, A. The Irruption of Polymers from Renewable Resources on the Scene of Macromolecular Science and Technology. *Green Chem.* **2011**, *13*, 1061–1083.
12. Gandini, A. Furans as Offspring of Sugars and Polysaccharides and Progenitors of a Family of Remarkable Polymers: A Review of Recent Progress. *Polym. Chem.* **2010**, *1*, 245–251.
13. Gandini, A.; Silvestre, A. J. D.; Neto, C. P.; Sousa, A. F.; Gomes, M. The Furan Counterpart of Poly(Ethylene Terephthalate): An Alternative Material Based on Renewable Resources. *J. Polym. Sci. Part Polym. Chem.* **2009**, *47*, 295–298.
14. Mialon, L.; Pemba, A. G.; Miller, S. A. Biorenewable Polyethylene Terephthalate Mimics Derived from Lignin and Acetic Acid. *Green Chem.* **2010**, *12*, 1704–1706.
15. Mialon, L.; Vanderhenst, R.; Pemba, A. G.; Miller, S. A. Polyalkylenehydroxybenzoates (PAHBs): Biorenewable Aromatic/Aliphatic Polyesters from Lignin. *Macromol. Rapid Commun.* **2011**, *32*, 1386–1392.
16. Kumar, N.; Pruthi, V. Potential Applications of Ferulic Acid from Natural Sources. *Biotechnol. Rep. Amsterdam Neth.* **2014**, *4*, 86–93.
17. Hatakeyama, H.; Hatakeyama, T. Lignin Structure, Properties, and Applications. In *Biopolymers: Lignin, Proteins, Bioactive Nanocomposites*; Abe, A., Dusek, K., Kobayashi, S., Eds.; Advances in Polymer Science; Springer: Berlin, Heidelberg, 2010; 1–63.
18. Takeshima, H.; Satoh, K.; Kamigaito, M. Scalable Synthesis of Bio-Based Functional Styrene: Protected Vinyl Catechol from Caffeic Acid and Controlled Radical and Anionic Polymerizations Thereof. *ACS Sustainable Chem. Eng.* **2018**, *6*, 13681–13686.
19. Takeshima, H.; Satoh, K.; Kamigaito, M. Bio-Based Functional Styrene Monomers Derived from Naturally Occurring Ferulic Acid for Poly(Vinylcatechol) and Poly(Vinylguaiacol) via Controlled Radical Polymerization. *Macromolecules* **2017**, *50*, 4206–4216.
20. Miyaji, H.; Satoh, K.; Kamigaito, M. Bio-Based Polyketones by Selective Ring-Opening Radical Polymerization of α-Pinene-Derived Pinocarvone. *Angew. Chem., Int. Ed.* **2016**, *55*, 1372–1376.
21. Tabata, Y.; Abe, H. Synthesis and Properties of Alternating Copolymers of 3-Hydroxybutyrate and Lactate Units with Different Stereocompositions. *Macromolecules* **2014**, *47*, 7354–7361.
22. Fujita, T.; Nguyen, H. D.; Ito, T.; Zhou, S.; Osada, L.; Tateyama, S.; Kaneko, T.; Takaya, N. Microbial Monomers Custom-Synthesized to Build True Bio-Derived Aromatic Polymers. *Appl. Microbiol. Biotechnol.* **2013**, *97*, 8887–8894.
23. Mu, W.; Liu, F.; Jia, J.; Chen, C.; Zhang, T.; Jiang, B. 3-Phenyllactic Acid Production by Substrate Feeding and p*H*-Control in Fed-Batch Fermentation of Lactobacillus Sp. SK007. *Bioresour. Technol.* **2009**, *100*, 5226–5229.

24. Niggeweg, R.; Michael, A. J.; Martin, C. Engineering Plants with Increased Levels of the Antioxidant Chlorogenic Acid. *Nat. Biotechnol.* **2004**, *22*, 746–754.
25. Kyndt, J. A.; Hurley, J. K.; Devreese, B.; Meyer, T. E.; Cusanovich, M. A.; Tollin, G.; Van Beeumen, J. J. Rhodobacter Capsulatus Photoactive Yellow Protein: Genetic Context, Spectral and Kinetics Characterization, and Mutagenesis. *Biochemistry* **2004**, *43*, 1809–1820.
26. Kort, R.; Phillips-Jones, M. K.; van Aalten, D. M. F.; Haker, A.; Hoffer, S. M.; Hellingwerf, K. J.; Crielaard, W. Sequence, Chromophore Extraction and 3-D Model of the Photoactive Yellow Protein from Rhodobacter Sphaeroides1The Nucleotide Sequences Reported in This Paper Have Been Submitted to the EMBL Nucleotide Sequence Database under Accession Numbers AJ002398 and X98889.1. *Biochim. Biophys. Acta BBA - Protein Struct. Mol. Enzymol.* **1998**, *1385*, 1–6.
27. Weng, W.; Markel, E. J.; Dekmezian, A. H. Synthesis of Long-Chain Branched Propylene Polymers via Macromonomer Incorporation. *Macromol. Rap. Commun.* **2001**, *22*, 1488–1492.
28. Johnson, O. H.; Green, D. E.; Pauli, R. The antibacterial action of derivatives and analogues of *p*-aminobenzoic acid. *J. Biol. Chem.* **1944**, *153*, 37–47.
29. Kaneko, T.; Tateyam, S.; Okajima, M.; Hojoon, S.; Takaya, N. Ultrahigh Heat-Resistant, Transparent Bioplastics from Exotic Amino Acid. *Mater. Today* **2016**, *3*, S21–S29.
30. Matsushita, Y.; Nakamura, A.; Aoki, D.; Yagami, S.; Fukushima, K. Bio-Based Polymer from Ferulic Acid by Electropolymerization. *BioResources* **2016**, *11*, 9789–9802.
31. Kaneko, T.; Thi, T. H.; Shi, D. J.; Akashi, M. Environmentally Degradable, High-Performance Thermoplastics from Phenolic Phytomonomers. *Nat. Mater.* **2006**, *5*, 966–970.
32. Tateyama, S.; Masuo, S.; Suvannasara, P.; Oka, Y.; Miyazato, A.; Yasaki, K.; Teerawatananond, T.; Muangsin, N.; Zhou, S.; Kawasaki, Y.; Zhu, L.; Zhou, Z.; Takaya, N.; Kaneko, T. Ultrastrong, Transparent Polytruxillamides Derived from Microbial Photodimers. *Macromolecules* **2016**, *49*, 3336–3342.
33. Kawasaki, Y.; Aniruddha, N.; Minakawa, H.; Masuo, S.; Kaneko, T.; Takaya, N. Novel Polycondensed Biopolyamide Generated from Biomass-Derived 4-Aminohydrocinnamic Acid. *Appl. Microbiol. Biotechnol.* **2018**, *102*, 631–639.
34. Tachibana, Y.; Masuda, T.; Funabashi, M.; Kunioka, M. Chemical Synthesis of Fully Biomass-Based Poly(Butylene Succinate) from Inedible-Biomass-Based Furfural and Evaluation of Its Biomass Carbon Ratio. *Biomacromolecules* **2010**, *11*, 2760–2765.
35. Wang, C.; Ming, W.; Yan, D.; Zhang, C.; Yang, M.; Liu, Y.; Zhang, Y.; Guo, B.; Wan, Y.; Xing, J. Novel Membrane-Based Biotechnological Alternative Process for Succinic Acid Production and Chemical Synthesis of Bio-Based Poly (Butylene Succinate). *Bioresour. Technol.* **2014**, *156*, 6–13.
36. Vardon, D. R.; Rorrer, N. A.; Salvachúa, D.; Settle, A. E.; Johnson, C. W.; Menart, M. J.; Cleveland, N. S.; Ciesielski, P. N.; Steirer, K. X.; Dorgan, J. R.; Beckham, G. T. Cis,Cis-Muconic Acid: Separation and Catalysis to Bio-Adipic Acid for Nylon-6,6 Polymerization. *Green Chem.* **2016**, *18*, 3397–3413.
37. Han, J. A. Bio-Based 'Green' Process for Catalytic Adipic Acid Production from Lignocellulosic Biomass Using Cellulose and Hemicellulose Derived γ-Valerolactone. *Energy Convers. Manag.* **2016**, *129*, 75–80.

38. Suzuki, H.; Ohnishi, Y.; Furusho, Y.; Sakuda, S.; Horinouchi, S. Novel Benzene Ring Biosynthesis from C(3) and C(4) Primary Metabolites by Two Enzymes. *J. Biol. Chem.* **2006**, *281*, 36944–36951.
39. Suzuki, H.; Ohnishi, Y.; Horinouchi, S. GriC and GriD Constitute a Carboxylic Acid Reductase Involved in Grixazone Biosynthesis in Streptomyces Griseus. *J. Antibiot. (Tokyo)* **2007**, *60*, 380–387.
40. Shozui, F.; Tajima, Y. Escherichia Coli Capable of Producing 3-Amino-4-Hydroxybenzoic Acid. US9587259B2, March 7, 2017.
41. Kawaguchi, H.; Ogino, C.; Kondo, A. Microbial Conversion of Biomass into Bio-Based Polymers. *Bioresour. Technol.* **2017**, *245*, 1664–1673.
42. Liu, R.; Liang, L.; Li, F.; Wu, M.; Chen, K.; Ma, J.; Jiang, M.; Wei, P.; Ouyang, P. Efficient Succinic Acid Production from Lignocellulosic Biomass by Simultaneous Utilization of Glucose and Xylose in Engineered Escherichia Coli. *Bioresour. Technol.* **2013**, *149*, 84–91.
43. Anand, J.; Palaniappan, S.; Sathyanarayana, D. N. Conducting Polyaniline Blends and Composites. *Prog. Polym. Sci.* **1998**, *23*, 993–1018.
44. Kulkarni, V. G. Polyanilines: Progress in Processing and Applications. In *Semiconducting Polymers*; ACS Symposium Series; American Chemical Society, 1999; Vol. 735, 174−183.
45. Armes, S. P.; Miller, J. F. Optimum Reaction Conditions for the Polymerization of Aniline in Aqueous Solution by Ammonium Persulphate. *Synth. Met.* **1988**, *22*, 385–393.
46. Kan, K.; Yamamoto, H.; Kaneko, D.; Tateyama, S.; Kaneko, T. Novel π-Conjugated Bio-Based Polymer, Poly(3-Amino-4-Hydroxybenzoic Acid), and Its Solvatochromism. *Pure Appl. Chem.* **2014**, *86*, 685–690.
47. Park, G. T.; Chang, J.-H.; Lim, A. R. Thermotropic Liquid Crystalline Polymers with Various Alkoxy Side Groups: Thermal Properties and Molecular Dynamics. *Polymers* **2019**, *11*, 992.
48. Jackson, W. J.; Kuhfuss, H. F. Liquid Crystal Polymers. I. Preparation and Properties of p-Hydroxybenzoic Acid Copolyesters. *J. Polym. Sci. Polym. Chem. Ed.* **1976**, *14*, 2043–2058.
49. Ophir, Z.; Ide, Y. Injection Molding of Thermotropic Liquid Crystal Polymers. *Polym. Eng. Sci.* **1983**, *23*, 792–796.
50. Uzman, M.; Kühnpast, K.; Springer, J. Orientational Behaviour of Polymer Blends Containing a Liquid-Crystalline Component. *Makromol. Chem.* **1989**, *190*, 3185–3194.
51. Lenz, R. W. Synthesis and Properties of Thermotropic Liquid Crystal Polymers with Main Chain Mesogenic Units. *Polym. J.* **1985**, *17*, 105–115.
52. Zeng, L.; Li, R.; Chen, P.; Xu, J.; Liu, P. Synthesis and Characterization of Thermotropic Liquid Crystalline Polyarylate with Ether Ether Ketone Segments in the Main Chain. *J. Appl. Polym. Sci.* **2016**, *133*, 43800.
53. Martínez-Gómez, A.; Encinar, M.; Fernández-Blázquez, J. P.; Rubio, R. G.; Pérez, E. Relationship Between Composition, Structure and Dynamics of Main-Chain Liquid Crystalline Polymers with Biphenyl Mesogens. In *Liquid Crystalline Polymers: Volume 1-Structure and Chemistry*; Thakur, V. K., Kessler, M. R., Eds.; Springer International Publishing: Cham, 2016; 453−476.
54. Damman, S. B.; Mercx, F. P. M.; Kootwijk-Damman, C. M. Liquid-Crystalline Main-Chain Polymers with a Poly(p-Phenylene Terephthalate) Backbone: 1. Synthesis, Characterization and Rheology of Polyesters with Alkoxy Side Chains. *Polymer* **1993**, *34*, 1891–1897.

55. Chang, J.-H.; Jo, B.-W. Blends of PBT with Rigid Thermotropic LCP Having Flexible Side Groups: Comparison of Their Tensile Properties with Semirigid Main-Chain TLCP Blends. *J. Appl. Polym. Sci.* **1996**, *60*, 939–946.
56. Chung, S.-J.; Kim, K.-K.; Jin, J.-I. Fluorescing Wholly Aromatic Polyesters Containing Diphenylanthracene Fluorophores. *Polymer* **1999**, *40*, 1943–1953.
57. Mori, T.; Kijima, M. Synthesis and Electroluminescence Properties of Carbazole-Containing 2,6-Naphthalene-Based Conjugated Polymers. *Eur. Polym. J.* **2009**, *45*, 1149–1157.
58. Lim, W. L.; Oo, C. W.; Kobayashi, T. Photoluminescent Enhancement of Polyesters via Photocrosslinking. *J. Appl. Polym. Sci.* **2015**, *132*, 41504.
59. Gross, R. A.; Kalra, B. Biodegradable Polymers for the Environment. *Science* **2002**, *297*, 803–807.
60. Ueda, M.; Sugita, H.; Sato, M. Synthesis of Poly(Benzoxazole)s by Direct Polycondensation of Dicarboxylic Acids with 3,3′-Dihydroxybenzidine Dihydrochloride Using Phosphorus Pentoxide/Methanesulfonic Acid as Condensing Agent and Solvent. *J. Polym. Sci. Part Polym. Chem.* **1986**, *24*, 1019–1026.
61. Imai, Y.; Taoka, I.; Uno, K.; Iwakura, Y. Polybenzoxazoles and Polybenzothiazoles. *Makromol. Chem.* **1965**, *83*, 167–178.
62. Ohta, Y.; Niiyama, T.; Yokoyama, A.; Yokozawa, T. Chain-Growth Condensation Polymerization Approach to Synthesis of Well-Defined Polybenzoxazole: Importance of Higher Reactivity of 3-Amino-4-Hydroxybenzoic Acid Ester Compared to 4-Amino-3-Hydroxybenzoic Acid Ester. *J. Polym. Sci. Part Polym. Chem.* **2014**, *52*, 1730–1736.
63. Moyer, W. W.; Cole, C.; Anyos, T. Aromatic Polybenzoxazoles. *J. Polym. Sci. A* **1965**, *3*, 2107–2121.
64. Zhou, C.; Wang, S.; Zhuang, Q.; Han, Z. Enhanced Conductivity in Polybenzoxazoles Doped with Carboxylated Multi-Walled Carbon Nanotubes. *Carbon* **2008**, *46*, 1232–1240.
65. Ali, M. A.; Shimosegawa, H.; Nag, A.; Takada, K.; Kaneko, T. Synthesis of Thermotropic Polybenzoxazole Using 3-Amino-4-Hydroxybenzoic Acid. *J. Polym. Res.* **2017**, *24*, 214.
66. Hsiao, S.-H.; Chiou, J.-H. Aromatic Polybenzoxazoles Containing Ether–Sulfone Linkages. *J. Polym. Sci. Part Polym. Chem.* **2001**, *39*, 2262–2270.
67. Hilborn, J. G.; Labadie, J. W.; Hedrick, J. L. Poly(Aryl Ether-Benzoxazoles). *Macromolecules* **1990**, *23*, 2854–2861.
68. Smith, J. G.; Connell, J. W.; Hergenrother, P. M. Chemistry and Properties of Poly(Arylene Ether Benzoxazole)s. *Polymer* **1992**, *33*, 1742–1747.
69. Miyazaki, T.; Hasegawa, M. Highly Tough and Highly Transparent Soluble Polybenzoxazoles. *High Perform. Polym.* **2007**, *19*, 243–269.
70. Oishi, Y.; Konnno, A.; Oravec, J.; Mori, K. Synthesis and Properties of Fluorine-Containing Polybenzoxazoles by In Situ Silylation Method. *J. Photopolym. Sci. Technol.* **2006**, *19*, 669–672.
71. Feig, V. R.; Tran, H.; Bao, Z. Biodegradable Polymeric Materials in Degradable Electronic Devices. *ACS Cent. Sci.* **2018**, *4*, 337–348.
72. Kohl, P. A. Low-Dielectric Constant Insulators for Future Integrated Circuits and Packages. *Annu. Rev. Chem. Biomol. Eng.* **2011**, *2*, 379–401.
73. Volksen, W.; Miller, R. D.; Dubois, G. Low Dielectric Constant Materials. *Chem. Rev.* **2010**, *110*, 56–110.

74. Maex, K.; Baklanov, M. R.; Shamiryan, D.; lacopi, F.; Brongersma, S. H.; Yanovitskaya, Z. S. Low Dielectric Constant Materials for Microelectronics. *J. Appl. Phys.* **2003**, *93*, 8793–8841.
75. Lin, Q.; Cohen, S. A.; Gignac, L.; Herbst, B.; Klaus, D.; Simonyi, E.; Hedrick, J.; Warlaumont, J.; Lee, H.-J.; Wu, W. Low Dielectric Constant Nanocomposite Thin Films Based on Silica Nanoparticle and Organic Thermosets. *J. Polym. Sci. Part B Polym. Phys.* **2007**, *45*, 1482–1493.
76. Ghosh, M. K.; Mittal, K. L. *Polyimides: Fundamentals and Applications*; Marcel Dekker: New York, 1996.
77. Baik, D.; Lee, W.; Park, Y. Interfacial Characterization of Polybenzoxazole/copper System. *Mol. Cryst. Liq. Cryst.* **2004**, *424*, 265–271.
78. Hasegawa, M.; Kobayashi, J.; Vladimirov, L. Solution-Processable Low-CTE Polybenzoxazoles. *J. Photopolym. Sci. Technol.* **2004**, *17*, 253–258.
79. Fukumaru, T.; Fujigaya, T.; Nakashima, N. Design and Preparation of Porous Polybenzoxazole Films Using the Tert-Butoxycarbonyl Group as a Pore Generator and Their Application for Patternable Low-k Materials. *Polym. Chem.* **2012**, *3*, 369–376.
80. MacDonald, S. A.; Ito, H.; Willson, C. G. Advances in the Design of Organic Resist Materials. *Microelectron. Eng.* **1983**, *1*, 269–293.
81. Wang, Y.; Yu, J.; Zhu, J.; Hu, Z. Hyperbranched Polybenzoxazoles Incorporated Polybenzoxazoles for High-Performance and Low-K Materials. *J. Polym. Sci. Part Polym. Chem.* **2016**, *54*, 1623–1632.
82. Fukukawa, K.; Shibasaki, Y.; Ueda, M. A Photosensitive Semi-Alicyclic Poly(Benzoxazole) with High Transparency and Low Dielectric Constant. *Macromolecules* **2004**, *37*, 8256–8261.
83. Kraytsberg, A.; Ein-Eli, Y. Review of Advanced Materials for Proton Exchange Membrane Fuel Cells. *Energy Fuels* **2014**, *28*, 7303–7330.
84. Asensio, J. A.; Borrós, S.; Gómez-Romero, P. Enhanced Conductivity in Polyanion-Containing Polybenzimidazoles. Improved Materials for Proton-Exchange Membranes and PEM Fuel Cells. *Electrochem. Commun.* **2003**, *5*, 967–972.
85. He, R.; Li, Q.; Bach, A.; Jensen, J. O.; Bjerrum, N. J. Physicochemical Properties of Phosphoric Acid Doped Polybenzimidazole Membranes for Fuel Cells. *J. Membr. Sci.* **2006**, *277*, 38–45.
86. Sannigrahi, A.; Arunbabu, D.; Sankar, R. M.; Jana, T. Tuning the Molecular Properties of Polybenzimidazole by Copolymerization. *J. Phys. Chem. B* **2007**, *111*, 12124–12132.
87. Asensio, J. A.; Borrós, S.; Gómez-Romero, P. Polymer Electrolyte Fuel Cells Based on Phosphoric Acid-Impregnated Poly(2,5-Benzimidazole) Membranes. *J. Electrochem. Soc.* **2004**, *151*, A304–A310.
88. Pahari, S.; Roy, S. Structural and Conformational Properties of Polybenzimidazoles in Melt and Phosphoric Acid Solution: A Polyelectrolyte Membrane for Fuel Cells. *RSC Adv.* **2016**, *6*, 8211–8221.
89. Omar, N.; Daowd, M.; Bossche, P. van den; Hegazy, O.; Smekens, J.; Coosemans, T.; Mierlo, J. van Rechargeable Energy Storage Systems for Plug-in Hybrid Electric Vehicles—Assessment of Electrical Characteristics. *Energies* **2012**, *5*, 2952–2988.
90. Tsuchiya, K.; Ishii, H.; Shibasaki, Y.; Ando, S.; Ueda, M. Synthesis of a Novel Poly(Binaphthylene Ether) with a Low Dielectric Constant. *Macromolecules* **2004**, *37*, 4794–4797.

91. Nag, A.; Ali, M. A.; Watanabe, M.; Singh, M.; Amornwachirabodee, K.; Kato, S.; Mitsumata, T.; Takada, K.; Kaneko, T. High-Performance Poly(Benzoxazole/Benzimidazole) Bio-Based Plastics with Ultra-Low Dielectric Constant from 3-Amino-4-Hydroxybenzoic Acid. *Polym. Degrad. Stab.* **2019**, *162*, 29–35.
92. Endo, K. Fluorinated Amorphous Carbon as a Low-Dielectric-Constant Interlayer Dielectric. *MRS Bull.* **1997**, *22*, 55–58.
93. Kloster, G.; Scherban, T.; Xu, G.; Blaine, J.; Sun, B.; Zhou, Y. Porosity Effects on Low-k Dielectric Film Strength and Interfacial Adhesion. In *Proceedings of the IEEE 2002 International Interconnect Technology Conference (Cat. No.02EX519)*; 2002; pp 242–244.
94. Vashishta, P.; Mundy, J. N.; Shenoy, G. K. *Fast Ion Transport in Solids: Electrodes and Electrolytes*; Elsevier North Holland, Inc.: United States, 1979.
95. O'Regan, B.; Grätzel, M. A Low-Cost, High-Efficiency Solar Cell Based on Dye-Sensitized Colloidal TiO 2 Films. *Nature* **1991**, *353*, 737–740.
96. Grätzel, M. Photoelectrochemical Cells. *Nature* **2001**, *414*, 338–344.
97. Kohle, O.; Grätzel, M.; Meyer, A. F.; Meyer, T. B. The Photovoltaic Stability of, Bis(Isothiocyanato)Rlutheniurn(II)-Bis-2,2′bipyridine-4,4′-Dicarboxylic Acid and Related Sensitizers. *Adv. Mater.* **1997**, *9*, 904–906.
98. Lin, D.; Liu, Y.; Cui, Y. Reviving the Lithium Metal Anode for High-Energy Batteries. *Nat. Nanotechnol.* **2017**, *12*, 194–206.
99. Tikekar, M. D.; Choudhury, S.; Tu, Z.; Archer, L. A. Design Principles for Electrolytes and Interfaces for Stable Lithium-Metal Batteries. *Nat. Energy* **2016**, *1*, 16114.
100. Liu, Y.; Lin, D.; Liang, Z.; Zhao, J.; Yan, K.; Cui, Y. Lithium-Coated Polymeric Matrix as a Minimum Volume-Change and Dendrite-Free Lithium Metal Anode. *Nat. Commun.* **2016**, *7*, 10992.
101. Zhang, H.; Eshetu, G. G.; Judez, X.; Li, C.; Rodriguez-Martínez, L. M.; Armand, M. Electrolyte Additives for Lithium Metal Anodes and Rechargeable Lithium Metal Batteries: Progress and Perspectives. *Angew. Chem., Int. Ed.* **2018**, *57*, 15002–15027.
102. Zhu, M.; Wu, J.; Wang, Y.; Song, M.; Long, L.; Siyal, S. H.; Yang, X.; Sui, G. Recent Advances in Gel Polymer Electrolyte for High-Performance Lithium Batteries. *J. Energy Chem.* **2019**, *37*, 126–142.
103. Wang, Y.; Zhong, W.-H. Development of Electrolytes towards Achieving Safe and High-Performance Energy-Storage Devices: A Review. *ChemElectroChem* **2015**, *2*, 22–36.
104. Bruce, P. G.; Scrosati, B.; Tarascon, J.-M. Nanomaterials for Rechargeable Lithium Batteries. *Angew. Chem., Int. Ed.* **2008**, *47*, 2930–2946.
105. Tarascon, J.-M.; Armand, M. Issues and Challenges Facing Rechargeable Lithium Batteries. *Nature* **2001**, *414*, 359–367.
106. Deng, K.; Wang, S.; Ren, S.; Han, D.; Xiao, M.; Meng, Y. A Novel Single-Ion-Conducting Polymer Electrolyte Derived from CO2-Based Multifunctional Polycarbonate. *ACS Appl. Mater. Interfaces* **2016**, *8*, 33642–33648.
107. Zhang, Y.; Lai, J.; Gong, Y.; Hu, Y.; Liu, J.; Sun, C.; Wang, Z. L. A Safe High-Performance All-Solid-State Lithium–Vanadium Battery with a Freestanding V2O5 Nanowire Composite Paper Cathode. *ACS Appl. Mater. Interfaces* **2016**, *8*, 34309–34316.

108. Hou, H.; Xu, Q.; Pang, Y.; Li, L.; Wang, J.; Zhang, C.; Sun, C. Efficient Storing Energy Harvested by Triboelectric Nanogenerators Using a Safe and Durable All-Solid-State Sodium-Ion Battery. *Adv. Sci.* **2017**, *4*, 1700072.
109. Mindemark, J.; Lacey, M. J.; Bowden, T.; Brandell, D. Beyond PEO—Alternative Host Materials for Li+-Conducting Solid Polymer Electrolytes. *Prog. Polym. Sci.* **2018**, *81*, 114–143.
110. Long, L.; Wang, S.; Xiao, M.; Meng, Y. Polymer Electrolytes for Lithium Polymer Batteries. *J. Mater. Chem. A* **2016**, *4*, 10038–10069.
111. Wakihara, M. Recent Developments in Lithium Ion Batteries. *Mater. Sci. Eng. R Rep.* **2001**, *33*, 109–134.
112. Matsumi, N.; Nakashiba, M.; Mizumo, T.; Ohno, H. Novel Polymer/Salt Hybrid Composed of Comblike Organoboron Polymer Electrolyte and Boron-Stabilized Imido Anion. *Macromolecules* **2005**, *38*, 2040–2042.
113. Zhang, H.; Li, C.; Piszcz, M.; Coya, E.; Rojo, T.; Rodriguez-Martinez, L. M.; Armand, M.; Zhou, Z. Single Lithium-Ion Conducting Solid Polymer Electrolytes: Advances and Perspectives. *Chem. Soc. Rev.* **2017**, *46*, 797–815.
114. Matsumi, N.; Sugai, K.; Ohno, H. Selective Ion Transport in Organoboron Polymer Electrolytes Bearing a Mesitylboron Unit. *Macromolecules* **2002**, *35*, 5731–5733.
115. Barthel, J.; Wühr, M.; Buestrich, R.; Gores, H. J. A New Class of Electrochemically and Thermally Stable Lithium Salts for Lithium Battery Electrolytes: I . Synthesis and Properties of Lithium Bis[1,2-benzenediolato(2-)-O,O′]Borate. *J. Electrochem. Soc.* **1995**, *142*, 2527–2531.
116. Barthel, J.; Buestrich, R.; Carl, E.; Gores, H. J. A New Class of Electrochemically and Thermally Stable Lithium Salts for Lithium Battery Electrolytes: II. Conductivity of Lithium Organoborates in Dimethoxyethane and Propylene Carbonate. *J. Electrochem. Soc.* **1996**, *143*, 3565–3571.
117. Barthel, J.; Buestrich, R.; Carl, E.; Gores, H. J. A New Class of Electrochemically and Thermally Stable Lithium Salts for Lithium Battery Electrolytes: III. Synthesis and Properties of Some Lithium Organoborates. *J. Electrochem. Soc.* **1996**, *143*, 3572–3575.
118. Barthel, J.; Schmid, A.; Gores, H. J. A New Class of Electrochemically and Thermally Stable Lithium Salts for Lithium Battery Electrolytes. V. Synthesis and Properties of Lithium Bis[2,3-Pyridinediolato(2−)-O,O′]Borate. *J. Electrochem. Soc.* **2000**, *147*, 21–24.
119. Matsumi, N.; Mizumo, T.; Ohno, H. Preparation of Comb-like Organoboron Polymer Electrolyte without Generation of Salt. *Chem. Lett.* **2004**, *33*, 372–373.
120. Qin, B.; Liu, Z.; Zheng, J.; Hu, P.; Ding, G.; Zhang, C.; Zhao, J.; Kong, D.; Cui, G. Single-Ion Dominantly Conducting Polyborates towards High Performance Electrolytes in Lithium Batteries. *J. Mater. Chem. A* **2015**, *3*, 7773–7779.
121. Ma, Q.; Zhang, H.; Zhou, C.; Zheng, L.; Cheng, P.; Nie, J.; Feng, W.; Hu, Y.-S.; Li, H.; Huang, X.; Chen, L.; Armand, M.; Zhou, Z. Single Lithium-Ion Conducting Polymer Electrolytes Based on a Super-Delocalized Polyanion. *Angew. Chem., Int. Ed.* **2016**, *55*, 2521–2525.
122. Feng, S.; Shi, D.; Liu, F.; Zheng, L.; Nie, J.; Feng, W.; Huang, X.; Armand, M.; Zhou, Z. Single Lithium-Ion Conducting Polymer Electrolytes Based on Poly[(4-Styrenesulfonyl)(Trifluoromethanesulfonyl)Imide] Anions. *Electrochim. Acta* **2013**, *93*, 254–263.

123. Bertasi, F.; Pagot, G.; Vezzù, K.; Negro, E.; Sideris, P. J.; Greenbaum, S. G.; Ohno, H.; Scrosati, B.; Di Noto, V. Exotic Solid State Ion Conductor from Fluorinated Titanium Oxide and Molten Metallic Lithium. *J. Power Sources* **2018**, *400*, 16–22.

124. Ayalneh Tiruye, G.; Muñoz-Torrero, D.; Palma, J.; Anderson, M.; Marcilla, R. All-Solid State Supercapacitors Operating at 3.5 V by Using Ionic Liquid Based Polymer Electrolytes. *J. Power Sources* **2015**, *279*, 472–480.

125. Ye, H.; Huang, J.; Xu, J. J.; Khalfan, A.; Greenbaum, S. G. Li Ion Conducting Polymer Gel Electrolytes Based on Ionic Liquid/PVDF-HFP Blends. *J. Electrochem. Soc.* **2007**, *154*, A1048–A1057.

126. Bhavsar, R. S.; Kumbharkar, S. C.; Rewar, A. S.; Kharul, U. K. Polybenzimidazole Based Film Forming Polymeric Ionic Liquids: Synthesis and Effects of Cation–Anion Variation on Their Physical Properties. *Polym. Chem.* **2014**, *5*, 4083–4096.

127. Jung, M.; Lee, W.; Nambi Krishnan, N.; Kim, S.; Gupta, G.; Komsiyska, L.; Harms, C.; Kwon, Y.; Henkensmeier, D. Porous-Nafion/PBI Composite Membranes and Nafion/PBI Blend Membranes for Vanadium Redox Flow Batteries. *Appl. Surf. Sci.* **2018**, *450*, 301–311.

128. Nawn, G.; Pace, G.; Lavina, S.; Vezzù, K.; Negro, E.; Bertasi, F.; Polizzi, S.; Di Noto, V. Interplay between Composition, Structure, and Properties of New H_3PO_4-Doped PBI4N–HfO2 Nanocomposite Membranes for High-Temperature Proton Exchange Membrane Fuel Cells. *Macromolecules* **2015**, *48*, 15–27.

129. Nawn, G.; Vezzù, K.; Bertasi, F.; Pagot, G.; Pace, G.; Conti, F.; Negro, E.; Di Noto, V. Electric Response and Conductivity Mechanism in H3PO4-Doped Polybenzimidazole-4N–HfO_2 Nanocomposite Membranes for High Temperature Fuel Cells. *Electrochim. Acta* **2017**, *228*, 562–574.

130. Xiao, L.; Zhang, H.; Jana, T.; Scanlon, E.; Chen, R.; Choe, E.-W.; Ramanathan, L. S.; Yu, S.; Benicewicz, B. C. Synthesis and Characterization of Pyridine-Based Polybenzimidazoles for High Temperature Polymer Electrolyte Membrane Fuel Cell Applications. *Fuel Cells* **2005**, *5*, 287–295.

131. Nag, A.; Ali, M. A.; Singh, A.; Vedarajan, R.; Matsumi, N.; Kaneko, T. N-Boronated Polybenzimidazole for Composite Electrolyte Design of Highly Ion Conducting Pseudo Solid-State Ion Gel Electrolytes with a High Li-Transference Number. *J. Mater. Chem. A* **2019**, *7*, 4459–4468.

132. Klongkan, S.; Pumchusak, J. Effects of Nano Alumina and Plasticizers on Morphology, Ionic Conductivity, Thermal and Mechanical Properties of PEO-$LiCF_3SO_3$ Solid Polymer Electrolyte. *Electrochim. Acta* **2015**, *161*, 171–176.

133. Chaurasia, K.; Shalu, S.; Gupta, K.; , A.; Verma, Y. L.; Singh, V. K.; Tripathi, A. K.; Saroj, A. L.; Singh, R. K. Role of Ionic Liquid [$BMIMPF_6$] in Modifying the Crystallization Kinetics Behavior of the Polymer Electrolyte PEO-$LiClO_4$. *RSC Adv.* **2015**, *5*, 8263–8277.

134. Wetjen, M.; Kim, G.-T.; Joost, M.; Winter, M.; Passerini, S. Temperature Dependence of Electrochemical Properties of Cross-Linked Poly(Ethylene Oxide)–Lithium Bis(Trifluoromethanesulfonyl)Imide–N-Butyl-N-Methylpyrrolidinium Bis(Trifluoromethanesulfonyl)Imide Solid Polymer Electrolytes for Lithium Batteries. *Electrochim. Acta* **2013**, *87*, 779–787.

135. Kim, G. T.; Appetecchi, G. B.; Carewska, M.; Joost, M.; Balducci, A.; Winter, M.; Passerini, S. UV Cross-Linked, Lithium-Conducting Ternary Polymer Electrolytes Containing Ionic Liquids. *J. Power Sources* **2010**, *195*, 6130–6137.
136. Rupp, B.; Schmuck, M.; Balducci, A.; Winter, M.; Kern, W. Polymer Electrolyte for Lithium Batteries Based on Photochemically Crosslinked Poly(Ethylene Oxide) and Ionic Liquid. *Eur. Polym. J.* **2008**, *44*, 2986–2990.
137. Choi, J.-W.; Cheruvally, G.; Kim, Y.-H.; Kim, J.-K.; Manuel, J.; Raghavan, P.; Ahn, J.-H.; Kim, K.-W.; Ahn, H.-J.; Choi, D. S.; Song, C. E. Poly(Ethylene Oxide)-Based Polymer Electrolyte Incorporating Room-Temperature Ionic Liquid for Lithium Batteries. *Solid State Ion.* **2007**, *178*, 1235–1241.
138. Kim, Y. H.; Cheruvally, G.; Choi, J. W.; Ahn, J. H.; Kim, K. W.; Ahn, H. J.; Choi, D. S.; Song, C. E. Electrochemical Properties of PEO-Based Polymer Electrolytes Blended with Different Room Temperature Ionic Liquids. *Macromol. Symp.* **2007**, *249–250*, 183–189.
139. Wang, M.; Zhao, F.; Dong, S. A Single Ionic Conductor Based on Nafion and Its Electrochemical Properties Used As Lithium Polymer Electrolyte. *J. Phys. Chem. B* **2004**, *108*, 1365–1370.
140. Kato, Y.; Yokoyama, S.; Yabe, T.; Ikuta, H.; Uchimoto, Y.; Wakihara, M. Ionic Conductivity and Transport Number of Lithium Ion in Polymer Electrolytes Containing PEG–Borate Ester. *Electrochim. Acta* **2004**, *50*, 281–284.
141. Sun, H.-Y.; Sohn, H.-J.; Yamamoto, O.; Takeda, Y.; Imanishi, N. Enhanced Lithium-Ion Transport in PEO-Based Composite Polymer Electrolytes with Ferroelectric $BaTiO_3$. *J. Electrochem. Soc.* **1999**, *146*, 1672–1676.
142. Smaran, K. S.; Joshi, P.; Vedarajan, R.; Matsumi, N. Optimisation of Potential Boundaries with Dynamic Electrochemical Impedance Spectroscopy for an Anodic Half-Cell Based on Organic–Inorganic Hybrid Electrolytes. *ChemElectroChem* **2015**, *2*, 1913–1916.
143. Patnaik, S. G.; Vedarajan, R.; Matsumi, N. BIAN Based Functional Diimine Polymer Binder for High Performance Li Ion Batteries. *J. Mater. Chem. A* **2017**, *5*, 17909–17919.

Green Sustainable Films and Bioplastics

Chapter 7

Xylose Utilization for Polyhydroxyalkanoate Biosynthesis

Richard D. Ashby

Eastern Regional Research Center, Agricultural Research Service, U.S. Departments of Agriculture, 600 East Mermaid Lane, Wyndmoor, Pennsylvania 19038, United States
Email: Rick.Ashby@usda.gov

The sustainable production of biochemicals is primarily driven by the costs associated with the specific biosynthetic process. To maximize the economics of any bio-based synthetic process, efforts are focused on the utilization of low-value, high-volume substrates to produce environmentally-benign, high-value products with a range of applications. One such substrate is xylose. Xylose is the main sugar associated with hemicellulose and is the second most abundant sugar in nature representing approximately 20-30% of the sugars present in terrestrial plant biomass. Polyhydroxyalkanoates (PHA) are bacterial polyesters that possess properties analogous to those demonstrated by a range of widely-utilized petroleum-based polymers with the added advantages of being biorenewable, biodegradable, and biocompatible. As such, xylose is currently recognized and evaluated as a potentially beneficial feedstock for the synthesis of many different products including PHA biopolymers. This chapter focuses on the utilization of xylose for PHA biosynthesis including membrane transport, and its integration into the central metabolic pathways associated with PHA production, its conversion to levulinic acid (a valuable co-substrate for copolymer synthesis) as well as identifying the different bacterial strains capable of PHA synthesis from xylose and the types of PHA produced.

Introduction

With the ongoing worldwide plastic disposal issues, many governments are implementing more stringent utilization and recycling requirements with the idea that fewer single-use applications will result in less plastic refuse. While this is a logical strategy, in reality these approaches will only be successful if they are combined with additional efforts designed to help in the environmental breakdown of discarded plastic waste. Currently it is estimated that less than 10% of all plastic waste is effectively reprocessed and that approximately 8 million metric tons of plastic enter the worlds waterways each year (*1*). This is a staggering amount of recalcitrant plastic that is entering the biosphere and the negative impact that these discarded wastes are having on dietary considerations, wildlife and planet aesthetics are immense. In addition to post-use repurposing efforts, alternative strategies are being explored and favorable results implemented in order to effectively breakdown discarded plastics. Currently, efforts are underway to investigate the possibility of discovering and/or designing unique enzymes that are capable of degrading large-volume single-use plastics. Recently it was reported that the bacterium *Ideonella sakaiensis* 201-F6 has the genetic capability to produce an enzyme that can degrade poly(ethylene terephthalate) (PET) which is the most common petroleum-based single-use plastic (*2*) and its replacement polyethylene-2,5-furandicarboxylate (*3*). While this discovery is groundbreaking, the utilization of controlled large-scale enzyme applications for natural plastic degradation is limited by slow reaction rates and dilution issues. A more realistic endeavor is to develop and utilize more environmentally-friendly plastic alternatives that can naturally degrade to carbon dioxide and water in microbially-active environments. This has been the subject of many scientific studies over the years but the single biggest drawback to this effort has been the prohibitive cost-to-produce of the bio-based polymers at production levels that would make them attractive substitutes.

Poly(hydroxyalkanoates) (PHA) belong to a class of bio-based polymers (biopolymers) that are typically derived from bacterial sources and are produced through fermentation processes. They are poly-β-esters and are naturally synthesized as carbon and energy reserves. PHA biopolymers have attracted much attention over the years because they have been demonstrated to be biorenewable, biocompatible and easily biodegradable. In addition, PHA can be produced with properties ranging from pliable elastomers (medium-chain PHA) to hard thermoplastics (short-chain PHA) (*4*). The properties of the short-chain (*sc*-) PHA biopolymers (typical monomeric chemical structures associated with *sc*-PHA are shown in Figure 1) have been favorably compared to those of polyethylene and polypropylene. Unfortunately, as with most other biopolymers, production costs are prohibitively higher when produced via fermentation compared to chemically-derived petroleum-based polymers. This cost increase is primarily due to the raw material costs which have driven efforts to engineer and diversify PHA biosynthesis (*5*) and to find low-value, high-volume materials that can be used by the individual bacterial strains to produce these valuable biopolymers more inexpensively. Some of the feedstocks investigated include crude glycerol (*6–8*), whey permeate (*9*), and soy molasses (*10*) among others.

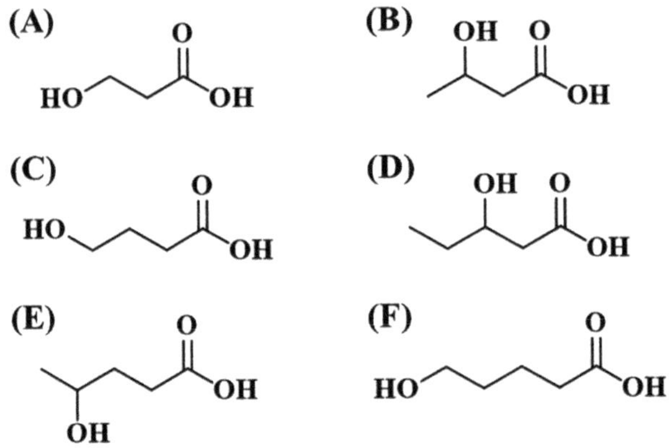

*Figure 1. Chemical structures for the typical monomers associated with sc-PHA biopolymers: 3-hydroxypropionic acid (3HP) (**A**), 3-hydroxybutyric acid (3HB) (**B**), 4-hydroxybutyric acid (4HB) (**C**), 3-hydroxyvaleric acid (3HV) (**D**), 4-hydroxyvaleric acid (4HV) (**E**), 5-hydroxyvaleric acid (5HV) (**F**).*

More recently, lignocellulosic materials have been examined as feedstocks for PHA biosynthesis. These materials have included forest biomass (*11*), wheat straw hydrolysate (*12*), sugar maple hemicellulosic hydrolysate (*13*), spruce sawdust hydrolysate (*14*), and tequila agave bagasse hydrolysate (*15*). Lignocellulose is composed of cellulose (a homopolymer of glucose units connected by β 1-4 glycosidic bonds; contents as high as 30-50 wt%), hemicellulose (a complex branched polysaccharide composed primarily of xylose, arabinose, mannose and galactose; contents typically ranging from 25-30 wt%), and lignin (comprised of phenolic polymers; contents normally reaching from 15-25 wt%). Lignocellulose is a very high-volume, low-value material that contains potentially fermentable carbohydrates. Unfortunately, the simple sugars associated with lignocellulose must be liberated prior to their utilization in fermentation processes. This liberation is typically done by acid hydrolysis which can present additional issues such as the formation of inhibitory compounds (*i.e.*, furfural, 5-hydroxymethyl furfural and organic acids) whose presence can impede the effectiveness of a fermentation. In addition, multiple unrelated simple sugars can cause carbon catabolite repression (CCR) which is a regulatory phenomenon in which a bacterium will initially utilize the most energy-efficient carbon source in a mixture (*16*). This leads to diauxic growth patterns and limits the effective utilization of secondary sugars and ultimately impedes the overall productivity of the fermentation. This is a common occurrence in lignocellulosic fermentations between glucose and xylose. Therefore, in an ideal situation either wildtype or genetically-altered bacterial strains must be obtained that can metabolize multiple sugars simultaneously to more efficiently utilize all of the sugar available and maintain the highest production levels. Although some research efforts have produced engineered bacterial strains capable of utilizing both glucose and xylose simultaneously (*17*), in reality these strains are not commonly known. As such, research into the viability of lignocellulose utilization for PHA biosynthesis has progressed by focusing on the individual sugars associated with lignocllulosic biomass. Glucose (the primary component of the cellulosic fraction of lignocellulose) is already well known as a viable substrate for the fermentative production of PHA biopolymers; however, xylose (the second most common sugar in nature) has only recently been identified as a plausible substrate for PHA biosynthesis. In this chapter the production of PHA from xylose will be discussed from the standpoint of the mechanisms involved in the bacterial membrane transport of xylose, the metabolic processes involved in the enzymatic formation of the precursors necessary for PHA biosynthesis, the chemical conversion of xylose (and other simple sugars) into levulinic acid (LevA; an important precursor molecule for PHA copolymer production), and lastly the successful utilization of xylose in the synthesis of *sc*-PHA biopolymers.

Membrane Transport

The first step in the assimilation process of any sugar molecule by bacteria or any other cell is the transport of that sugar across the cytoplasmic membrane. This process is vital in order to avail the sugar molecule to the enzymes necessary for energy and bio-intermediate generation. Bacteria possess a number of transport mechanisms that in many cases are sugar-specific (*18*). Each bacterial strain may possess multiple transporters designed to transfer different sugars across the cytoplasmic membrane. In *Escherichia coli* the transport mechanisms may include facilitated diffusion (glycerol), osmotic shock-sensitive active transport (maltose), proton-linked active transport (lactose), sodium cotransport (melibiose), and group translocation (glucose). Glucose is the most common lignocellulosic sugar. It's uptake is achieved through the phosphoenolpyruvate:sugar phosphotransferase system (PTS) which is a group translocation mechanism and is comprised of a series of soluble and membrane-bound enzymes that catalyze the transfer of a phosphate group from phosphoenolpyruvate to the sugar molecule (Figure 2A). The phosphorylation cascade begins with the transfer of a single phosphate group from phosphoenolpyruvate to a histidine residue on enzyme EI. The phosphate group is then sequentially transferred to the low-molecular-weight heat-stable protein HPr and then to the glucose-specific enzyme EIIA and finally to the membrane-associated enzyme EIIB which phosphorylates the glucose molecule as it passes the cytoplasmic membrane through the transmembrane EIIC protein channel. This reaction results in the formation of glucose-6-phosphate which is an intermediate in both the Embden-Meyerhof and Entner-Doudoroff Pathways (*19*).

As a result of CCR (described previously) glucose is typically the most efficient and preferred carbon source available for bacterial growth. As such, most chemoheterotrophic bacteria will utilize any glucose present prior to metabolizing secondary feedstocks. In contrast to the PTS transport mechanism demonstrated for glucose, xylose is transported through the cytoplasmic membrane either through the action of the major facilitator superfamily (MFS) of proteins (XylE protein; Figure 2B) or through an ABC-binding cassette transporter (XylFGH multi-component protein complex; Figure 2C) (*20*).

XylE

The XylE protein is a low affinity (K_m = 63-169 μM) D-xylose/proton symporter (an MFS protein) that is driven by the proton motive force (Figure 2B) and whose regulatory mechanism is poorly understood. The crystal structure of the protein has been elucidated and shows that XylE actually folds into 12 transmembrane domains, the first 6 comprising the N-domain with the last 6 making up the C-domain (*21*). In that study the crystal structures of XylE in both the inward-facing and outward-facing conformations with and without bound substrates were reported. These images showed the rocker-switch mechanism of the N- and C-domains during the transport process. Briefly, it was reported that the protonation of amino acid Asp27 (part of transmembrane region 1) is vital to binding the xylose substrate and inducing the switch from outward- to inward-facing conformation. This protonation of Asp27 is mediated by Arg133 (part of transmembrane region 4). Once protonation and substrate binding occur extracellularly, the substrate-binding site opens into the cytoplasm allowing the proton from Asp27 and the sugar to be released into the cytoplasm. Once this process is complete the interaction between Asp27 and Arg133 can be re-established and the protein conformation transitions back to outward-facing and the process begins again.

XylFGH

The XylFGH is an example of an ABC-binding cassette transporter. It is a multi-component, energy-intensive, high-affinity D-xylose transport complex that is composed of three distinct proteins, a substrate-binding protein (XylF), a membrane-spanning protein (XylH), and a nucleotide-binding protein (XylG) (Figure 2C) (*22*). This protein complex has a higher binding affinity (K_m = 0.2-4.0 μM) for D-xylose than the XylE proton symporter (described previously) so is thought to play a larger role in D-xylose transport across the bacterial cytoplasmic membrane (*23*). Transcriptional regulation of the *xylFGH* genes are coregulated along with the *xylAB* genes (which code for xylose catabolic enzymes) by cyclic AMP (cAMP) receptor protein (CRP) and the XylR transcriptional activator protein (*24*). The CRP-cAMP complex contributes to CCR and positively regulates the catabolic operons for secondary sugars such as xylose under glucose-limited conditions. In addition, recent work by Sun et al. (*25*) proposed the existence of an auxiliary regulator protein (XylFII) in *Clostridium beijerinckii* which is a transmembrane protein that does not participate in the sugar transport process itself, but interacts with and autophosphorylates (activates) another single transmembrane protein, histidine kinase (LytS), to sense the presence of D-xylose (*26*). The activated LytS gene product then triggers the response regulator YesN DNA binding protein which binds to the promoter region of the *xylFGH* gene and initiates gene expression and D-xylose uptake.

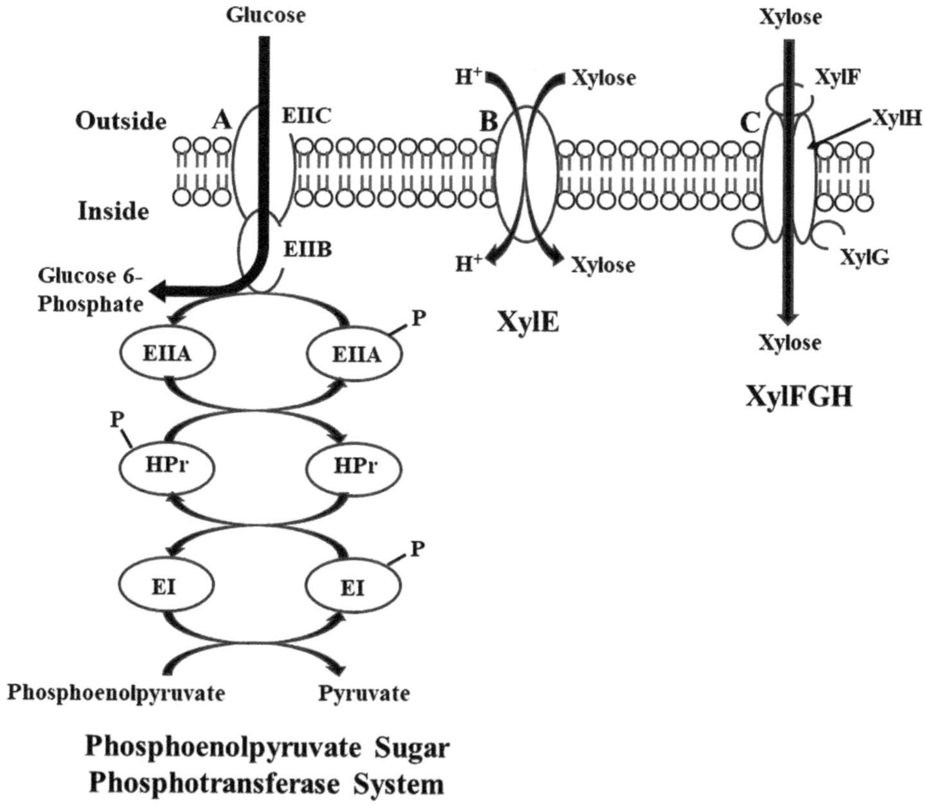

*Figure 2. Bacterial sugar transport mechanisms for glucose (group translocation); (**A**), xylose (proton symport mechanism; (**B**), and xylose (ABC-binding cassette transporter mechanism; (**C**).*

Metabolic Processes

The ultimate necessity for any bacteria to utilize xylose as a sugar precursor for *sc*-PHA biosynthesis is the ability to metabolically produce pyruvate and subsequently acetyl-CoA from which the poly-3-hydroxybutyrate (P3HB) precursor molecules can be synthesized. Once xylose is transported into the bacterial cytoplasm there are three commonly available enzymatic cascade pathways that a bacterium might utilize. These pathways are dictated by the genetic makeup of the specific bacterial strain (Figure 3).

Most bacteria that are capable of metabolizing xylose will intiate the xylose isomerase (XI) pathway. This pathway is diverse among prokaryotes and starts with the sequential conversion of xylose to xylulose. The xylulose is then phosphorylated to obtain xylulose-5-phosphate through the energy-requiring xylulose kinase reaction (*27*). In contrast, some enteric bacterial strains such as *E. coli* K12 have developed the ability to utilize the xylose reductase-xylitol dehydrogenase pathway (XR-XDH) to assimilate xylose. This pathway is typically found in *Saccharomyces cerevisiae* and catalyzes the enzymatic conversion of xylose into xylitol before proceeding to xylulose and finally xylulose-5-phosphate. While this pathway may be present in some bacteria, this pathway has been demonstrated to be comparatively inefficient compared to the XI pathway in prokaryotes (*28*). The end result of each of these enzymatic pathways is the generaton of xylulose-5-phosphate which is an important intermediate in the non-oxidative phase of the pentose phosphate pathway (*29*). In the pentose phosphate pathway xylulose-5-phosphate reacts with ribose-5-phosphate (derived from the enzymatic conversion of glucose-6-phosphate through substrate level phosphorylation) through a series of transketolase and transaldolase reactions to produce fructose-6-phosphate and glyceraldehyde-3-phosphate which, upon further substrate level phosphorylation, can be converted into pyruvate and subsequently to acetyl-CoA, a valuable intermediate in the P3HB biosynthetic pathway, through the pyruvate dehydrogenase complex.

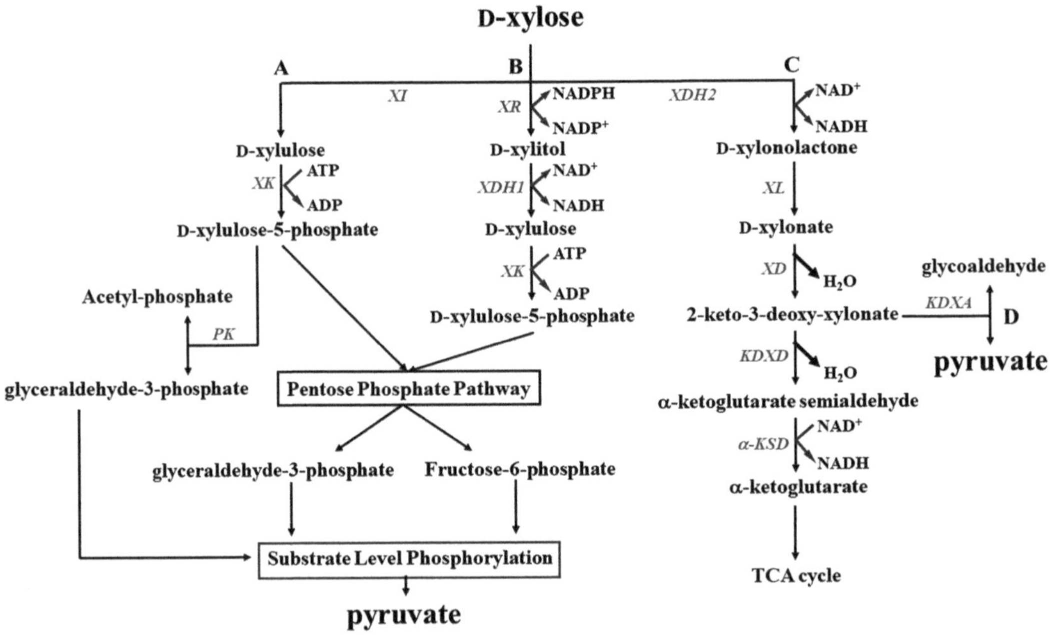

*Figure 3. Metabolic pathways for xylose assimilation; xylose isomerase pathway (**A**), xylose reductase – xylitol dehydrogenase (XR-XDH) pathway (**B**), α-ketoglutarate (Weimberg) pathway (**C**), Dahms pathway (**D**). Abbreviations: XI (D-xylose isomerase), XK (xylulose kinase), PK (phosphoketolase), XR (D-xylose reductase), XDH1 (xylitol dehydrogenase), XDH2 (D-xylose dehydrogenase), XL (xylonolactonase), XD (xylonate dehydratase), KDXA (2-keto-3-deoxy xylonate aldolase), KDXD (2-keto-3-deoxy xylonate dehydratase), α-KSD (α-ketoglutarate semialdehyde dehydrogenase).*

In some instances xylulose-5-phosphate can be cleaved into glyceraldehyde-3-phosphate and acetyl-phosphate through the action of the enzyme phosphoketolase in order to bypass a pyruvate decarboxylase reaction and increase the acetyl-CoA levels (*30, 31*). Although the phosphoketolase enzymatic reaction is more prevalent in yeasts and fungi than in bacteria, any PHA-producing strain of bacteria that is capable of undergoing this reaction will be benefitted by the increased acetyl-CoA levels which are important to the production of acetoacetyl-CoA, a valuable intermediate in the PHB biosynthetic process.

Lastly, an alternative xylose metabolizing pathway was discovered in *Pseudomonas fragi* (32) and more recently in *Caulobacter crescentus* (33). In this pathway xylose is initially converted into xylonolactone and xylonate by the sequential activity of xylose dehydrogenase and xylonolactonase.

Then, xylonate undergoes two rounds of dehydration in the presence of xylonate dehydratase and 2-keto-3-deoxyxylonate dehydratase to form 2-keto-3-deoxy-xylonate and finally α-ketoglutaric semialdehyde, respectively. Finally, the α-ketoglutaric semialdehyde is oxidized into α-ketoglutarate (an intermediate in the citric acid cycle) by α-ketoglutarate semialdehyde dehydrogenase (*33*). This enzymatic pathway has been termed the α-ketoglutarate pathway or Weimberg pathway. In addition, the 2-keto-3-deoxy-xylonate can be cleaved by an aldolase to yield pyruvate and glycoaldehyde which can then converted to glycolate through the action of a dehydratase. This pathway is known as the Dahms pathway (*34*). While this pathway is of lesser value in producing pyruvate for P3HB biosynthesis, the ability to synthesize glycolate using the Dahms pathway has been beneficial for the synthesis of other bio-based polymers from a metabolically engineered strain of *E. coli* (*35*). These biopolymers include poly(lactate-*co*-glycolate) and poly(lactate-*co*-glycolate-*co*-2-hydroxybutyrate (*36*).

Currently there are over 150 different monomers associated with PHA biopolymers (*37*) and historically, as a general rule, they are all categorized as either *sc*-PHA or medium-chain (*mc*-) PHA. Typically, wildtype bacterial strains that are capable of synthesizing PHA biopolymers produce one type or the other as few bacteria naturally possess the genetic makeup to polymerize both monomer types. In order to overcome this issue, genetic recombination efforts have been used to produce *sc*-/*mc*-PHA copolymers (*i.e.*, poly-3-hydroxybutyrate-*co*-3-hydroxyhexanoate; P3HB-3HHx, poly-3-hydroxybutyrate-*co*-3-hydroxyoctanoate; P3HB-3HO) with enhanced mechanical properties (*38, 39*). Of course, the ability to transport and metabolize xylose does not guarantee that any given bacterial strain can necessarily produce PHA biopolymers. The appropriate enzymes must be present to convert pyruvate into the requisite intermediates and then polymerize these intermediates into the mature polymer. While there are some bacterial strains that can produce *mc*-PHA from sugar substrates, normally the synthesis of *mc*-PHA is dependent upon the presence of triacylglycerol or free fatty acid substrates that can be easily formulated into mature *mc*- intermediates and polymerized into mature *mc*-PHA biopolymers. Those strains that do have the ability to make *mc*-PHA from sugar must synthesize the requisite *mc*- precursors *de novo* from the sugar substrate which can be an energy-intensive metabolic process that normally results in diminished product yields.

The most well-known member of the PHA family is poly-3-hydroxybutyrate (P3HB), an *sc*-PHA biopolymer which has been known for almost 100 years (*40*). This homopolymer is produced by a number of distinct bacterial strains many of which have unique processes to produce acetyl-CoA. For example, some bacterial strains can utilize glycerol as a primary feedstock for *sc*-PHA biosynthesis. Glycerol is typically transported through the cytoplasmic membrane by facilitated diffusion and enters substrate level phosphorylation at the glyceraldehyde-3-phosphate stage before being enzymatically converted to pyruvate and acetyl-CoA (*18*). A second possibility are those strains that can utilize fatty acids as substrates for *sc*-PHA biosynthesis. Upon β-oxidation, the fatty acid chain lengths are reduced by two carbon units through the liberation of acetyl-CoA. While this mechanism is vital for *mc*-PHA production from the standpoint of fatty acid chain length, the acetyl-CoA molecules that are released upon β-oxidation may also be utilized by certain bacterial strains to produce *sc*-PHA biopolymers (*41*).

Two of the most recognized P3HB producers are *Cupriavidus necator* (formerly *Ralstonia eutropha*) and *Azohydomonas lata* (formerly *Alcaligenes latus*). Each of these bacterial strains has been repeatedly demonstrated to produce *sc*-PHA from monosaccharides. As such, using the well-understood synthetic systems in these two wildtype PHA-producing strains it becomes evident how xylose can be further metabolized to produce *sc*-PHA and specifically P3HB. The biosynthetic pathway from pyruvate to P3HB is shown in Figure 4. First, pyruvate is enzymatically converted to acetyl-CoA through the action of the pyruvate dehydrogenase complex. This reaction is ubiquitous among cells for the introduction of acetyl-CoA into the citric acid cycle which is vital for energy and bio-intermediate production. The *sc*-PHA biosynthetic genes are found on the *phbCAB* operon. The PhbA gene product is a β-ketothiolase (EC 2.3.1.9) enzyme that catalyzes the condensation of two acetyl-CoA molecules into a single acetoacetyl-CoA. The PhbB gene product is an acetoacetyl-CoA reductase (EC 1.1.1.36) which catalyzes the conversion of acetoacetyl-CoA to R-3-hydroxybutyryl-CoA and lastly, the PhbC gene product is a PHA synthase (EC 2.3.1.41) that polymerizes the R-3-hydroxybutyryl-CoA intermediate into the mature polymer. Interestingly, studies in which the gene order within the operon was disrupted/rearranged resulted in ultra-high polymer molecular weights which can be additionally advantageous for property control (42).

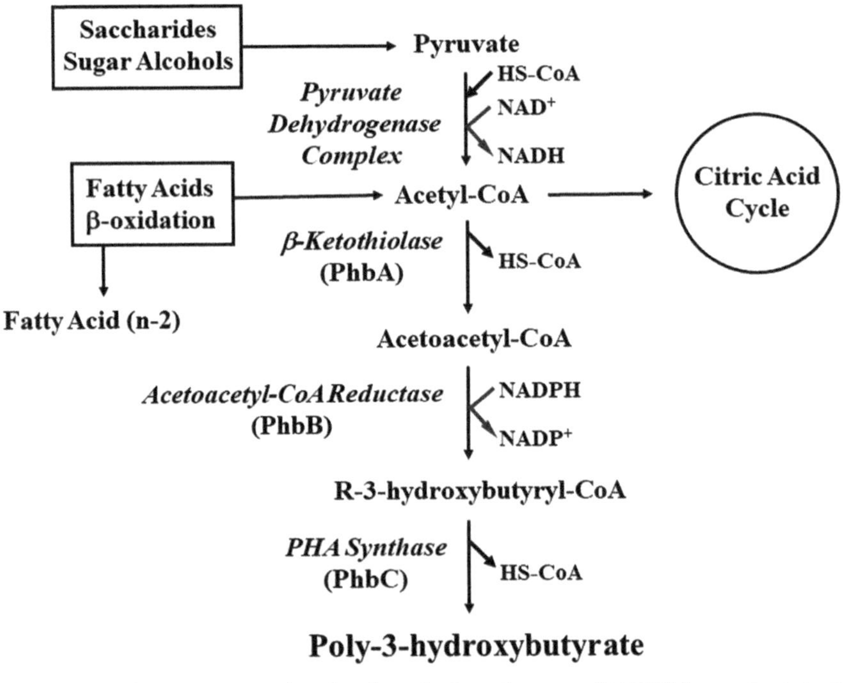

Figure 4. Metabolic pathways associated with poly-3-hydroxybutyrate (P3HB) biosynthesis in Cupriavidus necator from pyruvate to mature polymer.

Lignocellulosic Sugar Conversion to Levulinic Acid (LevA)

While P3HB is the most well-known and easily produced member of the *sc*-PHA family, its thermal and mechanical properties make widespread P3HB utilization less appealing. The high melting temperature of P3HB homopolymers (~180°C) complicates processing and the poor mechanical properties of P3HB (stiffness and brittleness) reduce the application potential for multi-functional uses. Historically, the solution to this dilemma has been the use of co-substrate materials that can result in *sc*-PHA copolymers with reduced processing temperatures and improved mechanical properties. Two of the most commonly used co-substrates are propionic acid and/or valeric acid which normally result in *sc*-PHA copolymers consisting of 3-hydroxybutyric acid (3HB) and 3-hydroxyvaleric acid (3HV) with improved pliability and reduced melting temperatures.

Another co-substrate that is finding increased use in *sc*-PHA biosynthesis is LevA. LevA (4-oxovaleric acid) is a value-added chemical that is structurally similar to valeric acid and can be produced from lignocellulosic biomass including from both 5-carbon (C5) and 6-carbon (C6) sugars (*43*). It has been classified among the top twelve bio-based platform chemicals by the United States Department of Energy (*44*) and in the year 2000 it was estimated that the cost of LevA could range to as low as 4 to 10 cents/lb (*45*). More recent estimates derived in 2015 place the cost of LevA at between 2 and 4 USD/lb (*46*) but because it is easily produced from plant biomass it remains an intriguing molecule for the fermentative production of *sc*-PHA copolymer synthesis.

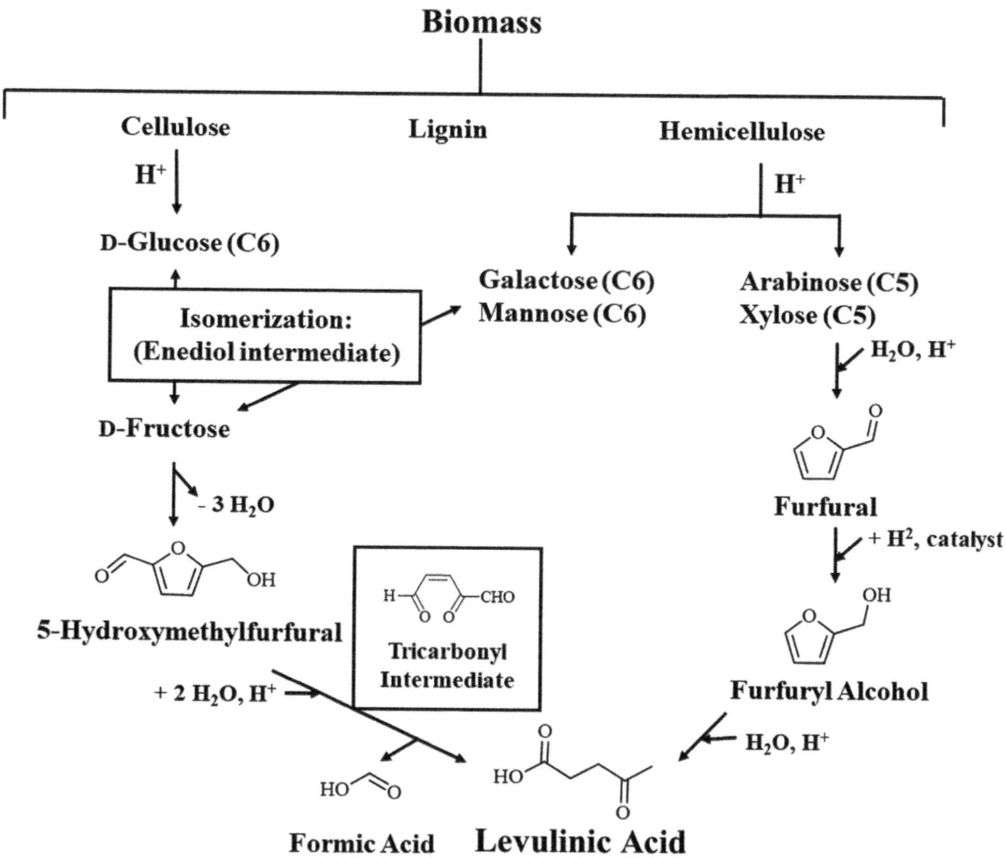

Figure 5. Chemical conversion pathways of biomass-derived C5 and C6 sugars into LevA. (for further details see reference (43)).

The pathways involved in LevA synthesis from lignocellulosic biomass are shown in Figure 5. Two different reaction conditions have been employed to facilitate monosaccharide generation from plant biomass. The first involves the use of dilute acid treatments, typically involving mineral acids (HCl or H_2SO_4), at high temperature and pressure while the second involves more concentrated acids at lower temperatures and normal pressures. Upon acid hydrolysis of lignocellulose both C5 and C6 sugars are liberated from the hemicellulosic and the cellulosic fractions, respectively. The conversion of the C6 sugars such as glucose, mannose and galactose to LevA is initiated through an isomerization reaction to produce fructose (*47, 48*) which is then dehydrated to 5-hydroxymethylfurfural (5-HMF) (*49*). Lastly, the 5-HMF is rehydrated through an unstable tricarbonyl intermediate to produce LevA and formic acid (*50*).

In contrast, C5 sugars such as xylose and arabinose are first converted to furfural (*51*) through a dehydration reaction with the subsequent hydrogenation of furfural to furfuryl alcohol and finally the acid-catalyzed hydrolysis of furfuryl alcohol to LevA (*52*). Unfortunately, from an economic perspective, both furfural and furfuryl alcohol are of themselves valuable chemicals. Furfural can be converted into solvents, polymers, fuels and other useful chemicals through catalytic reductions (*53, 54*), while furfuryl alcohol can be used for furan resins that can be utilized in thermoset polymer composites, cements, adhesives, and coatings (*55*). While LevA can certainly be produced from xylose and other C5 sugars, the valuable intermediates from the C5 pathway have focused the LevA synthetic process more towards the C6 pathway. In an effort to improve the economics involved in C5 sugar conversion to LevA, additional catalytic reactions have been studied including the use of modified zeolite (*56–59*) and metal chloride (*60, 61*) catalysts.

Xylose and Levulinic Acid Utilization for *sc*-PHA Biosynthesis

While xylose alone will typically result in P3HB biosynthesis, the inclusion of co-substrates will only result in copolymer production if the enzymatic mechanism is available for the conversion of the ancillary substrate to utilizable precursors. It is known that LevA can be utilized by certain bacterial strains to produce PHA biopolymers and although organic acids, including LevA, often inhibit cell growth at moderate concentrations, more recent studies have shown that some bacterial strains can withstand the detrimental effects of high concentrations of LevA in growth media (*62*). In small concentrations (typically less than 2 g/L) LevA can be utilized for cell growth and PHA biosynthesis (*63*). Studies have shown that the bacterium *Cupriavidus necator* has the capability to utilize LevA as a sole carbon source or co-substrate for cell growth and PHA copolymer biosynthesis (*63–65*). In 2011 a biosynthetic pathway was proposed for the synthesis of PHA copolymers from LevA (*66*). The reaction pathway is initiated by the conversion of LevA to its activated form (4-ketovaleryl-CoA) which is then split into acetyl-CoA and propionyl-CoA. The condensation of two acetyl-CoA molecules by a β-ketothiolase (PhaA) enzyme continues the reaction cascade towards the formation of R-3-hydroxybutyryl-CoA, the immediate precursor prior to 3HB polymerization (Figure 4). This proposed metabolic pathway also demonstrates the existence of a second β-ketothiolase enzyme (BktB) which is capable of condensing acetyl-CoA and propionyl-CoA to form 3-ketovaleryl-CoA, a 5-carbon intermediate which can then be enzymatically converted into 3-hydroxyvaleryl-CoA, the immediate precursor for 3HV incorporation into the *sc*-PHA biopolymer. This pathway describes the formation of the copolymer poly-3-hydroxybutyrate-*co*-3-hydroxyvalerate (P3HB-3HV). Interestingly, when LevA was used exclusively as the sole carbon source with *C. necator*, the terpolyester poly-3-hydroxybutyrate-*co*-3-hydroxyvalerate-*co*-4-hydroxyvalerate (P3HB-3HV-4HV) was formed with only a minor component (1-3%) of 4-hydroxyvalerate (4HV) (*63*). The metabolic pathway proposed in reference (*66*) did not provide for the conversion of LevA into 4HV due to the lack of an oxidoreductase enzyme to convert the levulinyl-CoA intermediate into 4-hydroxyvaleryl-CoA. This was problematic with respect to where the 4HV component of the terpolymer originated.

In 2017 another metabolic pathway was published for the synthesis of *sc*-PHA from LevA in *Pseudomonas putida* KT2440 (*67*). That study identified the *lvaABCDEFG* operon as coding for the enzymes involved in PHA precursor generation. Briefly, LevA is activated to levulinyl-CoA through the action of an acyl-CoA synthetase (LvaE gene product) and is then reduced to 4-hydroxyvaleryl-CoA by oxidoreductase (LvaD gene product). The 4-hydroxyvaleryl-CoA is then phosphorylated to 4-phosphovaleryl-CoA by a phosphotransferase enzyme from the LvaAB complex (LvaA gene product; the LvaB gene product has an unknown function but was shown to be essential for the enzymatic reaction to occur). The 4-phosphovaleryl-CoA is then dephosphorylated to produce pentenoyl-CoA that is subsequently hydrated to yield 3-hydroxyvaleryl-CoA through the action of a dehydrogenase enzyme (LvaC gene product). This pathway results in both the 3-hydroxyvaleryl-CoA and 4-hydroxyvaleryl-CoA precursors for *sc*-PHA copolymer biosynthesis. Of course, 3-hydroxyvaleryl-CoA can be further oxidized to acetyl-CoA and propionyl-CoA which can of themselves be utilized for 3HB and 3HV monomer synthesis. The LvaF and LvaG proteins are organic acid transport proteins and are not likely involved in the catabolism of LevA. This pathway accounts for all of the requisite -CoA precursors for P3HB-3HV-4HV biosynthesis.

Since organic acids including LevA can be toxic to bacteria at sufficiently high concentrations, in order to produce *sc*-PHA biopolymers from LevA alone, the substrate concentration in the media must remain at strain-specific levels that do not inhibit overall polymer production. In order to improve polymer production, research has focused on the utilization of mixed substrates including sugar and organic acids. Interestingly, when LevA was used in combination with sugar (glucose or fructose) *C. necator* consumed the LevA more efficiently than the available sugars (*66*). This was also evident in *Burkholderia sacchari* when grown in the presence of mixtures of glucose, xylose and LevA (*68*). In contrast, studies involving *Azohydromonas lata* demonstrated the utilization of the sugar component (glucose) in the mixed substrate prior to LevA in the production of P3HB-3HV-4HV terpolymers (*69*).

Individual studies on the utilization of xylose and LevA for PHA biosynthesis are shown in Table 1. In these studies PHA biosynthesis was reported by a number of different bacterial strains using xylose as the primary carbon source and, in some cases, organic acids as secondary carbon sources (*11, 68–78*).

As expected, numerous studies listed in Table 1 showed that xylose alone resulted in the formation of P3HB homopolymer regardless of the bacterial strain and culture conditions used. In contrast, the introduction of LevA into the growth media as a co-substrate induced the synthesis of either P3HB-3HV or P3HB-3HV-4HV copolymers whose overall compositions were dictated by the LevA concentration used in the fermentation process. Interestingly, reference (*68*) described the formation of P3HB-3HV block copolymers when xylose and LevA were used in combination. Typically, block copolymers are synthesized through the sequential feeding of two or more dissimilar feedstocks (*79, 80*) but *Burkholderia sacchari* DSM 17165 demonstrated the capability of synthesizing block copolymers comprised of 3HB and 3HV repeat units when both xylose and LevA were present at the onset of the fermentations. In that study it was demonstrated that the initial metabolic processes favored the utilization of LevA to produce block sequences of P3HV. Once the LevA concentration was reduced to basal levels, then the xylose was effectively utilized to produce sequences of P3HB which ultimately resulted in mature copolymers containing homopolymeric sequences of P3HB and P3HV (*68*). In contrast, *Azohydromonas lata* DSM 1122 could not grow on xylose but could utilize LevA in the presence of glucose to produce terpolymers composed of 3HB, 3HV, and 4HV where the 3HV component never exceeded 6 mol% and the 4HV component never exceeded 9 mol% (*69*). This indicated that while *A. lata* has the capability to assimilate LevA into both 3HV and 4HV monomers, the primary metabolic reactions involved glucose whose utilization was much more rapid than the LevA. When *A. lata* was grown in mixed-cultures with *B. sacchari* in the presence of mixtures of glucose, xylose and LevA, copolymer mixtures were produced that included both P3HB-3HV block copolymers and P3HB-3HV-4HV random terpolymers.

Table 1. Xylose Utilization for PHA Biopolymer Synthesis

	Bacterial Strain	*1°[a]*	*2°[b]*	*Type of PHA Prod.*	*Ref.*
1	*Pseudomonas cepacia* (ATCC 17759)	Xyl	-	P3HB	(*70*)
2	*Pseudomonas pseudoflava* (ATCC 33668)	Xyl	-	P3HB	(*71*)
3	*Burkholderia cepacia*	Xyl	LevA	P3HB-3HV	(*11*)
4	*Bacillus megaterium* (J-65)	Xyl	-	P3HB	(*72*)
5	*Burkholderia sacchari* (LFM900 recomb.)	Xyl	-	P3HB	(*73*)
6	**A**) *Burkholderia sacchari* (IPT101)	Xyl	-	P3HB	(*74*)
	B) Multiple soil isolates identified through 16S DNA comp. including: *Burkholderia cepacia, Bacillus megaterium, Bacillus cereus*	Xyl	-	P3HB	
7	*Pseudomonas putida* (KT 2440) pSLM1 recomb.	Xyl	-	P3HB	(*75*)
		Xyl	OA	*mc*-PHA	
8	*Enterobacter* (sp. TF) recomb.	Xyl	-	Primarily P3HB; some C10 monomers in 1 strain	(*76*)
9	*Burkholderia sacchari* (DSM 17165)	Xyl	-	P3HB	(*77*)
		Glu/Xyl	-	P3HB	
10	*Burkholderia sacchari* (LMG 19450) var. recomb.	Xyl	-	P3HB	(*78*)
11	*Burkholderia sacchari* (DSM 17165)	Xyl	-	P3HB	(*68*)
		Glu/Xyl	-	P3HB	
		Xyl	LevA	P3HB-3HV block copolymers	
		Glu/Xyl	LevA	P3HB-3HV block copolymers	
12	*Azohydromonas lata* (DSM 1122)	Glu	-	P3HB	(*69*)
		Glu	LevA	P3HB-3HV-4HV Terpolymers	
		Xyl	-	-	
	Azohydromonas lata (DSM 1122) +	Glu	-	P3HB	
	Burkholderia sacchari (DSM 17165)	Xyl	-	P3HB	
	mixed culture	Glu/Xyl	LevA	P3HB-3HV-4HV and P3HB-3HV block copolymer mixtures	

[a] 1° refers to the primary carbon source (Glu = glucose; Xyl = xylose). [b] 2° refers to the secondary carbon source (LevA = levulinic acid; OA = octanoic acid).

Conclusions

Polyhydroxyalkanoates (PHA) have been demonstrated to possess properties analogous to many of the petroleum-based polymers with the added benefits of renewability and degradability. Unfortunately, the widespread use of PHA biopolymers has been thwarted due to their high production costs. In an effort to reduce their cost-to-produce (via fermentation) many studies have been performed that focused on the utilization of cheap substrates for the production of PHA biopolymers. One such substrate is lignocellulosic biomass. Unfortunately, the use of lignocellulosic feedstocks presents two major bottlenecks. The first involves the need for pretreatment (i.e., acid/ base hydrolysis, enzymatic hydrolysis) to effectively liberate the individual monosaccharides from the cellulose and hemicellulose fractions so that they can be efficiently used by the producing bacterial strains. Additionally, acid hydrolysis typically results in the formation of inhibitory compounds that must be neutralized prior to fermentation success. Regrettably, these processes impart additional cost to the production process which may usurp the economic benefits of using lignocellulosic feedstocks for PHA production in the first place. The second bottleneck involves the actual chemical composition of the lignocellulosic biomass. Upon pretreatment various pentose and hexose sugars are present for fermentation processes. Unfortunately, bacterial strains will initially utilize the sugars that are the most energy-efficient before moving to the less energy-favorable feedstocks. This limits the effective utilization of the secondary sugars, which in the case of lignocellulosic feedstocks may be considerable, and results in diauxic growth patterns which restricts product synthesis. In addition, the presence of phenolic compounds typically found in lignin may be inhibitory to some producing strains. While additional work is needed to overcome the previously-mentioned bottlenecks for a cost-effective process that results in maximum sugar utilization and PHA yields from lignocellulosic materials, currently there is great interest in PHA production from xylose, the most common sugar component of hemicellulose. While only a few naturally-producing bacterial strains have been demonstrated to be capable of metabolizing xylose for PHA biosynthesis, the fact that this substrate has been shown to be a successful feedstock for PHA biosynthesis is beneficial in the overall effort to reduce the cost-to-produce of these valuable biopolymers. In addition, the ability to convert both 5-carbon and 6-carbon sugars into LevA has aided in the overall production of PHA biopolymers by improving their physical and mechanical properties. By utilizing xylose in combination with LevA with the appropriate bacterial strains and culture conditions, *sc*-PHA copolymers have been produced from pentose sugars and their chemically-derived products thereby increasing the possibility of a more cost-effective process to make these valuable bio-based polymers from plant biomass.

Acknowledgments

Mention of trade names or commercial products in this publication is solely for the purpose of providing specific information and does not imply recommendation or endorsement by the U.S. Department of Agriculture (USDA). USDA is an equal opportunity provider and employer.

References

1. Jambeck, J. R.; Geyer, R.; Wilcox, C.; Siegler, T. R.; Perryman, M.; Andrady, A.; Narayan, R.; Law, K. L. Plastic waste input from land into the ocean. *Science* **2015**, *347*, 768–771.

2. Yoshida, S.; Hiraga, K.; Takehana, T.; Taniguchi, I.; Yamaji, H.; Maeda, Y.; Toyohara, K.; Miyamoto, K.; Kimura, Y.; Oda, K. A bacterium that degrades and assimilates poly(ethylene terephthalate). *Science* **2016**, *351*, 1196–1199.
3. Austin, H. P.; Allen, M. D.; Donohoe, B. S.; Rorrer, N. A.; Kearns, F. L.; Silveira, R. L.; Pollard, B. C.; Dominick, G.; Duman, R.; El Omari, K.; Mylhaylyk, V.; Wagner, A.; Michener, W. E.; Amore, A.; Skaf, M. S.; Crowley, M. F.; Thorne, A. W.; Johnson, C. W.; Woodcock, H. L.; McGeehan, J. E.; Beckham, G. E. Characterization and engineering of a plastic-degrading aromatic polyesterase. *Proc. Natl. Acad. Sci. USA* **2018**, *115*, E4350–E4357.
4. Madison, L. L.; Huisman, G. W. Metabolic engineering of poly(3-hydroxyalkanoates): DNA to plastic. *Microbiol. Mol. Biol. Rev.* **1999**, *63*, 21–53.
5. Zheng, Y.; Chen, J.-C.; Ma, Y.-M.; Chen, G.-Q. Engineering biosynthesis of polyhydroxyalkanoates (PHA) for diversity and cost reduction. *Metabol. Eng.* **2020**, *58*, 82–93.
6. Ashby, R. D.; Solaiman, D. K. Y.; Foglia, T. A. Bacterial poly(hydroxyalkanoate) polymer production from the biodiesel co-product stream. *J. Polym. Environ.* **2004**, *12*, 105–112.
7. Cavalheiro, J. M. B. T.; de Almeida, M. C. M. D.; Grandfils, C.; da Fonseca, M. M. R. Poly(3-hydroxybutyrate) production by *Cupriavidus necator* using waste glycerol. *Proc. Biochem.* **2009**, *44*, 509–515.
8. Zhu, C.; Nomura, C. T.; Perrota, J.; Stipanovic, A. J.; Nakas, J. P. Production and characterization of poly-3-hydroxybutyrate from biodiesel-glycerol by *Burkholderia cepacia* ATCC 17759. *Biotechnol. Prog.* **2010**, *26*, 424–430.
9. Amaro, T. M. M. M.; Rosa, D.; Comi, G.; Lacumin, L. Prospects for the use of whey for polyhydroxyalkanoate (PHA) production. *Front. Microbiol.* **2019**, *10*article 992.
10. Solaiman, D. K. Y.; Ashby, R. D.; Hotchkiss, A. T.; Foglia, T. A. Biosynthesis of medium-chain-length poly(hydroxyalkanoates) from soy molasses. *Biotechnol. Lett.* **2006**, *28*, 157–162.
11. Keenan, T. M.; Nakas, J. P.; Tanenbaum, S. W. Polyhydroxyalkanoate copolymers from forest biomass. *J. Ind. Microbiol. Biotechnol.* **2006**, *33*, 616–626.
12. Cesário, M. T.; Raposo, R. S.; de Almeida, M. C. M. D.; van Keulen, F.; Ferreira, B. S.; Telo, J. P.; da Fonseca, M. M. R. Production of poly(3-hydroxybutyrate-co-4-hydroxybutyrate) by *Burkholderia sacchari* using wheat straw hydrolysates and gamma-butyrolactone. *Int. J. Biol. Macromol.* **2014**, *71*, 59–67.
13. Pan, W.; Perrotta, J. A.; Stipanovic, A. J.; Nomura, C. T.; Nakas, J. P. Production of polyhydroxyalkanoates by *Burkholderia cepacia* ATCC 17759 using a detoxified sugar maple hemicellulosic hydrolysate. *J. Ind. Microbiol. Biotechnol.* **2012**, *39*, 459–469.
14. Kucera, D.; Benesova, P.; Ladicky, P.; Pekar, M.; Sedlacek, P.; Obruca, S. Production of polyhydroxyalkanoates using hydrolyzates of spruce sawdust: Comparison of hydrolyzates detoxification by application of overliming, active carbon, and lignite. *Bioengineering* **2017**, *4*, article 53.
15. González-Garcia, Y.; Grieve, J.; Meza-Contreras, J. C.; Clifton-García, B.; Silva-Guzman, J. A. Tequila agave bagasse hydrolysate for the production of polyhydroxybutyrate by *Burkholderia sacchari*. *Bioengineering* **2019**, *6*, article 115.
16. Görke, B.; Stülke, J. Carbon catabolite repression in bacteria: Many ways to make the most out of nutrients. *Nat. Rev. Microbiol.* **2008**, *6*, 613–624.

17. Kim, S. M.; Choi, B. Y.; Ryu, Y. S.; Jung, S. H.; Park, J. M.; Kim, G. H.; Lee, S. K. Simultaneous utilization of glucose and xylose via novel mechanisms in engineered *Escherichia coli*. *Metabol. Eng.* **2015**, *30*, 141–148.
18. Dills, S. S.; Apperson, A.; Schmidt, M. R.; Saier, M. H., Jr. Carbohydrate Transport in Bacteria. *Microbiol. Rev.* **1980**, *44*, 385–418.
19. Lengeler, J. W.; Drews, G.; Schlegel, H. G. *Biology of Prokaryotes*; Blackwell Science: Stuttgart, Germany, 1999, pp 83–84.
20. Zhao, Z.; Xian, M.; Liu, M.; Zhao, G. Biochemical routes for uptake and conversion of xylose by microorganisms. *Biotechnol. Biofuels* **2020**, *13*, article 21.
21. Wisedchaisri, G.; Park, M.-S.; Iadanza, M. G.; Zheng, H.; Gonen, T. Proton-coupled sugar transport in the prototypical major facilitator superfamily protein XylE. *Nat. Commun.* **2014**, *5*, article 4521.
22. Davidson, A. L.; Maloney, P. C. ABC transporters: How small machines do a big job. *Trends Microbiol.* **2007**, *15*, 448–455.
23. Davis, E. O.; Henderson, P. J. The cloning and DNA sequence of the gene *xylE* for xylose-proton symport in *Escherichia coli* K12. *J. Biol. Chem.* **1987**, *262*, 13928–13932.
24. Song, S.; Park, C. Organization and regulation of the D-xylose operons in *Escherichia coli* K-12: XylR acts as a transcriptional activator. *J. Bacteriol.* **1997**, *179*, 7025–7032.
25. Sun, Z.; Chen, Y.; Yang, C.; Yang, S.; Gu, Y.; Jiang, W. A novel three-component system-based regulatory model for D-xylose sensing and transport in *Clostridium beijerinckii*. *Mol. Microbiol.* **2015**, *95*, 576–589.
26. Buelow, D. R.; Raivio, T. L. Three (and more) component regulatory systems – auxiliary regulators of bacterial histidine kinases. *Mol. Microbiol.* **2010**, *75*, 547–566.
27. Mishra, P.; Singh, A. Microbial pentose utilization. *Adv. Appl. Microbiol.* **1993**, *39*, 91–152.
28. Doten, R. C.; Mortlock, R. P. Inducible xylitol dehydrogenases in enteric bacteria. *J. Bacteriol.* **1985**, *162*, 845–848.
29. Scalcinati, G.; Otero, J. M.; Van Vleet, J. R.; Jefferies, T. W.; Olsson, L.; Nielsen, J. Evolutionary engineering of *Saccharomyces cerevisiae* for efficient aerobic xylose consumption. *FEMS Yeast Res.* **2012**, *12*, 582–597.
30. Liu, L.; Zhang, L.; Tang, W.; Gu, Y.; Hua, Q.; Yang, S.; Jiang, W.; Yang, C. Phosphoketolase pathway for xylose catabolism in *Clostridium acetobutylicum* revealed by ^{13}C metabolic flux analysis. *J. Bacteriol.* **2012**, *194*, 5413–5422.
31. Tanaka, K.; Komiyama, A.; Sonomoto, K.; Ishizaki, A.; Hall, S. J.; Stanbury, P. F. Two different pathways for D-xylose metabolism and the effect of xylose concentration on the yield coefficient of L-lactate in mixed-acid fermentation by the lactic acid bacterium *Lactococcus lactis* IO-1. *Appl. Microbiol. Biotechnol.* **2002**, *60*, 160–167.
32. Weimberg, R. Pentose oxidation by *Pseudomonas fragi*. *J. Biol. Chem.* **1961**, *236*, 629–635.
33. Stephens, C.; Christen, B.; Fuchs, T.; Sundaram, V.; Watanabe, K.; Jenal, U. Genetic analysis of a novel pathway for D-xylose metabolism in *Caulobacter crescentus*. *J. Bacteriol.* **2007**, *189*, 2181–2185.
34. Dahms, A. S. 3-deoxy-D-petulosonic acid aldolase and its role in a new pathway of D-xylose degradation. *Biochem. Biophys. Res. Commun.* **1974**, *60*, 1433–1439.

35. Choi, S. Y.; Park, S. J.; Kim, W. J.; Yang, J. E.; Lee, H.; Shin, J.; Lee, S. Y. One-step fermentative production of poly(lactate-*co*-glycolate) from carbohydrates in *Escherichia coli*. *Nat. Biotechnol.* **2016**, *34*, 435–440.
36. Choi, S. Y.; Kim, W. J.; Yu, S. J.; Park, S. J.; Im, S. G.; Lee, S. Y. Engineering the xylose-catabolizing Dahms pathway for production of poly(D-lactate-co-glycolate) and poly(D-lactate-co-glycolate-co-D-2-hydroxybutyrate) in *Escherichia coli*. *Microb. Biotechnol.* **2017**, *10*, 1353–1364.
37. Steinbüchel, A.; Valentin, H. E. Diversity of bacterial polyhydroxyalkanoic acids. *FEMS Microbiol. Lett.* **1995**, *128*, 219–228.
38. Nomura, C. T.; Tanaka, T.; Gan, Z.; Kuwabara, K.; Abe, H.; Takase, K.; Taguchi, K.; Doi, Y. Effective enhancement of short-chain-length—medium-chain-length polyhydroxyalkanoate copolymer production by coexpression of genetically engineered 3-ketoacyl-acyl-carrier-protein synthase III (*fabH*) and polyhydroxyalkanoate synthesis genes. *Biomacromolecules* **2004**, *5*, 1457–1464.
39. Green, P. R.; Kemper, J.; Schechtman, L.; Guo, L.; Satkowski, M.; Fiedler, S.; Steinbüchel, A.; Rehm, B. H. A. Formation of short chain length/medium chain length polyhydroxyalkanoate copolymers by fatty acid b-oxidation inhibited *Ralstonia eutropha*. *Biomacromolecules* **2002**, *3*, 208–213.
40. Lemoigne, M. Produits de deshydration et de polymerization de l'acide b-oxobutyrique. *Bull. Soc. Chim. Biol.* **1926**, *8*, 770–782.
41. Chen, G-Q.; Jiang, X-R. Engineering bacteria for enhanced polyhydroxyalkanoates (PHA) biosynthesis. *Synth. Syst. Biotechnol.* **2017**, *2*, 192–197.
42. Hiroe, A.; Tsuge, K.; Nomura, C. T.; Itaya, M.; Tsuge, T. Rearrangement of gene order in the *phaCAB* operon leads to effective production of ultrahigh-molecular-weight poly[(R)-3-hydroxybutyrate] in genetically engineered *Escherichia coli*. *Appl. Environ. Microbiol.* **2012**, *78*, 3177–3184.
43. Antonetti, C.; Licursi, D.; Fulignati, S.; Valentini, G.; Galletti, A. M. R. New frontiers in the catalytic synthesis of levulinic acid: From sugars to raw and waste biomass as starting feedstock. *Catalysts* **2016**, *6*, article 196.
44. Werpy, T.; Petersen, G.; Aden, A.; Bozell, J.; Holladay, J.; White, J.; Manheim, A.; Eliot, D.; Lasure, L.; Jones, S.; Gerber, M.; Ibsen, K.; Lumberg, L.; Kelley, S. Results of screening for potential candidates from sugars and synthesis gas. In *Top Value Added Chemicals from Biomass*; Werpy, T., Petersen, G., Eds.; Office of Energy Efficiency and Renewable Energy, US Department of Energy: Washington, D.C., USA, 2004; Volume 1, pp 1—67.
45. Bozell, J. J.; Moens, L.; Elliott, D. C.; Wang, Y.; Neuenscwander, G. G.; Fitzpatrick, S. W. Production of levulinic acid and its use as a platform chemical for derived products. *Res. Conserv. Recycl.* **2000**, *28*, 227–239.
46. Grand View Research. *Levulinic Acid: Market trends and trend analysis*; Grand View Research Inc., 2017, https://www.grandviewresearch.com/industry-analysis/levulinic-acid-market. Accessed 2020-07-15.
47. Amarasekara, A. S.; Williams, L. D.; Ebede, C. C. Mechanism of the dehydration of D-fructose to 5-hydroxymethylfurfural in dimethyl sulfoxide at 150°C: An NMR study. *Carbohydr. Res.* **2008**, *343*, 3021–3024.

48. Guan, J.; Cao, Q.; Guo, X.; Mu, X. The mechanism of glucose conversion to 5-hydroxymethylfurfural catalyzed my metal chlorides in ionic liquid: A theoretical study. *Comput. Theor. Chem.* **2011**, *963*, 453–462.
49. Van Putten, R. J.; van der Waal, J. C.; de Jong, E.; Rasrenda, C. B.; Heeres, H. J.; de Vries, J. G. Hydroxymethylfurfural, a versatile platform chemical made from renewable resources. *Chem. Rev.* **2013**, *113*, 1499–1597.
50. Horvat, J.; Klaić, B.; Metelko, B.; Šunjić, V. Mechanism of levulinic acid formation in acid catalysed hydrolysis of 2-hydroxymethylfurane and 5-hydroxymethylfurane-2-carbaldehyde. *Croat. Chem. Acta* **1986**, *59*, 429–438.
51. Zeitsch, K. J. *The chemistry and technology of furfural and its many by-products*, 1st ed.; Elsevier: Amsterdam, The Netherlands, 2000.
52. Van de Graaf, W. D.; Lange, J. P. A process for the conversion of furfuryl alcohol into levulinic acid or alkyl levulinate. U.S. Patent Appl. # 2007/0049771 A1, 2007.
53. Dalvand, K. Economics of biofuels: Market potential of furfural and its derivatives. *Biomass Bioeng.* **2018**, *115*, 56–63.
54. Chen, S.; Wojcieszak, R.; Dumeignil, F.; Marceau, E.; Royer, S. How catalysts and experimental conditions determine the selective hydroconversion of furfural and 5-hydroxymethylfurfural. *Chem. Rev.* **2018**, *118*, 11023–11117.
55. Brydson, J. A. Furan Resins. In *Plastics Materials*, 7th ed.; Brydosn, J. A., Ed.; Butterworth-Heineman: Oxford, 1999: pp 810—813.
56. Jeong, H.; Park, S.-Y.; Ryu, G.-H.; Choi, J.-H.; Kim, J.-H.; Choi, W.-S.; Lee, S. M.; Choi, J. W.; Choi, I.-G. Catalytic conversion of hemicellulosic sugars derived from biomass to levulinic acid. *Catal. Commun.* **2018**, *117*, 19–25.
57. Ya'aini, N.; Saidina Amin, N. A.; Asmadi, M. Optimization of levulinic acid from lignocellulosic biomass using a new hybrid catalyst. *Bioresour. Technol.* **2012**, *116*, 58–65.
58. Ya'aini, N.; Saidina Amin, N. A.; Endud, S. Characterization and performance of hybrid catalysts for levulinic acid production from glucose. *Micropor. Mesopor. Mater.* **2013**, *171*, 14–23.
59. Ramli, N. A. S.; Saidina Amin, N. A. Kinetic study of glucose conversion to levulinic acid over Fe/HY zeolite catalyst. *Chem. Eng. J.* **2016**, *283*, 150–159.
60. Wang, C.; Zhang, Q.; Chen, Y.; Zhang, X.; Xu, F. Highly efficient conversion of xylose residues to levulinic acid over $FeCl_3$ catalyst in green salt solutions. *ACS Sustainable Chem. Eng.* **2018**, *6*, 3154–3161.
61. Peng, L.; Lin, L.; Zhang, J.; Zhuang, J.; Zhang, B.; Gong, Y. Catalytic conversion of cellulose to levulinic acid by metal chlorides. *Molecules* **2010**, *15*, 5258–5272.
62. Habe, H.; Sato, S.; Morita, T.; Fukuoka, T.; Kirimura, K.; Kitamoto, D. Isolation and characterization of bacterial strains with the ability to utilize high concentrations of levulinic acid, a platform chemical from inedible biomass. *Biosci. Biotechnol. Biochem.* **2015**, *79*, 1552–1555.
63. Yu, J.; Chen, L. X. L.; Sato, S. Biopolyester synthesis and protein regulations in *Ralstonia eutropha* on levulinic acid and its derivatives from biomass refining. *J. Biobased Mater. Bioenergy* **2009**, *3*, 113–122.

64. Chung, S. H.; Choi, G. G.; Kim, H. W.; Rhee, Y. H. Effect of levulinic acid on the production of poly(3-hydroxybutyrate-co-3-hydroxyvalerate) by *Ralstonia eutropha* KHB-8862. *J. Microbiol.* **2001**, *39*, 79–82.
65. Lee, S. E.; Li, Q. X.; Yu, J. Diverse protein regulations on PHA formation in *Ralstonia eutropha* on short chain organic acids. *Int. J. Biol. Sci.* **2009**, *5*, 215–225.
66. Jaremko, M.; Yu, J. The initial metabolic conversion of levulinic acid in *Cupriavidus necator*. *J. Biotechnol.* **2011**, *155*, 293–298.
67. Rand, J. M.; Pisithkul, T.; Clark, R. L.; Thiede, J. M.; Mehrer, C. R.; Agnew, D. E.; Campbell, C. E.; Markley, A. L.; Price, M. N.; Ray, J.; Wetmore, K. M.; Suh, Y.; Arkin, A. P.; Deutschbauer, A. M.; Amador-Noguez, D.; Pfleger, B. F. A metabolic pathway for catabolizing levulinic acid in bacteria. *Nat. Microbiol.* **2017**, *2*, 1624–1634.
68. Ashby, R. D.; Solaiman, D. K. Y.; Nuñez, A.; Strahan, G. D.; Johnston, D. B. *Burkholderia sacchari* DSM 17165: A source of compositionally-tunable block-copolymeric short-chain poly(hydroxyalkanoates) from xylose and levulinic acid. *Bioresour. Technol.* **2018**, *253*, 333–342.
69. Ashby, R. D.; Solaiman, D. K. Y.; Strahan, G. D. The use of *Azohydromonas lata* DSM 1122 to produce 4-hydroxyvalerate-containing polyhydroxyalkanoate terpolyesters, and unique polymer blends from mixed-cultures with *Burkholderia sacchari* DSM 17165. *J. Polym. Environ.* **2019**, *27*, 198–209.
70. Young, F. K.; Kastner, J. R.; May, S. W. Microbial production of poly-b-hydroxybutyric acid from D-xylose and lactose by *Pseudomonas cepacia*. *Appl. Environ. Microbiol.* **1994**, *60*, 4195–4198.
71. Bertrand, J. L.; Ramsay, B. A.; Ramsay, J. A.; Cavarie, C. Biosynthesis of poly-b-hydroxyalkanoates from pentoses by *Pseudomonas pseudoflava*. *Appl. Environ. Microbiol.* **1990**, *56*, 3133–3138.
72. Jun, H.-K.; Jin, Y.-H.; Kim, H. N.; Kim, Y.-T.; Kim, S. W.; Baik, H. S. Production of poly-3-hydroxybutyrate from xylose by *Bacillus megaterium* J-65. *J. Life Sci.* **2008**, *18*, 1625–1630.
73. Lopes, M. S. G.; Gomez, J. G. C.; Silva, L. F. Cloning and overexpression of the xylose isomerase gene from *Burkholderia sacchari* and production of polyhydroxybutyrate from xylose. *Can. J. Microbiol.* **2009**, *55*, 1012–1015.
74. Lopes, M. S. G.; Rocha, R. C. S.; Zanotto, S. P.; Gomez, J. G. C.; da Silva, L. F. Screening of bacteria to produce polyhydroxyalkanoates from xylose. *World J. Microbiol. Biotechnol.* **2009**, *25*, 1751–1756.
75. Le Meur, S.; Zinn, M.; Egli, T.; Thöny-Meyer, L.; Ren, Q. Production of medium-chain-length polyhydroxyalkanoates by sequential feeding of xylose and octanoic acid in engineered *Pseudomonas putida* KT2440. *BMC Biotechnol.* **2012**, *12*, article 53.
76. Tanaka, K.; Mori, S.; Matsumoto, K.; Yamamoto, K.; Wakida, I.; Goto, S.; Mastusaki, H. Growth characteristics of newly isolated xylose-assimilating bacterium and accumulation of green plastic, polyhydroxyalkanoate in the genetic engineered strain. *Int. J. Biol. Eng.* **2016**, *6*, 1–6.
77. Raposo, R. S.; de Almeida, C. M. D.; de Oliveira, M. M. A.; da Fonseca, M. M.; Cesário, M. T. A *Burkholderia sacchari* cell factory: Production of poly-3-hydroxybutyrate, xylitol and xylonic acid from xylose-rich sugar mixtures. *New Biotechnol.* **2017**, *34*, 12–22.

78. Guamán, L. P.; Barba-Ostria, C.; Zhang, F.; Oliveira-Filho, E. R.; Gomez, J. G. C.; Silva, L. F. Engineering xylose metabolism for production of polyhydroxybutyrate in the non-model bacterium *Burkholderia sacchari*. *Microb. Cell Fact.* **2018**, *17*, article 74.
79. Pederson, E. N.; McChalicher, C. W. J.; Srienc, F. Bacterial synthesis of PHA block copolymers. *Biomacromolecules* **2006**, *7*, 1904–1911.
80. Li, S. Y.; Dong, C. L.; Wang, S. Y.; Ye, H. M.; Chen, G. Q. Microbial production of polyhydroxyalkanoate block copolymer by recombinant *Pseudomonas putida*. *Appl. Microbiol. Biotechnol.* **2011**, *90*, 659–669.

Chapter 8

Polyhydroxyalkanoates for Biodegradable Mulch Films Applications

Yuchuan Tian and James H. Wang*

Sinopec Shanghai Research Institute of Petrochemical Technology, Shanghai 201028, China

***Email: JamesWang.sshy@sinopec.com**

The wide applications of non-biodegradable polymer mulch films are causing a major environmental problem on agricultural sustainability around the world. It becomes urgent and important to develop biodegradable mulch films to tackle this issue. Among the biodegradable polymers, polyhydroxyalkanoates are the most structurally diverse and easily biodegradable in composting, soil or aquatic environment. However, polyhydroxyalkanoates also face challenges in processability, availability, and cost. This chapter provides a brief review on PHA biosynthesis, properties, blends, degradation, and recent development of mulch films from PHA.

Introduction

Mulch films have played an important role in modern agriculture to improve crop yield by preventing water evaporation, increasing soil temperature, and eliminating weed growth, etc. The consumption of mulch films has grown substantially in the last decade. The global market is projected to reach $4.07 billion by 2020 (*1*). However, commercial mulch films are made from non-degradable polymers such as polyolefins (2), ethylene-vinyl acetate copolymer (EVA), and polyvinyl chloride (PVC) (3). Since these polymers are not degradable and mulch films cannot be easily collected after harvesting, the accumulation of film residues in soil over time has caused a serious sustainability problems due to the pollution of film fragments and their adverse effect on crop growth (Figure 1) (*1*).

To solve this problem, one approach is to develop biodegradable mulch films which can be naturally biodegraded within a crop planting cycle. This has led to significant research efforts in creating fully functional biodegradable mulch films by both industrial and academic scientists. The polymers investigated for biodegradable mulch films include polybutylene adipate terephthalate (PBAT), polylactic acid (PLA), polybutylene succinate (PBS), polybutylene succinate adipate (PBSA), poly-3-hydroxybutyrate-*co*-3-hydroxyvalerate (PHBV), polycaprolactone (PCL), and blends of these polymers.

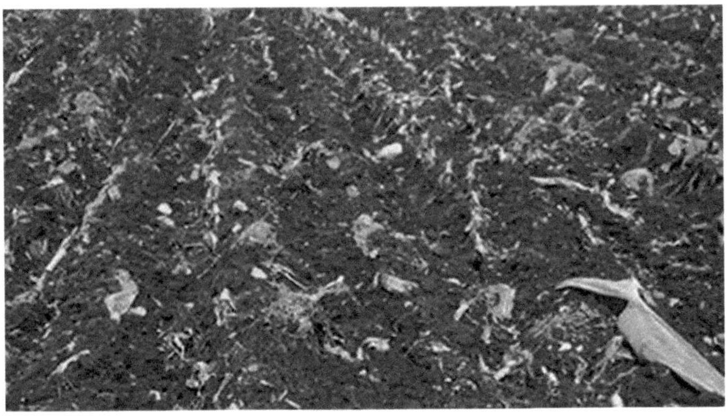

Figure 1. Soil pollution caused by non-biodegradable mulch films.

Polyhydroxyalkanoates (PHA) are accumulated intracellularly in cells by various microorganisms (*4–7*), a general structure is shown in Figure 2. PHA can be biosynthesized from renewable carbon resources as feedstocks such as sugars, vegetable oils by microbial fermentation (*8–10*). PHA materials are 100% bio-based and the most easily biodegradable polymers in composting, soil or aquatic environment. The properties of PHA can be tailored by changing comonomer structures and copolymer compositions. However, substantial challenges still remain for PHA to become practically applicable in commercial uses. For instance, PHA is prone to thermomechanical degradation during film extrusion (*8*). Therefore, it's necessary to improve the thermal and mechanical performance of PHA, mainly by blending with other polymers, nanofillers, and structural design (*11–15*).

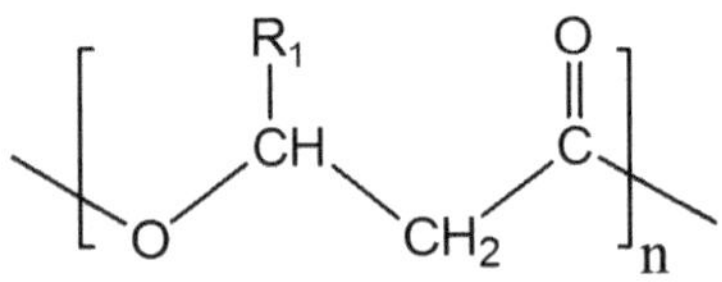

a

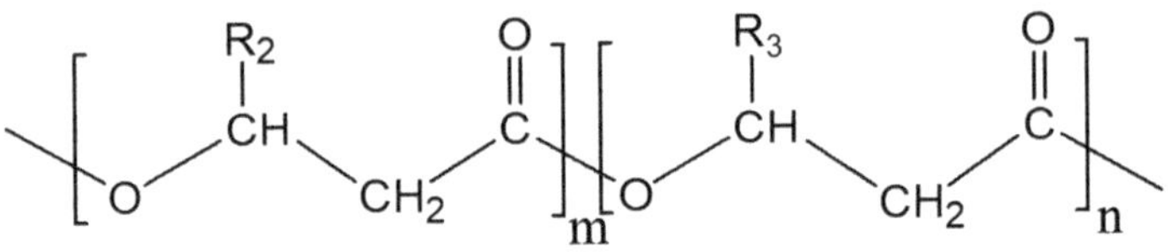

b

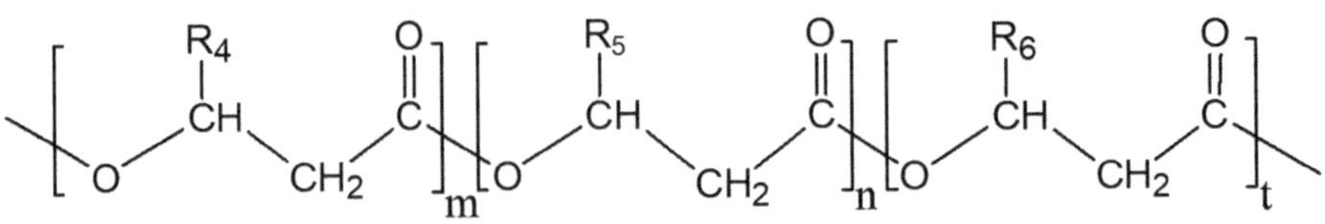

c

Figure 2. General structures of PHA: (a) a homopolymer, (b) a copolymer, and (c) a terpolymer. R_1~R_6: C_1~C_{16} alkyl groups, $R_2 \neq R_3$; $R_4 \neq R_5 \neq R_6$.

PHA Biosynthesis

Biosynthesis of PHA usually consists of the following stages: strain development, shake flask fermentation, lab and pilot scale fermentation, and industrial scale production (*16*). Effective microbial production of PHA depends on several factors including cell density, PHA% in cell dry weight, bacterial growth rate, time to reach final cell density, substrate to product transformation efficiency, cost of substrates, and purification.

The supply of substrates and conversion from the substrates to monomers are the two main steps involved in the biosynthesis of PHA (*4*, *17*). Generally, glucose and/or fatty acids are commonly used as substrates. For example, producing PHA consisting of monomers other than 3-hydroxybutyrate (3HB) needs fatty acids as precursors for forming other short-chain-length (scl) or medium-chain-length (mcl) monomers (*18–20*). There are three well-known PHA biosynthesis pathways (*4*).

Pathway I

PHA synthesis by *Cupriavidus necator*. This is the well known PHA biosynthetic pathways. In this pathway, 3HB monomer is generated by condensation of two acetyl-CoA molecules, then acetoacetyl-CoA reductase acts on acetoacetyl-CoA to form 3-hydroxybutyryl-CoA. Finally, PHA synthase catalyzes polymerization via esterification of 3-hydroxybutyryl-CoA into poly-3-hydroxybutyrate (PHB).

Pathway II

The fatty acid β-oxidation pathway. The substrates that can be polymerized by the PHA synthases of *Pseudomonas* are involved in this pathway. In *Aeromonas caviae*, for example, the β-oxidation intermediate, trans-2-enoyl-CoA is converted to (R)-hydroxyacyl-CoA by a (R)-specific enoyl-CoA hydratase (*21*, *22*).

Pathway III

PHA synthesis from glucose (*23*). The intermediates were obtained from the fatty acid de novo biosynthetic pathway. Pathway III is of significant importance because it helps to generate monomers for PHA biosynthesis from structurally unrelated, simple, and inexpensive carbon sources such as glucose, sucrose, and fructose.

Various substrates and PHA synthase activities have led to rather unstable PHA structures and molecular weights during PHA biosynthesis (*24*). Therefore, numerous studies have been conducted in order to fine tune the chemical structures of PHA chains (*25–28*). Notably, researchers clearly demonstrated that β-oxidation deficient mutants allowed controllable PHA structures, as supported by several other studies (*27*). For example, recombinant *Pseudomonas putida* KT2442 with its β-oxidation mutant *Pseudomonas putida KTQQ20* was constructed as a platform for PHA synthesis with controllable monomer contents and polymer compositions (*25*). Homopolymers of C5 (3-hydroxyvalerate, 3HV) to C14 (3-hydroxytetradecanoate) were synthesized by β-oxidation deficient *P. entomophila* LAC23 fed with different fatty acids as precursors, respectively (*29*).

Similarly, block copolymers of poly-3-hydroxypropinate-*b*-poly-4- hydroxybutyrate (P3HP-*b*-P4HB) can be synthesized using recombinant E. coli S17-1 by feeding precursors with a pre-defined ratio (*28*, *30*). In order to overcome the challenge of high cost of PHA production, next generation of industrial biotechnology is based on recombinant *H. bluephagenesis* grown under un-sterile and continuous conditions to reduce the complexity and the cost of PHA production (*31*).

Properties of PHA

The properties of PHA are highly dependent on the exact chemical structures of PHA. As shown in Figure 2, PHA can be a homopolymer of only one hydroxyalkanoate, a copolymer of two hydroxyalkanoates, or a terpolymer of three hydroxyalkanoates, etc. The simplest PHA homopolymer from biosynthesis is poly-3-hydroxybutyrate, since it has high crystallinity and low thermal stability (*6*).

Crystallinity

The crystallinity of PHA can be determined by X-ray diffraction (XRD) (*32*). Degree of crystallinity is calculated from the crystalline and amorphous areas on XRD diffractogram using Lorentzian and Gauss functions (*33*).

The degree of crystallinity of PHA ranges from 0% to 70%, i.e. varying from amorphous to highly crystalline structures (*34, 35*). The crystalline structure of PHBV is determined by its 3-hydroxyvalerate (HV) content. According to a study by Doi (*36*), the degree of crystallinity of PHBV increased from 55% to 70% as molar fraction of HV increased from 0% to 95%. While for poly-3-hydroxybutyrate-co-4-hydroxybutyrate (P3HB4HB), the degree of crystallinity decreased from 55% to 14% as 4HB amount increased from 0 to 49%. With increasing 4HB content, the crystallinity of P3HB4HB decreased while the radius of spherulites increased (*37*).

Thermal Properties

The thermal properties of PHA were studied by analyzing their glass transition temperature (T_g), melting temperature (T_m), melting enthalpy (ΔH_m) and other properties from DSC scans. T_m of PHA is usually less than 180°C (*38*). Kunioka (*36*) studied the thermal degradation of P3HB-*b*-4HB, P3HB-*co*-3HV, and PHB. Based on their DSC results, the two PHA copolymers were unstable at a temperature above 180°C.

For PHB, its T_m was 177°C which is the highest among the three PHA materials, however, the molecular weight of PHB was drastically reduced at 180°C (*36*). Therefore, PHB has an extremely narrow temperature window for thermoplastic processing, which has caused significant difficulty in melt processing PHB. The PHA copolymers had a wider processing window than PHB, but it is still quite narrow with comparison to other biodegradable polymers such as PLA. As a result, PHA is usually blended with other biodegradable polymers or nanomaterials to improve melt processability.

PHA Blends

The processability of PHA can be effectively improved by blending with other biodegradable polymers and/or plasticizers. The biodegradable polymers that PHA can be melt blended with include PLA (*39–41*) PBS (*42*), PCL (*43*), lignin (*44*), polypropylene carbonate (PPC) (*45*), PBAT (*46, 47, 48*), and PBAT/starch (*49*).

The properties of PHA blends from literature are summarized in Table 1. Basically, most of the PHA blends have low tensile strength except for the PLA/PHA 85/15 blends. The PHA blends also have low elongation at break with only the PHBV/PCL/starch blend as an exception.

Table 1. Properties of PHA Blends

<table>
<tr><th>Blend Compositions</th><th>Methods</th><th>Tensile Strength/ MPa</th><th>Tensile Modulus/ GPa</th><th>Elongation at Break/%</th><th>T_c/°C</th><th>T_m/°C</th><th>T_g/°C</th><th>ΔH_m/J·g^{-1}</th><th>Ref. No.</th></tr>
<tr><td rowspan="2">PHBV (45%)/PBAT (55%)</td><td rowspan="2">Injection molding</td><td rowspan="2">21</td><td rowspan="2">1.2</td><td rowspan="2">6.9</td><td rowspan="2">N</td><td>142.4</td><td>-</td><td rowspan="2">28.6</td><td rowspan="2">(47) & (48)</td></tr>
<tr><td>156.9</td><td>-</td></tr>
<tr><td>PHBV (45%)/PBAT (55%)+40% Swiss grass</td><td>Injection molding</td><td>18</td><td>2.3</td><td>-</td><td>-</td><td>-</td><td>-</td><td>-</td><td>(78)</td></tr>
<tr><td>PHBV (40%)/PCL (36%)+14.4% Starch</td><td>Cast film extrusion</td><td>15</td><td>0.45</td><td>405</td><td>-</td><td>-</td><td>-</td><td>-</td><td>(48)</td></tr>
<tr><td rowspan="2">PHB (80%)/ PBAT (20%)</td><td rowspan="2">Twin screw extrusion</td><td rowspan="2">-</td><td rowspan="2">-</td><td rowspan="2">-</td><td>126</td><td rowspan="2">176</td><td>9</td><td>-</td><td rowspan="2">(54)</td></tr>
<tr><td>78</td><td>-34</td><td>-</td></tr>
<tr><td rowspan="2">PLA (75%)/PHA (25%)</td><td rowspan="2">Injection molding</td><td rowspan="2">16</td><td rowspan="2">1.3</td><td rowspan="2">7.1</td><td rowspan="2">115</td><td rowspan="2">173</td><td>1.7</td><td rowspan="2">13</td><td rowspan="4">(41)</td></tr>
<tr><td>62</td></tr>
<tr><td rowspan="2">PLA(75%)/ PHA (25%)+7% Lapol®</td><td rowspan="2">Injection molding</td><td rowspan="2">15</td><td rowspan="2">1.1</td><td rowspan="2">15</td><td rowspan="2">120</td><td rowspan="2">173</td><td>0.6</td><td rowspan="2">24</td></tr>
<tr><td>58</td></tr>
<tr><td>PLA</td><td>Solution cast films</td><td>-</td><td>-</td><td>-</td><td>120</td><td>179</td><td>59</td><td>51</td><td>(51)</td></tr>
<tr><td>PLA (95%)/ PHB (5%)</td><td>Solution cast films</td><td>-</td><td>-</td><td>-</td><td>120</td><td>179</td><td>53</td><td>64</td><td></td></tr>
<tr><td rowspan="2">PLA (30%)/PBAT (70%)+PHBV - MCPA (15%)</td><td rowspan="2">Hot pressed films</td><td rowspan="2">-</td><td rowspan="2">-</td><td rowspan="2">-</td><td rowspan="2">99</td><td rowspan="2">149</td><td>-27.4</td><td rowspan="2">-</td><td rowspan="2">(77)</td></tr>
<tr><td>36.9</td></tr>
<tr><td>PLA (85%)/PHA (15%)</td><td>Twin screw extrusion</td><td>27.1</td><td>1.5</td><td>48</td><td>-</td><td>-</td><td>-</td><td>-</td><td>(52)</td></tr>
</table>

"N": Not detectable; "-" Not provided in the literature.

PHA and PLA Blends

PLA is a kind of broadly used biodegradable polymer with properties similar to polystyrene (PS) and polyethylene terephthalate (PET). PHA was blended with PLA to improve mechanical properties (*50, 51*). Ohkoshi et al. (*52*) studied the miscibility and phase structure of binary blends of PLA with PHB via wide angle X-ray scattering (WAXS) and DSC. They found that PHB and PLA were miscible when PHB content is less than 50 wt.%. In contrast, the blends of PHB and a high molecular weight PLA exhibited two glass transition temperatures indicating immiscibility. Increasing PHB content in the blends facilitates the growth of spherulites within PLA and further promotes crystallization. This was attributed to enhanced chain mobility of high molecular weight PLA by low molecular weight PHB with a relative low viscosity.

Botta et al. (*53*) reported melt processing of PLA and a medium-chain-length (mcl-) PHA blends, and the PHA were copolymers of C6, C8, C10, and C12 hydroxyalkanoates. They studied the rheological, mechanical, and other properties of the blends. The results of dynamic mechanical analysis indicated that PHA functioned as not only a plasticizer but also a lubricant for PLA chains both in melt and in solid state. A decrease in viscosity and an increase in ductility of the blends were observed, while the glass transition temperature was only slightly reduced. Therefore, as compared to neat PLA, the PHA/PLA blends exhibited better films processability.

Abdelwahab et al. (*41*) studied the effects of a bio-renewable plasticizer (Lapol® 108) on PHB, PLA, and PLA/PHB blends. As expected, the melting temperature and thermal stability of the PLA and PHB blends did not change with increased plasticizer content, however, the ductility of the PLA/PHB blends was substantially improved with addition of the plasticizer.

PHA and PBAT Blends

PBAT is a biodegradable aliphatic aromatic copolyester which has excellent tensile strength, flexibility, and toughness (*54*). Javadi et al. blended PHBV with PBAT through microcellular and conventional injection molding (*47*). The PHBV/PBAT blend specimens showed an extraordinary improvement of specific toughness, up to two orders of magnitude for solid and microcellular specimens as compared to neat PHBV. The PHBV/PBAT blends also showed >500% increase in strain-at-break versus neat PHBV.

Larsson et al. (*55*) blended PHB with PBAT in a twin screw extruder. Dicumyl peroxide (DCP) was added to the PBAT and PHB blends as a free radical initiator. Before melt processing, PHB was pretreated by extraction with chloroform and washing with an aqueous hydrochloric acid solution, the thermal stability of PHA was improved after such pretreatment. The results suggested that blending PHB with PBAT is an efficient approach to improve processability and other properties of PHB when PBAT is the continuous phase in the blends. Adding a free radical initiator such as DCP to the blends provided the needed compatibility between PHA and PBAT.

In order to achieve the required performance of biodegradable mulch films, the film thickness typically needs to be less than 10 μm, tensile strength is at least 25 MPa and elongation at break is usually >300%. However, as the literature data shown in Table 1, PHA blends with the properties meeting the requirements of mulch films are rare.

Wang et al. at Sinopec (*56–58*) has systematically investigated melt processing and properties of a new PHA block copolymer: poly-3-hydroxybutyrate-*b*-4-hydroxybutyrate (P3HB-*b*-4HB) by blending it with PBAT (Figure 3). Through an extensive research on the blend compositions and processing conditions (*57*), ultra-thin, high performance biodegradable mulch films with excellent strength and toughness were developed from the PHA block copolymer (*57*). Amazingly, an extremely low film thickness of <5 μm was successfully achieved. The films had a tensile strength higher than 30 MPa and an elongation at break greater than 600%.

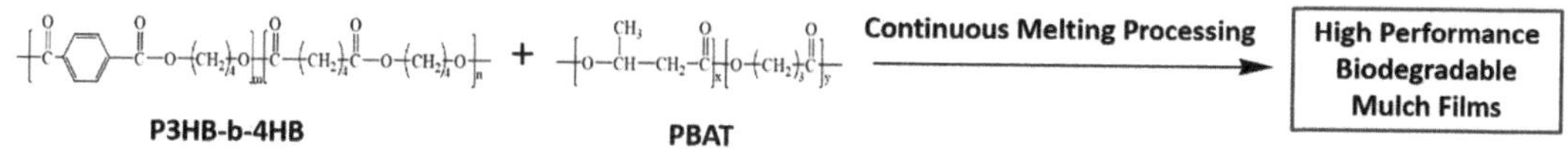

Figure 3. Biodegradable mulch films from P3HB-b-4HB and PBAT blends.

PHA Blends with Other Polymers

Starch is a low-cost, abundant natural polymer separated from a variety of crops such as corn, potatoes, cassava, etc. It became a convenient material to blend with other biodegradable materials to reduce cost. Parulekar et al. (*49*) prepared PHBV/PBAT/starch blends by extrusion. The resulting material had a tensile strength of 17 MPa and an elongation at break of 300% at 70% bio-based content. They found that the presence of starch had improved aging properties of the blends. The secondary crystallization of PHBV component was inhibited by miscible starch phase and by disruptive and intensive mixing during melt processing.

Blending PHA with natural fibers is another approach to improve the mechanical properties of PHA. When PHBV was reinforced with straw fibers (30% w/w), the rate of crystallization is increased while the biodegradation rate is less affected. The PHBV fiber composites can be used in agricultural mulching and transplantation (*59*). PHB blends with polycaprolactone were reported to have improved toughness (*60*). PHB and PCL are not miscible but the blends exhibited improvement in heat stability and impact strength.

Degradation of PHA

There are two main degradation mechanisms for biodegradable mulch films in use: (1) photodegradation, and (2) biodegradation. The extent of photodegradation determines the service life of mulch films when used in the fields, while biodegradation is responsible for complete decomposition of mulch films in soil. When exposed to intensive sunlight, the biodegradable mulch films tend to become brittle, then form cracks, tears, and holes before they finally disintegrate into small flakes via enzymatic hydrolysis (*61*).

Photodegradation of PHA

In comparison to biodegradation, the studies on photodegradation of PHA-related films are relatively limited. For biodegradable mulch films are placed on soil, the photodegradation process is predominant and it determines if mulch films can meet the service life requirement. In general, the useful life of unmodified biodegradable polymer films in field is too short to meet the average time required for crop farming (*61*). Therefore, significant efforts are needed to improve the stability of PHA-based mulch films under sunlight exposure. Adding different UV stabilizers and UV absorbers can efficiently prevent photo degradation under UV exposure (*62–64*).

Paola et al. (*65*) studied the photo-oxidative weathering of PHBV/PBAT blend films, neat films and films with added natural and organo-modified halloysite nanofillers. The photodegradation of PHBV/PBAT films was tested under accelerated aging test in laboratory. Adding nanofillers improved the tensile properties of the blend films. However, the presence of nanofillers was not able to slow down the degradation process under accelerated aging.

Katarína et al. (*66*) studied natural aging of PLA/PHB mulch films with 3 wt.% carbon black (CB) after application for one season. They investigated 3 different conditions: (a) buried in soil, (b) exposed to sunlight, and (c) shadowed by plants. The plasticized PLA/PHB/CB film buried in soil lost its tensile properties faster than the film shadowed by plants or exposed to sunlight. The deterioration of mechanical properties during degradation resulted in significant decrease in ductility.

Recently, Wang et al. at Sinopec has studied the degradation process of biodegradable PHA/PBAT mulch films under Xeon accelerated aging test (*67*). By adding different types of stabilizers, the mulch films showed significantly improved retention of mechanical performance after aging. Based on the extensive evaluation results by XPS, FTIR, DSC and XRD, detailed degradation mechanisms of the PHA/PBAT films were proposed. The results suggested that PHA/PBAT films underwent both photo-oxidation and Norrish type I/II photodegradation under accelerated aging test. As shown in Figure 4, the photo-oxidation process of PHA/PBAT blend films involves a classical hydrogen abstraction on the polymeric backbone usually at the tertiary carbon, leading to the formation of macroradicals first, and this eventually cause chain scissions at ester bonds.

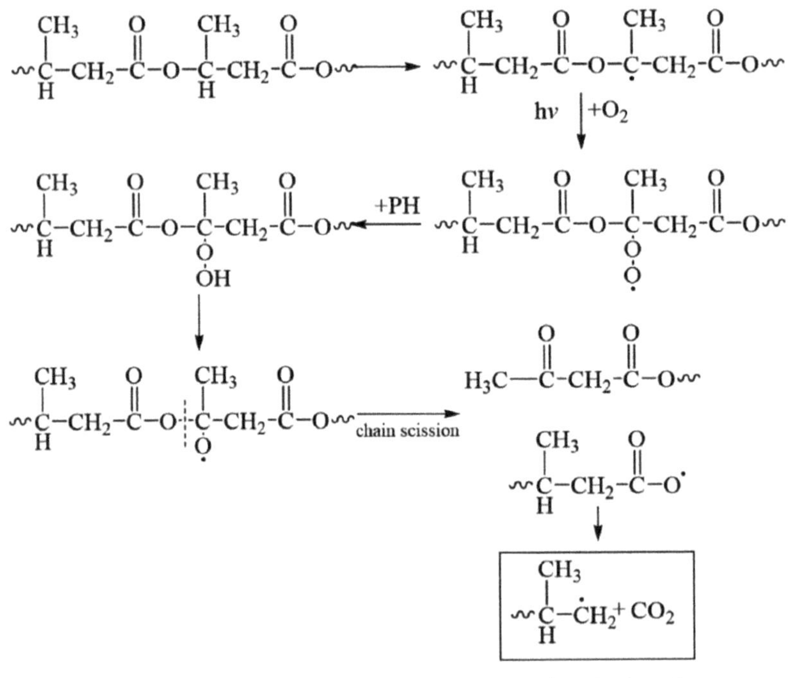

Figure 4. A mechanistic scheme of photo-oxidative reactions of the 4 hydroxybutyrate structure in P3HB4HB.

Biodegradation of PHA

Biodegradation is the decomposition of materials initiated by bacteria, fungi or through other biological means whether aerobically or anaerobically. In short, biodegradable polymers are those which can degraded in biological environments through enzymatic and non-enzymatic hydrolysis or biologically catalyzed chain scission reactions, but not through thermal oxidation, photolysis, or hydrolysis (*68*).

Theoretically, PHA can be completely mineralized to CO_2 and H_2O in aerobic systems (*69*). However, the biodegradation process of PHA or other biodegradable materials is rather complicated, and it can be influenced by not only the inherent properties of biodegradable films but also by external factors (*70*). For instance, the biodegradation mechanism varies as the sample thickness is changed (*68*). High temperature usually increases degradation rate of materials (*71*).

Mergaert et al. (72) studied the degradation of PHBV in soil at different temperatures, pH, and types of degrading organisms. PHBV tended to degrade faster at a lower pH and a higher temperature. Based on the results of molecular weight change in different environments, Mergaert proposed that the degradation in soil seems to occur on at least two levels: (i) in the bulk of samples, the molecular weight is decreased by simple hydrolysis without affecting the sample weight, and (ii) at the surface of the samples, the polymer is degraded by enzymes to low molecular weight fragments that are rapidly metabolized by microbial cells, resulting in weight loss of the polymer samples.

Similarly, Sintim et al. (73) studied the degradation of PLA/PHA mulch films in compost and in soil. They found that the degradation rate of PLA/PHA mulch films is in close relationship with the conditions and environment. The films degraded faster in compost, but degraded more slowly in soil or a cold environment. It was assumed that the high glass transition temperature (T_g : 49.2°C) of PLA/PHA blend (74) inhibited the mobility of polymer chains at ambient temperatures, making them less accessible for biodegradation (75).

Application of PHA in Biodegradable Mulch Films

Due to the combination of factors including limited availability, difficult to process, high cost, and high biodegradation rate than needed, reported research on PHA-related mulch films are rather limited (76). Procter & Gamble once produced Nodax™ PHA which is poly-3-hydroxybutyrate-*co*-3-hydroxyhexanoate) (PHBHx) (77). Nodax™ PHA could potentially be used for developing biodegradable mulch films, but the PHA products were discontinued even before they became readily available on the market.

Figure 5. Field applications of biodegradable PHA/PBAT mulch films: (a) peanuts, (b) corn, (c) strawberry, (d) rice.

Metabolix Inc. and ADM collaborated on Mirel™ PHA resin development and commercialization (78). The PHA was produced by fermentation using natural sugars as carbon source, the effort was also terminated before mass production began.

As discussed in previous sections, Sinopec group has developed a series of high performance PHA/PBAT mulch films. The biodegradable mulch films were successfully used for several crops as shown in Figure 5. Through a systematic investigation on using PHA/PBAT mulch films for corn, we found that the service life of the PHA/PBAT mulch films can meet the requirement for corn. After 60 days, the PHA/PBAT mulch films began to degrade in soil. Moreover, the PHA/PBAT films can keep a similar temperature under the films as non-biodegradable mulch films to help seed germination and plant growth. As compared with bare field without mulch films, the growth of corn under PHA/PBAT mulch films was significantly improved in both plant height and stem diameter as shown in Table 2.

Table 2. Field Study Results of PHA/PBAT Mulch Films for Corn Cultivation

Mulch Film Type	*30 Days*			*60 Days*		
	Plant Height/cm	*Stem Diameter/cm*	*Temperature/°C*	*Plant Height/cm*	*Stem Diameter/ cm*	*Temperature/°C*
PHA/PBAT Film	47.6	1.6	15.5	182.1	2.9	28.2
PE Film	47.8	1.5	15.8	179.0	3.0	28.5
Bare Soil	44.9	1.3	14.2	170.5	2.7	27.6

Another potential application of PHA films in agriculture is a pesticide/ herbicide delivery system by covalently bonding pesticide or herbicide onto PHA-based films through transesterification. Kwiecien et al. reported a method to prepare bioactive and biodegradable PLA-PBAT based mulch films containing a 2-methyl-4-chlorophenoxyacetic acid (MCPA)-PHBV conjugate by melt transesterification (79). The results showed that when these films were used under glass house conditions, they were able to successfully suppress broadleaf weed species by controlled release of the herbicide, while growth of crop species was essentially not affected.

Conclusions

Polyhydroxyalkanoates (PHA) can potentially be applicable polymers to address the pollution issue caused by non-biodegradable mulch films due to their easy biodegradation feature. However, developing a successful PHA mulch film has to overcome several challenges. PHA has a narrow thermal processing window, low mechanical performance, and a short service life in field. It is important to understand the degradation mechanisms of PHA in order to extend the service life of PHA mulch films. Ultra-thin, high performance PHA/PBAT mulch films were recently developed and shown to be suitable for a number of crops, the results demonstrated that it is possible to use PHA in biodegradable mulch films through innovative material development and extensive field testing.

Acknowledgments

This research was supported by the National Key Research and Development Program of China (Grant No. 2016YFB0302500).

References

1. Mulch Films Market by Type (Clear/Transparent, Black, Colored, and Degradable), Application (Agricultural Farms, and Horticulture), Element (LLDPE, LDPE, HDPE, EVA, PLA, and PHA), & by Region - Global Trends and Forecast to 2020. 2020. https://www.marketsandmarkets.com/Market-Reports/mulch-films-market-220908278.html.
2. Halley, P.; Rutgers, R.; Coombs, S.; Kettels, J.; Gralton, J.; Christie, G.; Jenkins, M.; Beh, H.; Griffin, K.; Jayasekara, R. Developing biodegradable mulch films from starch-based polymers. *Starch-Stärke* **2001**, *53* (8), 362–367.
3. Jayasekara, R.; Harding, I.; Bowater, I.; Lonergan, G. Biodegradability of a selected range of polymers and polymer blends and standard methods for assessment of biodegradation. *J. Polym. Environ.* **2005**, *13* (3), 231–251.
4. Philip, S.; Keshavarz, T.; Roy, I. Polyhydroxyalkanoates: biodegradable polymers with a range of applications. *J. Chem. Technol. Biotechnol.: Int. Res. Process, Environ. Clean Technol.* **2007**, *82* (3), 233–247.
5. Ong, S. Y.; Chee, J. Y.; Sudesh, K. *Degradation of polyhydroxyalkanoate (PHA): a review*; 2017.
6. Bugnicourt, E.; Cinelli, P.; Lazzeri, A.; Alvarez, V. A. Polyhydroxyalkanoate (PHA): Review of synthesis, characteristics, processing and potential applications in packaging. *eXPRESS Polym. Lett.* **2014**, *8* (11), 791–808.
7. Chen, G. G.-Q., *Plastics from bacteria: natural functions and applications*; Springer Science & Business Media: 2009; Vol. 14.
8. Wang, S.; Chen, W.; Xiang, H.; Yang, J.; Zhou, Z.; Zhu, M. Modification and potential application of short-chain-length polyhydroxyalkanoate (SCL-PHA). *Polymers* **2016**, *8* (8), 273.
9. Chen, G.-Q.; Chen, X.; Wu, F.; Chen, J. Polyhydroxyalkanoates (PHA) Toward Cost Competitiveness and Functionality. *Adv. Ind. Eng. Polym. Res.* **2019**, *3* (1), 1–7.
10. Bordes, P.; Hablot, E.; Pollet, E.; Averous, L. Effect of clay organomodifiers on degradation of polyhydroxyalkanoates. *Polym. Degrad. Stab.* **2009**, *94* (5), 789–796.
11. Buzarovska, A.; Grozdanov, A.; Avella, M.; Gentile, G.; Errico, M. Poly (hydroxybutyrate-*co*-hydroxyvalerate)/titanium dioxide nanocomposites: A degradation study. *J. Appl. Polym. Sci.* **2009**, *114* (5), 3118–3124.
12. Sridhar, V.; Lee, I.; Chun, H.; Park, H. Graphene reinforced biodegradable poly (3-hydroxybutyrate-*co*-4-hydroxybutyrate) nano-composites. *eXPRESS Polym. Lett.* **2013**, *7* (4), 320–328.
13. Wang, S.; Xiang, H.; Wang, R.; Peng, C.; Zhou, Z.; Zhu, M. Morphology and properties of renewable poly (3-hydroxybutyrate-*co*-3-hydroxyvalerate) blends with thermoplastic polyurethane. *Polym. Eng. Sci.* **2014**, *54* (5), 1113–1119.
14. Li, Z.-J.; Shi, Z.-Y.; Jian, J.; Guo, Y.-Y.; Wu, Q.; Chen, G.-Q. Production of poly (3-hydroxybutyrate-*co*-4-hydroxybutyrate) from unrelated carbon sources by metabolically engineered Escherichia coli. *Metabolic Eng.* **2010**, *12* (4), 352–359.
15. Povolo, S.; Romanelli, M. G.; Basaglia, M.; Ilieva, V. I.; Corti, A.; Morelli, A.; Chiellini, E.; Casella, S. Polyhydroxyalkanoate biosynthesis by Hydrogenophaga pseudoflava DSM1034 from structurally unrelated carbon sources. *New Biotechnol.* **2013**, *30* (6), 629–634.

16. Chen, G.-Q. A microbial polyhydroxyalkanoates (PHA) based bio-and materials industry. *Chem. Soc. Rev.* **2009**, *38* (8), 2434–2446.
17. Choi, J.; Lee, S. Y. Factors affecting the economics of polyhydroxyalkanoate production by bacterial fermentation. *Appl. Microbiol. Biotechnol.* **1999**, *51* (1), 13–21.
18. Andreeßen, B.; Lange, A. B.; Robenek, H.; Steinbüchel, A. Conversion of glycerol to poly (3-hydroxypropionate) in recombinant Escherichia coli. *Appl. Environ. Microbiol.* **2010**, *76* (2), 622–626.
19. Cavalheiro, J. M.; Raposo, R. S.; de Almeida, M. C. M.; Cesário, M. T.; Sevrin, C.; Grandfils, C.; Da Fonseca, M. Effect of cultivation parameters on the production of poly (3-hydroxybutyrate-*co*-4-hydroxybutyrate) and poly (3-hydroxybutyrate-4-hydroxybutyrate- 3-hydroxyvalerate) by Cupriavidus necator using waste glycerol. *Bioresource Technol.* **2012**, *111*, 391–397.
20. Ling, C.; Qiao, G. Q.; Shuai, B. W.; Song, K. N.; Yao, W. X.; Jiang, X. R.; Chen, G. Q. Engineering self-flocculating Halomonas campaniensis for wastewaterless open and continuous fermentation. *Biotechnol. Bioeng.* **2019**, *116* (4), 805–815.
21. Fukui, T.; Doi, Y. Cloning and analysis of the poly (3-hydroxybutyrate-co-3-hydroxyhexanoate) biosynthesis genes of Aeromonas caviae. *J. Bacteriol.* **1997**, *179* (15), 4821–4830.
22. Fukui, T.; Shiomi, N.; Doi, Y. Expression and characterization of (R)-specific enoyl coenzyme A hydratase involved in polyhydroxyalkanoate biosynthesis by Aeromonas caviae. *J. Bacteriol.* **1998**, *180* (3), 667–673.
23. Huijberts, G.; Eggink, G.; De Waard, P.; Huisman, G. W.; Witholt, B. Pseudomonas putida KT2442 cultivated on glucose accumulates poly(3-hydroxyalkanoates) consisting of saturated and unsaturated monomers. *Appl. Environ. Microbiol.* **1992**, *58* (2), 536–544.
24. Chen, G.-Q.; Chen, X.-Y.; Wu, F.-Q.; Chen, J.-C. Polyhydroxyalkanoates (PHA) toward cost competitiveness and functionality. *Adv. Ind. Eng. Polym. Res.* **2020**, *3* (1), 1–7.
25. Tripathi, L.; Wu, L.-P.; Dechuan, M.; Chen, J.; Wu, Q.; Chen, G.-Q. Pseudomonas putida KT2442 as a platform for the biosynthesis of polyhydroxyalkanoates with adjustable monomer contents and compositions. *Bioresource Technol.* **2013**, *142*, 225–231.
26. Tripathi, L.; Wu, L.-P.; Chen, J.; Chen, G.-Q. Synthesis of Diblock copolymer poly-3-hydroxybutyrate-block-poly-3-hydroxyhexanoate [PHB-b-PHHx] by a β-oxidation weakened Pseudomonas putida KT2442. *Microb. Cell Fact.* **2012**, *11* (1), 44.
27. Li, S.; Cai, L.; Wu, L.; Zeng, G.; Chen, J.; Wu, Q.; Chen, G.-Q. Microbial synthesis of functional homo-, random, and block polyhydroxyalkanoates by β-oxidation deleted Pseudomonas entomophila. *Biomacromolecules* **2014**, *15* (6), 2310–2319.
28. Tripathi, L.; Wu, L.-P.; Meng, D.; Chen, J.; Chen, G.-Q. Biosynthesis and characterization of diblock copolymer of P(3-hydroxypropionate)-*block*-P(4-hydroxybutyrate) from recombinant Escherichia coli. *Biomacromolecules* **2013**, *14* (3), 862–870.
29. Wang, Y.; Chung, A.; Chen, G. Q. Synthesis of Medium-Chain-Length Polyhydroxyalkanoate Homopolymers, Random Copolymers, and Block Copolymers by an Engineered Strain of Pseudomonas entomophila. *Adv. Healthcare Mater.* **2017**, *6* (7), 1601017.
30. Meng, D.-C.; Shi, Z.-Y.; Wu, L.-P.; Zhou, Q.; Wu, Q.; Chen, J.-C.; Chen, G.-Q. Production and characterization of poly (3-hydroxypropionate-*co*-4-hydroxybutyrate) with fully

controllable structures by recombinant Escherichia coli containing an engineered pathway. *Metab. Eng.* **2012,** *14* (4), 317–324.

31. Chen, Y.; Chen, X.-Y.; Du, H.-T.; Zhang, X.; Ma, Y.-M.; Chen, J.-C.; Ye, J.-W.; Jiang, X.-R.; Chen, G.-Q. Chromosome engineering of the TCA cycle in *Halomonas* bluephagenesis for production of copolymers of 3-hydroxybutyrate and 3-hydroxyvalerate (PHBV). *Metab. Eng.* **2019,** *54,* 69–82.
32. Dufresne, A.; Kellerhals, M. B.; Witholt, B. Transcrystallization in Mcl-PHAs/cellulose whiskers composites. *Macromolecules* **1999,** *32* (22), 7396–7401.
33. Sánchez, R. J.; Schripsema, J.; da Silva, L. F.; Taciro, M. K.; Pradella, J. G.; Gomez, J. G. C. Medium-chain-length polyhydroxyalkanoic acids (PHAmcl) produced by Pseudomonas putida IPT 046 from renewable sources. *Eur. Polym. J.* **2003,** *39* (7), 1385–1394.
34. Rai, R.; Keshavarz, T.; Roether, J.; Boccaccini, A. R.; Roy, I. Medium chain length polyhydroxyalkanoates, promising new biomedical materials for the future. *Mater. Sci. Eng., R: Rep.* **2011,** *72* (3), 29–47.
35. Chanprateep, S. Current trends in biodegradable polyhydroxyalkanoates. *J. Biosci. Bioeng.* **2010,** *110* (6), 621–632.
36. Kunioka, M.; Tamaki, A.; Doi, Y. Crystalline and thermal properties of bacterial copolyesters: poly (3-hydroxybutyrate-*co*-3-hydroxyvalerate) and poly (3-hydroxybutyrate-*co*-4-hydroxybutyrate). *Macromolecules* **1989,** *22* (2), 694–697.
37. Lu, X.; Wen, X.; Yang, D. Isothermal crystallization kinetics and morphology of biodegradable poly(3-hydroxybutyrate-*co*-4-hydroxybutyrate). *J. Mater. Sci.* **2011,** *46* (5), 1281–1288.
38. Chen, G.-Q., Introduction of Bacterial Plastics PHA, PLA, PBS, PE, PTT, and PPP. In *Plastics from bacteria*; Springer: 2010; pp 1–16.
39. Armentano, I.; Fortunati, E.; Burgos, N.; Dominici, F.; Luzi, F.; Fiori, S.; Jiménez, A.; Yoon, K.; Ahn, J.; Kang, S. Processing and characterization of plasticized PLA/PHB blends for biodegradable multiphase systems. *eXPRESS Polym. Lett.* **2015,** *9* (7), 583–596.
40. Arrieta, M. P.; Fortunati, E.; Dominici, F.; Rayón, E.; López, J.; Kenny, J. M. Multifunctional PLA–PHB/cellulose nanocrystal films: Processing, structural and thermal properties. *Carbohydr. Polym.* **2014,** *107,* 16–24.
41. Abdelwahab, M. A.; Flynn, A.; Chiou, B.-S.; Imam, S.; Orts, W.; Chiellini, E. Thermal, mechanical and morphological characterization of plasticized PLA–PHB blends. *Polym. Degrad. Stab.* **2012,** *97* (9), 1822–1828.
42. Ma, P.; Hristova-Bogaerds, D. G.; Zhang, Y.; Lemstra, P. J. Enhancement in crystallization kinetics of the bacterially synthesized poly(β-hydroxybutyrate) by poly(butylene succinate). *Polym. Bull.* **2014,** *71* (4), 907–923.
43. Gassner, F.; Owen, A. Physical properties of poly(β-hydroxybutyrate)-poly(ε-caprolactone) blends. *Polymer* **1994,** *35* (10), 2233–2236.
44. Mousavioun, P.; Halley, P. J.; Doherty, W. O. Thermophysical properties and rheology of PHB/lignin blends. *Ind. Crops Prod.* **2013,** *50,* 270–275.
45. Li, J.; Lai, M.; Liu, J. Control and development of crystallinity and morphology in poly (β-hydroxybutyrate-co-β-hydroxyvalerate)/poly (propylene carbonate) blends. *J. Appl. Polym. Sci.* **2005,** *98* (3), 1427–1436.

46. Dacko, P.; Kowalczuk, M.; Janeczek, H.; Sobota, M. In *Physical properties of the biodegradable polymer compositions containing natural polyesters and their synthetic analogues*; Macromolecular symposia; Wiley Online Library: 2006; pp 209–216.
47. Javadi, A.; Kramschuster, A. J.; Pilla, S.; Lee, J.; Gong, S.; Turng, L. S. Processing and characterization of microcellular PHBV/PBAT blends. *Polym. Eng. Sci.* **2010**, *50* (7), 1440–1448.
48. Nagarajan, V.; Misra, M.; Mohanty, A. K. New engineered biocomposites from poly (3-hydroxybutyrate-*co*-3-hydroxyvalerate)(PHBV)/poly (butylene adipate-*co*-terephthalate)(PBAT) blends and switchgrass: Fabrication and performance evaluation. *Ind. Crops Prod.* **2013**, *42*, 461–468.
49. Parulekar, Y.; Mohanty, A. K. Extruded biodegradable cast films from polyhydroxyalkanoate and thermoplastic starch blends: fabrication and characterization. *Macromol. Mater. Eng.* **2007**, *292* (12), 1218–1228.
50. Takagi, Y.; Yasuda, R.; Yamaoka, M.; Yamane, T. Morphologies and mechanical properties of polylactide blends with medium chain length poly(3-hydroxyalkanoate) and chemically modified poly (3-hydroxyalkanoate). *J. Appl. Polym. Sci.* **2004**, *93* (5), 2363–2369.
51. Scaffaro, R.; Dintcheva, N. T.; Marino, R.; La Mantia, F. Processing and properties of biopolymer/polyhydroxyalkanoates blends. *J. Polym. Environ.* **2012**, *20* (2), 267–272.
52. Ohkoshi, I.; Abe, H.; Doi, Y. Miscibility and solid-state structures for blends of poly [(S)-lactide] with atactic poly [(R, S)-3-hydroxybutyrate]. *Polymer* **2000**, *41* (15), 5985–5992.
53. Botta, L.; Mistretta, M.; Palermo, S.; Fragalà, M.; Pappalardo, F. Characterization and processability of blends of polylactide acid with a new biodegradable medium-chain-length polyhydroxyalkanoate. *J. Polym. Environ.* **2015**, *23* (4), 478–486.
54. Raquez, J. M.; Nabar, Y.; Narayan, R.; Dubois, P. Novel high-performance talc/poly [(butylene adipate)-co-terephthalate] hybrid materials. *Macromol. Mater. Eng.* **2008**, *293* (4), 310–320.
55. Larsson, M.; Markbo, O.; Jannasch, P. Melt processability and thermomechanical properties of blends based on polyhydroxyalkanoates and poly (butylene adipate-*co*-terephthalate). *RSC Adv.* **2016**, *6* (50), 44354–44363.
56. Wang, J. H.; Kuang, J.; Jia, J., Modified Polyhydroxyalkanoate Materials and Their Application as Biodegradable Films and the Preparation Methods. Patent No. CN111205602A, 2020.
57. Wang, J. H.; Wang, Z.; Kuang, J.; Zhou, B.; Jia, Q. G.-Q. Ultrathin, High Strength and High Toughness Biodegradable Materials and Films, and Their Preparation Methods and Applications. Patent No. CN111205604A, 2020.
58. Wang, J. H.; Wang, Z.; Kuang, J.; Zhou, B.; Jia, Q. High Melt Strength Polyhydroxyalkanoate Blends and Their Preparation and Applications. Patent No. CN111205605A, 2020.
59. Avella, M.; La Rota, G.; Martuscelli, E.; Raimo, M.; Sadocco, P.; Elegir, G.; Riva, R. Poly (3-hydroxybutyrate-*co*-3-hydroxyvalerate) and wheat straw fibre composites: thermal, mechanical properties and biodegradation behaviour. *J. Mater. Sci.* **2000**, *35* (4), 829–836.
60. Urakami, T.; Imagawa, S.; Harada, M.; Iwamoto, A.; Tokiwa, Y. Development of biodegradable plastic-poly-β-hydroxybutyrate/polycaprolactone blend polymer. *Kobunshi ronbunshu (Tokyo)* **2000**, *57* (5), 263–270.

61. Kijchavengkul, T.; Auras, R.; Rubino, M.; Ngouajio, M.; Fernandez, R. T. Assessment of aliphatic–aromatic copolyester biodegradable mulch films. Part I: Field study. *Chemosphere* **2008**, *71* (5), 942–953.
62. Avar, L.; Bechtold, K. Studies on the interaction of photoreactive light stabilizers and UV-absorbers. *Prog. Org. Coat.* **1999**, *35* (1-4), 11–17.
63. Zhu, S.; Chen, Y.; Tang, Y.; Li, Q.; Zhong, B.; Zeng, X.; Xie, D.; Jia, Z.; Jia, D. A novel nanosilica-supported ultraviolet absorber for the preparation of robust biodegradable plastic film with high ultraviolet aging resistance. *Polym. Compos.* **2019**, *40* (10), 4154–4161.
64. Souza, P. M. S.; Morales, A. R.; Sanchez, E. M. S.; Mei, L. H. I. Study of PBAT Photostabilization with Ultraviolet Absorber in Combination with Hindered Amine Light Stabilizer and Vitamin E, Aiming Mulching Film Application. *J. Polym. Environ.* **2018**, *26* (8), 3422–3436.
65. Scarfato, P.; Acierno, D.; Russo, P. Photooxidative weathering of biodegradable nanocomposite films containing halloysite. *Polym. Compos.* **2015**, *36* (6), 1169–1175.
66. Mosnáčková, K.; Šlosár, M.; Kollár, J.; Janigová, I.; Šišková, A.; Chmela, Š.; Sikorska, W.; Perďochová, D.; Gálisová, I.; Alexy, P. Ageing of plasticized poly (lactic acid)/poly (3-hydroxybutyrate)/carbon black mulching films during one season of sweet pepper production. *Eur. Polym. J.* **2019**, *114*, 81–89.
67. Ikada, Y.; Tsuji, H. Biodegradable polyesters for medical and ecological applications. *Macromol. Rapid Commun.* **2000**, *21* (3), 117–132.
68. Tian, Y.; Wang, J. H., Understanding the Degradation Process of Polyhydroxyalkanoate (PHA)/Poly(Butylene Adipate-*co*-Terephthalate) (PBAT) Biocomposite Mulch Films and the Effects of Stabilizers Under Accelerated Aging Test. 259th American Chemical Society National Meeting, Mar. 22–26, 2020, an online oral presentation, https://doi.org/10.1021/scimeetings.0c02391.
69. Krupp, L. R.; Jewell, W. J. Biodegradability of modified plastic films in controlled biological environments. *Environ. Sci. Technol.* **1992**, *26* (1), 193–198.
70. Haider, T. P.; Völker, C.; Kramm, J.; Landfester, K.; Wurm, F. R. Plastics of the future? The impact of biodegradable polymers on the environment and on society. *Angew. Chem., Int. Ed.* **2019**, *58* (1), 50–62.
71. Shen, J.; Burgess, D. J. Accelerated *in-vitro* release testing methods for extended-release parenteral dosage forms. *J. Pharm. Pharmacol.* **2012**, *64* (7), 986–996.
72. Mergaert, J.; Webb, A.; Anderson, C.; Wouters, A.; Swings, J. Microbial degradation of poly(3-hydroxybutyrate) and poly(3-hydroxybutyrate-*co*-3-hydroxyvalerate) in soils. *Appl. Environ. Microbiol.* **1993**, *59* (10), 3233–3238.
73. Sintim, H. Y.; Bary, A. I.; Hayes, D. G.; Wadsworth, L. C.; Anunciado, M. B.; English, M. E.; Bandopadhyay, S.; Schaeffer, S. M.; DeBruyn, J. M.; Miles, C. A. *In situ* degradation of biodegradable plastic mulch films in compost and agricultural soils. *Sci. Total Environ.* **2020**, 138668.
74. Hayes, D. G.; Wadsworth, L. C.; Sintim, H. Y.; Flury, M.; English, M.; Schaeffer, S.; Saxton, A. M. Effect of diverse weathering conditions on the physicochemical properties of biodegradable plastic mulches. *Polym. Test.* **2017**, *62*, 454–467.

75. Lyu, S.; Untereker, D. Degradability of polymers for implantable biomedical devices. *Int. J. Mol. Sci.* **2009**, *10* (9), 4033–4065.
76. Niaounakis, M. *Biopolymers: Applications and Trends*; Elsevier: Waltham, USA, 2015; p 407
77. Hassan, M. K.; Abou-Hussein, R.; Zhang, X.; Mark, J. E.; Noda, I. Biodegradable copolymers of 3-hydroxybutyrate-co-3-hydroxyhexanoate (Nodax™), including recent improvements in their mechanical properties. *Mol. Cryst. Liq. Cryst.* **2006**, *447* (1), 341–362.
78. Amelia, T. S. M.; Govindasamy, S.; Tamothran, A. M.; Vigneswari, S.; Bhubalan, K. Applications of PHA in Agriculture. In *Biotechnological Applications of Polyhydroxyalkanoates*; Springer: 2019; pp 347–361.
79. Kwiecien, I.; Adamus, G.; Jiang, G.; Radecka, I.; Baldwin, T. C.; Khan, H. R.; Johnston, B.; Pennetta, V.; Hill, D.; Bretz, I. Biodegradable PBAT/PLA blend with bioactive MCPA-PHBV conjugate suppresses weed growth. *Biomacromolecules* **2018**, *19* (2), 511–520.

Chapter 9

Physical Properties and Structures of Novel Bio-Based Films and Fibers Derived from Paramylon Esters

Hongyi Gan and Tadahisa Iwata*

Science of Polymeric Materials, Department of Biomaterial Sciences, Graduate School of Agricultural and Life Sciences, The University of Tokyo, 1-1-1 Yayoi, Bunkyo-ku, Tokyo 113-8657, Japan

***Email: atiwata@g.ecc.u-tokyo.ac.jp**

Paramylon is a β-1,3-glucan that is photoproduced by microalgae. It is a promising raw material for bio-based plastics. This chapter describes the manufacture of fully substituted paramylon ester derivatives with alkyl side-chains of various lengths. The thermal and mechanical properties of the paramylon ester derivatives can be controlled over wide range by adjusting the lengths of the side-chains. The manufactured paramylon esters had excellent thermal stability compared with paramylon esters with short side-chain, and compared with general petroleum-based plastics. The molecular and crystal structures of paramylon propionate, paramylon butyrate, and paramylon valerate were investigated using wide-angle X-ray diffraction. Melt-spun fibers of paramylon esters were also successfully produced at various take-up rates. The relationship between the crystallinity and tensile properties of the fibers was also investigated.

Introduction

Plastics manufactured from petroleum-based polymers—which are lightweight, inexpensive, and easy to produce—have improved people's comfort and made their lives more convenient. However, the constant increase in the volume of plastics being fabricated has resulted in serious environmental problems, especially climate change and waste pollution. Consequently, bioplastics—which are derived from sustainable and renewable natural sources, and are biodegradable—have attracted attention as potential alternatives to petroleum-based plastics (*1*).

Among such biodegradable bioplastics manufactured from biomass resources, polylactic acid (PLA) and polyhydroxyalkanoates (PHAs) have been studied extensively. PLA is a thermoplastic aliphatic polyester derived from fermented plant starch, obtained from such sources as corn, sugar cane, or sugar beet pulp. It has a glass transition temperature (T_g) of 60 °C and a melting temperature (T_m) of 175 °C (*2*). PLA-based materials such as films, fibers, and bulk polymers have been developed. PHAs are a series of polyesters produced in nature by numerous microorganisms by fermenting sugars or lipids. More than 150 kinds of monomers from this family can be combined to produce materials with various properties (*3*). PHAs are also highly biocompatible, which suggests their potential for a variety of applications in the medical field.

Moreover, polysaccharide derivatives are also considered to be useful bio based polymers with high potential. Polysaccharides—which constitute the most abundant natural polymer resource on the planet—are mainly composed of glucose units bonded by α or β glycosidic linkages. Because they are already polymerized, polysaccharides do not need to be manufactured from monomers. However, their thermoplasticity and solubility in organic solvents are restricted by strong inter- and intra-hydrogen bonds between the hydroxyl groups of the monosaccharide units. The hydroxyl groups are usually converted to functional ester groups by esterification to suppress hydrogen bonding, which improves solubility and thermoplasticity.

The cellulose contained in plant cell walls and plant fibers is the most abundant polysaccharide resource, and is commonly used. Cellulose has a molecular structure comprised of β-1,4-glucans without branches. Cellulose triacetate (CTA) is an important cellulose ester derivative that is mainly used in optical films, cigarette filters, and fiber materials (*4*). However, despite their excellent properties, a plasticizer is required when heat-molding CTA-based materials owing to their poor thermal fluidity (*5*).

In nature, in addition to cellulose, there are many kinds of polysaccharides with various characteristic structures. In recent years, several unique properties of β-1,3-glucan polymers derived from paramylon or curdlan have been discovered, including high thermal fluidity after esterification. Curdlan is an extracellular microbe-derived polysaccharide that was first discovered and studied in 1964 (*6*). Paramylon is a storage polysaccharide that is photosynthesized by the microalga *Euglena* (*7*). Curdlan and paramylon are both linear β-1,3-glucan polymers without branches; the molecular weight of paramylon is approximately one-fifth that of curdlan. Curdlan and paramylon both have characteristic six-fold helical mainchain structures (*8–11*), which differ from the two-fold helical structure of cellulose.

Fully substituted curdlan esters with alkyl side-chains of various lengths have been prepared in a carboxylic acid/trifluoroacetic anhydride (TFAA) heterogeneous reaction system (*12*, *13*). Fundamental physical properties such as thermal, mechanical, and crystalline properties were controlled by changing the lengths of the side-chains. Branched ester groups have been introduced into curdlan to obtain more favorable physical properties (*14*, *15*). Owing to the denser packing structure of their molecular chains, the curdlan branched esters display higher melting temperatures and tensile strengths than curdlan linear esters.

Paramylon mixed esters containing both short and long alkyl side-chains have been prepared (*16–18*). Precise control of physical properties was achieved by adjusting the types and ratios of the carboxylic acids used. Owing to their high thermoplasticity, the paramylon mixed esters could be processed into injection molded specimens as well as self-standing films.

However, other than as mentioned above, paramylon homoesters have not been manufactured. Paramylon esters with lower molecular weights than those of curdlan esters are supposed to provide higher thermoplasticity. Furthermore, paramylon homoesters with short side-chain lengths may have higher crystallinity than those of mixed esters, which could facilitate structural analysis.

Accordingly, paramylon, which is a β-1,3-glucan, has been modified in several ways. Eight kinds of fully substituted paramylon esters with various alkyl side-chain lengths were prepared, and their thermal and mechanical properties were investigated (*19*). Crystal structure analysis was then carried out (*20*). The substituted paramylon esters were also melt-spun into fibers to investigate their thermal processing behavior (*21*).

Materials and Methods

Materials

Paramylon was purchased from the Euglena Co. (Tokyo, Japan). TFAA and acetic, propionic, butyric, valeric, hexanoic, octanoic, decanoic, and lauric acids were purchased from Wako Pure Chemicals (Tokyo, Japan), and were used without further purification.

Esterification of Paramylon

Figure 1 represents the synthesis of the paramylon esters. A premixed solution of TFAA (80 mL) and the corresponding fatty acid (40 mL) was stirred at 50 °C in a flask for 5 min, then pre-dried paramylon was immediately added. The solution was stirred at 50 °C for 1 h. After cooling to room temperature, the solution was poured into a mixture of water and methanol (2.0 L). The precipitate was further collected by filtration, dissolved in chloroform, and reprecipitated in a mixture of water and methanol. Finally, the precipitate was collected, and dried at reduced pressure for 1 day.

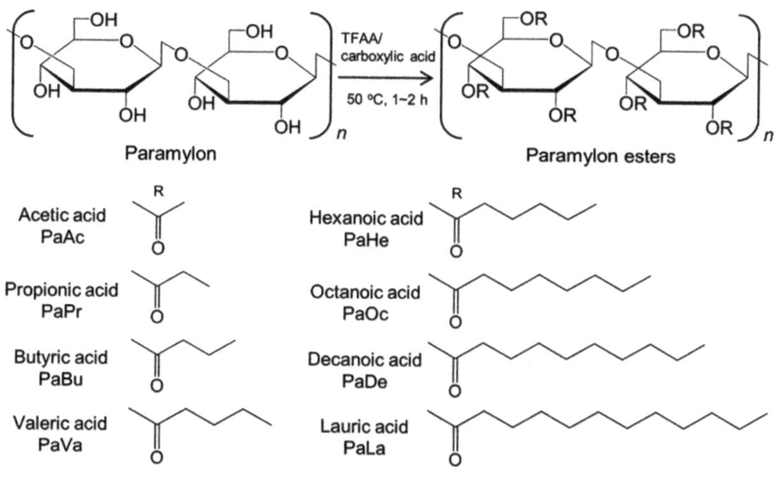

Figure 1. Synthesis of the paramylon esters.

Nuclear Magnetic Resonance (NMR) Measurements

Proton NMR (^{1}H-NMR) spectra were recorded using a JNM-A500 FT-NMR system (500 MHz; JEOL Ltd.) at 25 °C. Deuterated chloroform ($CDCl_3$) was used as the solvent for the paramylon esters except for PaAc (TFA), wherein the solution concentration was set to 10 mg/mL. Chemical shifts (δ in ppm) were referenced to the resonance of tetramethylsilane (TMS; δ = 0) for $CDCl_3$. The coupling constants (*J*) are reported in Hz. The degree of substitution (DS) of the paramylon esters was calculated from the ratio of the integrated area of the methyl protons of the acyl group to the ring protons of glucose: DS = ([CH3]/3)/([ring-H]/7)

Gel Permeation Chromatography (GPC) Measurements

The number- and weight-average molecular weights (M_n and M_w) and polydispersity index values (M_w/M_n) were estimated by GPC (RID-20A refractive index detector, Shimadzu) in chloroform at 40 °C. Shodex columns (K-806M, K-802) were used, and the flow rate was 0.8 mL/min. A calibration curve was constructed using polystyrene (PS) standards (Shodex).

Thermogravimetric Analysis

Thermogravimetric analysis (TGA) was performed using an STA 6000 instrument (Perkin Elmer) under nitrogen gas supplied at a flow rate of 20 mg/s. A sample weighing approximately 10 mg was packed into an aluminum pan. Thermograms were obtained from 30 °C to 450 °C at a heating rate of 10 °C/min.

Differential Scanning Calorimetry (DSC) Measurement

Samples for DSC measurements were prepared by solvent casting using chloroform. The paramylon esters (0.25 g) were dissolved in chloroform (10 mL) and poured into a polytetrafluoroethylene (PTFE) petri dish (diameter 4 cm) after filtration through cotton. The solvent was then evaporated in air at room temperature. The obtained films were dried *in vacuo* for 1 day to remove the solvent.

DSC thermograms were recorded on a DSC8500 system (Perkin Elmer) under nitrogen gas supplied at a flow rate of 20 mg/s. The measurements were performed using 2–6 mg samples on a DSC pan. The samples were first heated from 30 °C to 250 °C at a rate of 100 °C/min, cooled to -70 °C at a rate of 200 °C/min, and scanned while heating from -70 °C to 250 °C at a rate of 10 °C/min.

Tensile Tests on Cast Films and Melt-Quenched Films

Solvent-cast films were prepared using the same method described in the above Section.

To prepare the melt-quenched films, powders of the paramylon esters were placed between PTFE or polyimide sheets and pressed at 30 °C to 50 °C above their melting temperatures, then quenched in ice water or at room temperature. A film thickness of approximately 100 μm was produced by maintaining the pressure between 3 and 15 MPa. The melt-quenched films are amorphous due to X-ray measurement.

The tensile tests were performed at room temperature (controlled at 25 °C) using an EZ-test machine (Shimadzu, Japan). The crosshead speed was 20 mm/min, and the initial gauge length was 10 mm. Approximately five paramylon ester film specimens (length 30 mm, width 3 mm) were used for each measurement, and the data were averaged for each film.

Preparation of Stretched and Annealed Films of Paramylon Esters

The melt-quenched paramylon ester films were cut into pieces (length 20 mm, width 5 mm). The cut films were fixed in a stretching machine and stretched in an oven at 10 °C to 20 °C above their glass transition temperatures. The films were then annealed in an oven below their T_m values for 1 h. The paramylon ester films were all stretched to four times their initial length. Stretched and annealed films with a high degree of orientation (approximately 95%) were obtained, as confirmed by their X-ray fiber diagrams.

Wide-Angle X-ray Diffraction (WAXD) at SPring-8

X-ray fiber diagrams of the stretched and annealed films were obtained by WAXD measurements using beamlines BL45XU and BL03XU at the SPring-8 large synchrotron radiation facility (Harima, Japan) at a wavelength of 0.1000 nm. Each WAXD diagram was recorded with a PILATUS detector at a sample-to-detector distance of 266 mm. A vacuum path was set up between the sample holder and the detector. Each two-dimensional (2-D) X-ray fiber diagram was analyzed using the free software Fit 2-D and ImageJ.

Melt Spinning

Melt spinning was performed using a system comprising a melt indexer (IMC-19F8; Imoto, Japan) and a customized spinning apparatus (Figure 2). Paramylon ester powders were used for melt spinning. The melt indexer had a die with a diameter of 1.0 mm and an L/D ratio of 2. The polymer was first placed in the indexer (which was set to the processing temperature), melted for 3 min, then extruded at a rate of 0.5 mm/s (the descent speed of the extruded bar). The initial part of the extruded polymer was quenched in air and pasted on a roller, which was part of the spinning apparatus. The distance between the die and the roller was set at 40 mm. The take-up rate of the spinning apparatus could be varied from 0 to 180 m/min. In the present study, four take-up rates were established: 18, 36, 108, and 180 m/min.

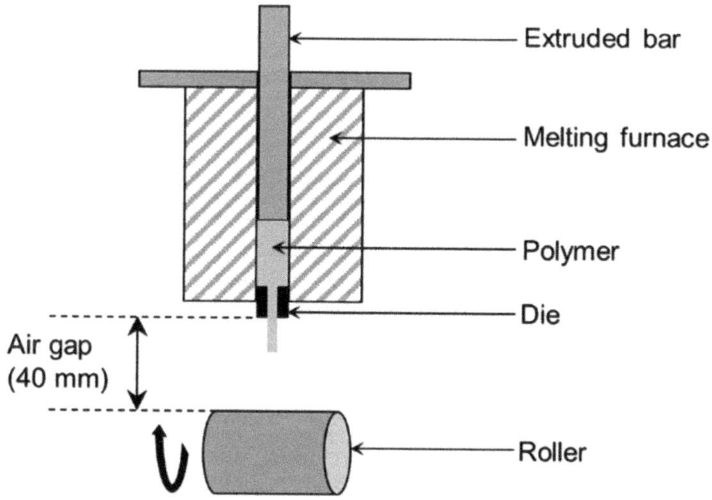

Figure 2. Schematic diagram of the melt spinning apparatus.

Annealing Treatment of the Melt-Spun Fibers

To investigate the influence of crystallinity, PaPr fibers were fixed and further annealed for 30 min at 150 °C in a thermostatic oven.

Wide-Angle X-ray Diffraction of the Melt-Spun Fibers

Two-dimensional WAXD measurements of the fibers were obtained using a MicroMax-007HF system (Rigaku MicroMax-007HF) operated at 40 kV and 30 mA with Cu Kα radiation ($\lambda = 0.15418$ nm). Each X-ray fiber pattern was recorded on an imaging plate (Fujifilm Corp.; 2540 × 2540 pixels, 50 × 50 μm^2/pixel), and read using a RAXIA-D*i* system (Rigaku Corp.). The sample-to-detector distance was set at approximately 83 mm. The sample holder and the detector were set in a vacuum chamber. Silicon powder was used as a standard sample, and the measurements were obtained at room temperature. Two-dimensional fiber pattern analysis and conversion to a one-dimensional powder pattern were performed using 2DP software (Rigaku Corp.). Some of the measurements were also carried out using beamlines BL40B2 and BL03XU at the SPring-8 large synchrotron radiation facility (Harima, Japan) at a wavelength of 0.1000 nm. The degree of orientation was calculated using the formula: **F = (180-Wh)/180**, in which "F" and "Wh" represent the degree of orientation and the half width of the orientation peak from the intensity–azimuthal profile, respectively.

Tensile Tests on Melt-Spun Fibers

The tensile tests were performed at room temperature using an EZ-test machine (Shimadzu, Japan) equipped with a 10 N load cell. The crosshead tension speed was 10 mm/min, and the initial gauge length was 10 mm. The fibers were assumed to be cylindrical, and their diameters were measured using a polarized optical microscope (POM). Approximately five melt-spun fiber specimens were used for each measurement, and the data were averaged for each fiber.

Results and Discussions

Esterification of Paramylon

Eight types of paramylon esters with various side-chain lengths were successfully prepared using a carboxylic acid/TFAA reaction system. The degree of substitution (DS) according to the ^{1}H-NMR analysis was 3, indicating that all of the hydroxyl groups in the glucose unit were completely substituted by the corresponding carboxyl groups.

Table 1. Thermal and Mechanical Properties of the Paramylon and Curdlan Esters

Name	$M_w \times 10^5$	M_w/M_n	*DS*	T_g (°C)	T_m (°C)	σ_t (MPa)^a	ε_b (%)^b
PaAc	–[c]	–[c]	3	168	281	–[f]	–[f]
PaPr	3.4	1.8	3	112	221	31.4±3.4[g]	62±6[g]
PaBu	5.8	1.8	3	76	207	20.6±1.2	87±25
PaVa	8.1	2.1	3	–[d]	196	7.9±0.7	61±28
PaHe	8.5	2.1	3	–[d]	114	7.5±0.6	446±18
PaOc	8.9	1.9	3	–[d]	–[d]	2.7±0.3	415±66
PaDe	9.2	1.9	3	–[d]	–[d]	2.8±0.6	328±48
PaLa	9.6	1.8	3	–[d]	–[d]	2.7±0.5	529±108
CDAc*	–[c]	–[c]	3	171	287	21.2±4.0[g]	12±3[g]
CDPr*	6.0	1.8	3	110	213	23.7±4.2[g]	46±20[g]
CDBu*	8.9	1.9	3	74	190	15.4±1.9	26±5
CDVa*	13.0	2.0	3	51[e]	168	12.4±2.3	79±9
CDHe*	14.0	2.0	3	69[e]	167	14.7±1.0	150±28
CDOc*	14.0	2.1	3	50[e]	–[d]	6.0±2.0	112±43
CDDe*	16.0	2.1	3	55[e]	–[d]	4.6±0.4	144±11
CDLa*	16.0	2.3	3	47[e]	–[d]	1.7±0.3	172±35

[a] *Tensile strength (maximal stress value).* [b] *Elongation at break.* [c] *Not measured due to its insolubility in the GPC solution (CHCl$_3$).* [d] *Not detected by DSC measurement.* [e] *Determined by DMA (tan δ).* [f] *Not measured because PaAc could not form self-standing films.* [g] *Melt-quench film as sample.* * Marubayashi et al. *Carbohydr. Polym.* **2014**, *103*, 427–433.

The weight-average molecular weights and polydispersity values of the paramylon esters ranged from 3.0×10^5 to 9.6×10^5, and 1.8 to 2.6, respectively (see Table 1), suggesting that the conditions were adequate for the manufacture of the plastic materials. The weight-average molecular weights of the curdlan esters ranged from 6.0×10^5 to 16.0×10^5, i.e., twice those of the paramylon esters. The GPC measurements revealed that the degree of polymerization (DP) of each paramylon ester was approximately 1500, i.e., slightly lower than that of paramylon itself (approximately 1800 before esterification). Thus, the mainchain decomposed during the reaction owing to the use of TFAA. However, the decrease did not influence esterification.

Thermal Properties of the Paramylon Esters

According the TGA results, the $T_{d\_5\%}$ value ($T_{d\_5\%}$ represents the temperature when 5% mass reduction occurs) of each paramylon ester is above 300 °C, which was approximately 50 °C higher than that of paramylon itself, suggesting that esterification improves the thermal stability. However, no such tendency was apparent following adjustment of the alkyl side-chain lengths.

The thermal behaviors of the paramylon esters were analyzed using DSC thermal diagrams, as shown in Figure 3. As revealed by the first heating scans, the paramylon esters with C2–C6 side-chains produced endothermic peaks, whereas those with C8–C12 side-chains lacked a heat transition. Therefore, the paramylon esters with C2–C6 side-chains were crystalline polymers, and those with C8–C12 side-chains were amorphous. Furthermore, PaPr had two melting temperatures in the first run, which indicates polymorphism, as reported by Marubayashi *et al.* for curdlan propionate (*13*).

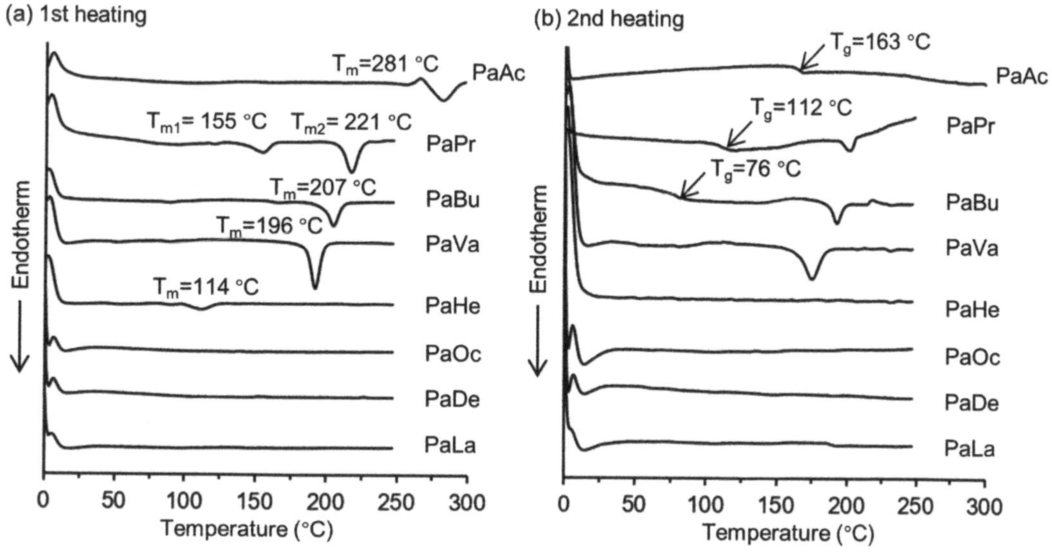

Figure 3. Differential scanning calorimetry (dsc) thermal diagrams of the solvent-cast films.

As shown in Figure 4, compared with general petroleum-based plastics, the T_m of PaAc (281 °C) was higher than that of polyethylene terephthalate (PET, $T_m = 269$ °C). Furthermore, PaPr, PaBu, and PaVa had T_m values of 221 °C, 207 °C, and 196 °C, respectively, which were higher than that of polypropylene (PP, $T_m = 175$ °C). Considering that the T_m values of PaPr, PaBu, PaVa, and PaHe were significantly lower than the thermal degradation temperature at approximately 300 °C, these ester derivatives have a wide process window for thermoforming. With regard to the T_g values in the second heating scan, PaAc, PaPr, and PaBu produced peaks at 168 °C, 112 °C, and 76 °C, respectively. In particular, PaAc and PaPr had glass transition temperatures above 100 °C, indicating that they are more thermally stable than PET, PP, or polyethylene (PE).

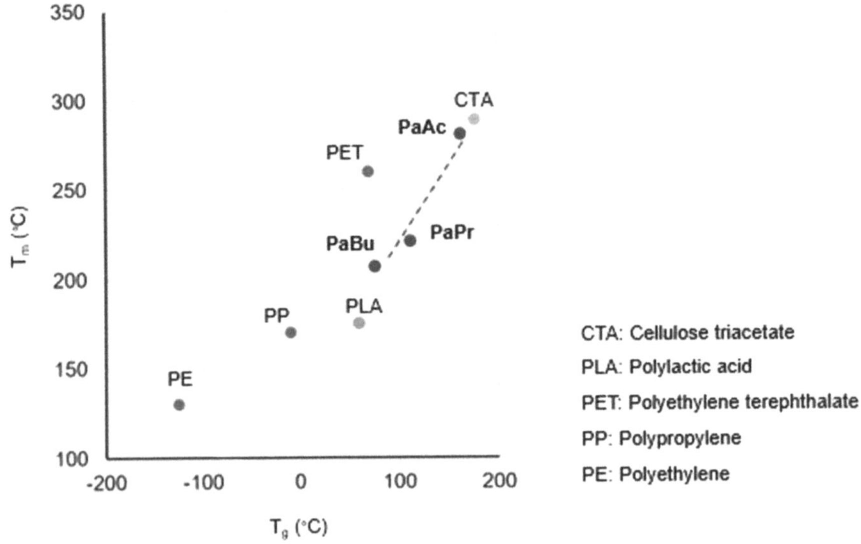

Figure 4. Comparison of the thermal properties of paramylon esters with general plastics.

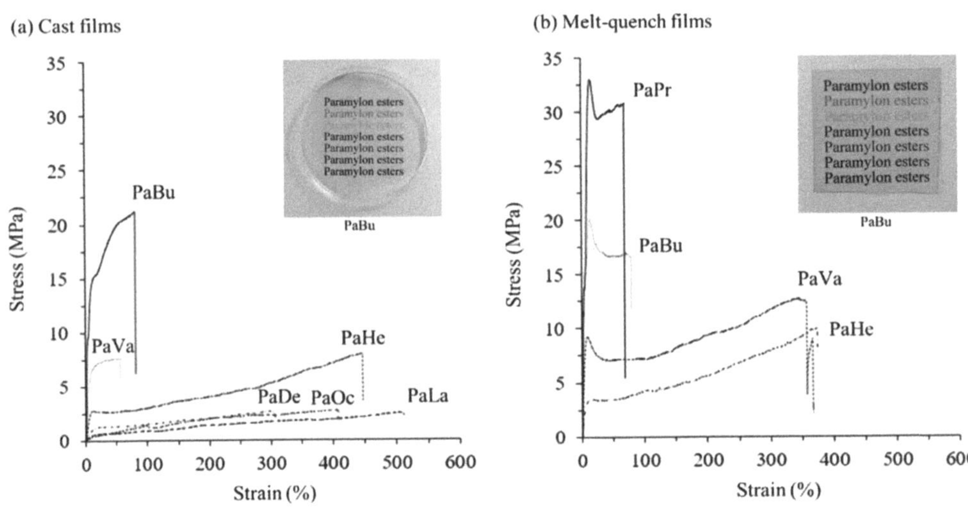

Figure 5. Stress-strain curves of (a) cast films and (b) melt-quenched films. Photograph of pabu films (embedded picture).

Tensile Tests on the Cast Films and Melt-Quenched Films of the Paramylon Esters

Figure 5a and Figure 5b show the stress–strain curves of the cast films and melt-quenched films of the paramylon esters, respectively. The cast films of PaAc and PaPr could not be set to the instrument owing to their brittleness. With increasing alkyl side-chain lengths, the tensile strength and Young's modulus values decreased, whereas the elongation at break tended to increase. Table 2 summarizes the tensile strength, elongation at break, and Young's modulus values of the solvent-cast and melt-quenched films. Except for PaPr and PaVa, there were only slight differences in the tensile strengths and elongation at break values of the cast and melt-quencheded films with corresponding ester groups. In the case of PaPr, the melt-quencheded film had a high tensile strength of 30 MPa, but the cast film was very brittle. Furthermore, the melt-quencheded film of PaVa had a high elongation at break of approximately 350%, whereas that of the cast film was approximately 60%. These differences seem to have resulted from the difference in crystallinity between the cast and melt-quenched films.

Table 2. Tensile Properties of the Cast and Melt-Quenched Films

	Cast films			*Melt-quenched films*		
Name	σ_t *(MPa)*[a]	ε_b *(%)*[b]	*E (MPa)*[c]	σ_t *(MPa)*[a]	ε_b *(%)*[b]	*E (MPa)*[c]
PaPr	—[d]	—[d]	—[d]	31.4±3.4	62±6	597±52
PaBu	20.6±1.2	87±25	352±54	16.8±2.9	51±31	355±55
PaVa	7.9±0.7	61±28	133±19	13.8±2.2	364±51	198±22
PaHe	7.5±0.6	446±18	47±3	8.9±1.6	374±44	49±13
PaOc	2.7±0.3	415±66	3±1	—[e]	—[e]	—[e]
PaDe	2.8±0.6	328±48	12±1	—[e]	—[e]	—[e]
PaLa	2.7±0.5	529±108	5±1	—[e]	—[e]	—[e]

[a] *Tensile strength (maximal stress value).* [b] *Elongation at break.* [c] *Young's modulus.* [d] *Not measured due to its brittleness.* [e] *Melt-quench film could not be formed.*

Comparison of the Physical Properties of the Paramylon and Curdlan Esters

Considering the influence of molecular weight on β-1,3-glucan esters, Table 1 presents the thermal and mechanical properties of the paramylon and curdlan esters. With regard to thermal properties, the esters with C2–C4 alkyl side-chains had almost the same T_g and T_m values. The T_m values of the valeric (C5) and hexanoic (C6) derivatives differed owing the different form of aggregation in the cast films. Almost all the ester derivatives which have the same alkyl side-chain lengths had similar tensile strengths. Furthermore, the paramylon esters with C6–C12 alkyl side-chains had higher elongation at break values than the curdlan esters. Therefore, although the molecular weights of the paramylon esters were half those of the curdlan esters, their thermal and mechanical properties were not affected.

Molecular and Crystal Structures of PaPr, PaBu, and PaVa

The 2-D X-ray fiber diagrams of PaPr, PaBu, and PaVa are shown in Figure 6. As reported by Marubayashi *et al.* (*13*), the X-ray fiber diagram of PaPr has a strong reflection on the meridian of the fifth layer lines, which indicates that PaPr has a five-fold helical conformation of molecular chains in the crystal, according to the dependence of the structure factors of helical polymers on the layer line order (*22*). PaBu and PaVa have similar X-ray fiber diagrams, which indicates that they also have five-fold helical conformations of molecular chains along the molecular axis.

The crystal lattice parameters of curdlan propionate have already been calculated and refined as a pseudohexagonal unit cell, which belongs to the P1 space group (*13*). Accordingly, the crystal lattice constant of PaPr was calculated using the formula for a hexagonal cell. The X-ray fiber diagram of PaPr yielded a total of 15 reflections, which were extracted from the first quadrant and indexed to the hexagonal system with $a = b = 1.181$ nm and c (the fiber axis) = 1.847 nm, which produced a calculated density of 1.24 g/cm^3 (Table 3). These lattice parameters of PaPr are the same as those of CDPr, suggesting that the molecular weight does not affect the crystal structure.

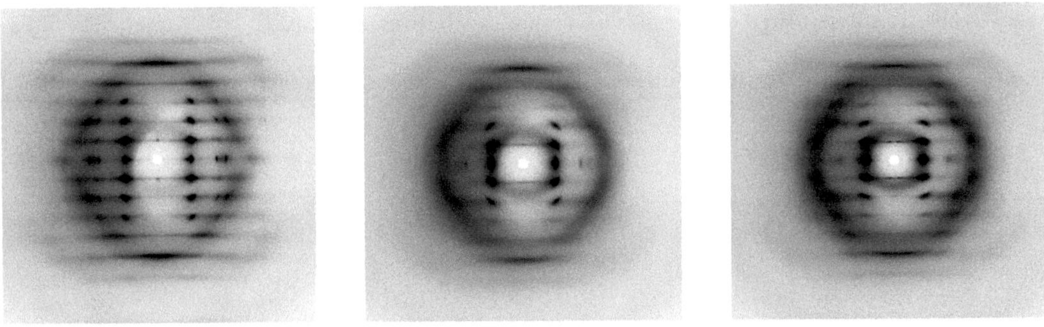

Figure 6. 2-D X-ray fiber diagrams of the paramylon esters.

For PaBu and PaVa, considering that the molecular conformations are the same as that of PaPr but the alkyl side-chain lengths are different, they appear to have the same crystal system as PaPr. PaBu and PaVa both have 12 reflections in the first quadrant, which are indexed to a hexagonal crystal system. The small range of errors between the calculated and observed *d*-spacings indicates that PaBu and PaVa both belong to hexagonal crystal systems. The crystal lattice constants and densities of the unit cells of PaPr, PaBu, and PaVa are summarized in Table 3. Above all, the crystal constants of PaPr ($a = b = 1.181$ nm and $c = 1.847$ nm), PaBu ($a = b = 1.251$ nm, and $c = 1.907$ nm), and PaVa ($a = b = 1.384$ nm, and $c = 1.957$ nm) were successfully indexed. With increasing side-chain length, all the *a*, *b*, and *c* (fiber axis) lattice parameters increased, which indicates that interactions of the side-chains affect both inter- and inner-molecular chain distances. The densities of the stretched and annealed films of PaPr, PaBu, and PaVa obtained by the sink-float method were 1.22, 1.16, and 1.01 g/cm^3, respectively. Based on the crystal lattice constants, the densities of the crystal unit cells, and the five-fold helical structures of the molecular chains, the molecular chain numbers in the unit cells were all calculated to be 1, which means that the unit cells contained five pyranose residues. Accordingly, the densities of the crystal unit cells of PaPr, PaBu, and PaVa were calculated to be 1.24, 1.21, and 1.07 g/cm^3, respectively.

Table 3. Crystal Lattice Constants of PaPr, PaBu, and PaVa

	PaPr	*PaBu*	*PaVa*
a, b (nm)	1.181	1.251	1.384
c (nm)	1.847	1.907	1.957
ρ_{obs}[a] (g/cm^3)	1.22	1.16	1.01
ρ_{cal}[b] (g/cm^3)	1.24	1.21	1.07
Helix	5	5	5
Chain number	1	1	1

[a] Density of crystal lattice (observed value). [b] Density of crystal lattice (calculated value).

Melt Spinning of PaPr, PaBu, and PaVa

PaPr, PaBu, and PaVa all had good melt spinnability, and Figure 7 shows images of the obtained melt-spun fibers. Continuous melt spinning was achieved at take-up rates of 18, 36, 108, and 180 m/min. According to the optical microscopy investigations, the average diameters of the PaPr melt-spun fibers at take-up rates of 18, 36, 108, and 180 m/min were 144, 142, 96, and 52 μm, respectively. It was also worth noting that the melt spinning of PaPr, PaBu, and PaVa did not require any additives. Particularly in the case of PaPr, which had relatively high glass transition and melting temperatures, the melt spinning of the paramylon esters provides a good reference for further applications.

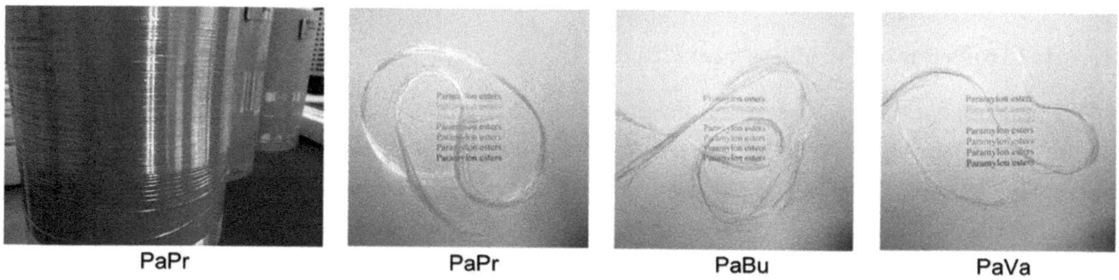

Figure 7. Images of the PaPr, PaBu, and PaVa melt-spun fibers.

Orientation and Crystallinity of PaPr Fibers

Figure 8a shows 2-D fiber diffraction patterns of the PaPr fibers at various take-up rates obtained by WAXD. The fibers manufactured at the lower take-up rates of 18 and 36 m/min produced amorphous patterns. In contrast, the fibers manufactured at the relatively high take-up rates of 108 and 180 m/min clearly show patterns of crystallization and crystal orientation. Considering that higher take-up rates during melt spinning correspond to greater loads applied to the melting resin, shearing occurs more readily among the crystals, which induces crystallization.

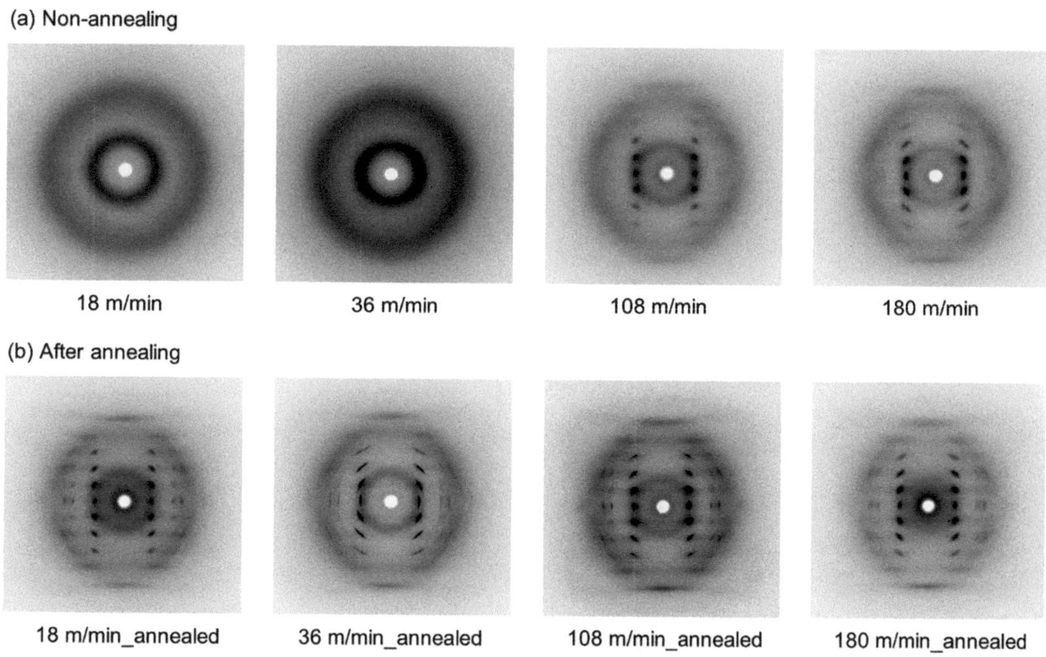

Figure 8. 2-D X-ray fiber diffraction patterns of paper fibers (a) before and (b) after annealing.

Figure 8b shows the 2-D fiber diffraction patterns of the fibers formed at various take-up rates after annealing. The degree of orientation of the (100) crystal plane (the numbers represent the Miller indices) in equatorial line and the degree of crystallization values calculated from the WAXD results are presented in Table 4. The annealed fibers produced at low take-up rates of 18 and 36 m/min suggest the promotion of orientation and crystallization. Moreover, all the (100) crystal planes of the PaPr fibers mentioned in the present study were well oriented, with degree of orientation values above 90%. In addition, the calculated degree of crystallization in each annealed fiber was approximately 50%, which was much higher than that in each non-annealed fiber. Therefore, a higher take-up rate during melt spinning and annealing treatment promotes further crystallization.

Tensile Properties of the Melt-Spun Fibers

The results of the tensile tests on the PaBu and PaVa fibers at a take-up rate of 108 m/min, and on the PaPr fibers at various take-up rates are presented in Table 4. The tensile strength of the PaPr fibers was 138 MPa, which was much higher than the tensile strengths of the PaBu (28 MPa) and PaVa (58 MPa) fibers. Considering that increasing the length of the side-chain reduces the crystallinity of paramylon esters, the differences in the tensile properties of the fibers were expected.

The stress–strain curves of the non-annealed and annealed PaPr fibers manufactured at take-up rates of 18, 36, 108, and 180 m/min are presented in Figure 9a and Figure 9b, respectively. Table 4 summarizes the tensile properties of the fibers. The tensile strengths of the non-annealed PaPr fibers manufactured at take-up rates of 18, 36, 108, and 180 m/min were 106, 112, 138, and 157 MPa, respectively. The Young's moduli of the fibers manufactured at take-up rates of 18, 36, 108, and 180 m/min were 1.12, 1.16, 1.57, and 1.94 GPa, respectively. The highest tensile strengths and Young's moduli corresponded to fibers manufactured at a spinning rate of 180 m/min. Therefore, the tensile strengths and Young's moduli of the fibers tended to increase with increasing take-up rates, whereas the elongation at break decreased. The tensile strengths of the annealed fibers manufactured at take-up rates of 18, 36, 108, and 180 m/min were 59, 103, 151, and 200 MPa, respectively. There was an improvement in tensile strength, especially in the fibers manufactured at high take-up rates of 108 and 180 m/min.

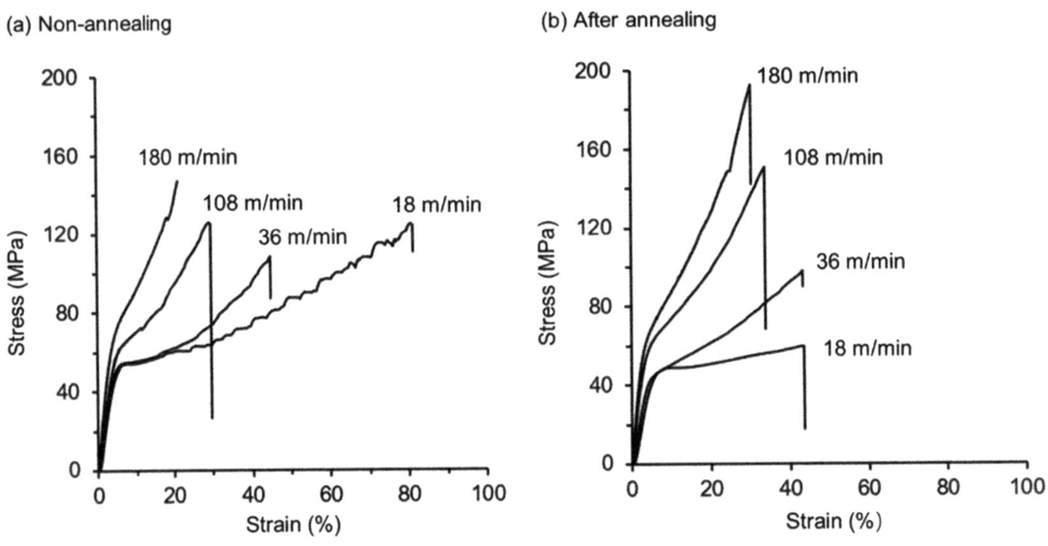

Figure 9. Stress-strain curves of the papr fibers (a) before and (b) after annealing.

Above all, the extent of molecular chain alignment, and the degree of crystallization in the PaPr fibers had major influences on their tensile properties. Firstly, considering that the shear force increased with increasing take-up rate during melt spinning, the molecular chains of the fibers tended to align more. Therefore, fibers with highly aligned molecular chains obtained at higher take-up rates had greater tensile strengths. Secondly, annealing treatment accelerated crystallization, which also improved the tensile strength of the fibers.

Table 4. Tensile and Crystalline Properties of the Melt-Spun Fibers

	Tensile strength (MPa)	*Elongation at break (%)*	*Youngs modulus (GPa)*	*Degree of orientation (%)*	*Degree of crystallization (%)*
PaPr_18 m/min	106 ± 21	100 ± 31	1.12 ± 0.19	–	–
PaPr_36 m/min	112 ± 13	58 ± 25	1.16 ± 0.18	–	–
PaPr_108 m/min	138 ± 13	31 ± 6	1.57 ± 0.21	95	22
PaPr_180 m/min	157 ± 8	34 ± 10	1.94 ± 0.20	92	23
PaPr_18 m/min _annealed	59 ± 11	52 ± 11	0.95 ± 0.11	97	54
PaPr_36 m/min _annealed	103 ± 9	52 ± 8	1.19 ± 0.09	93	48
PaPr_108 m/min _annealed	151 ± 12	35 ± 3	1.53 ± 0.19	97	45
PaPr_180 m/min _annealed	200 ± 14	34 ± 4	1.64 ± 0.30	95	58
PaBu_108 m/min	28 ± 3	54 ± 17	0.48 ± 0.13	–[a]	–[a]
PaVa_108 m/min	58 ± 14	25 ± 6	0.31 ± 0.06	–[a]	–[a]

Conclusions

Paramylon—which is a β-1,3-glucan and one of the most abundant natural resources—has been investigated as a prospective biomass. A series of paramylon esters with various alkyl side-chain lengths was prepared. It was possible to control the thermal properties, mechanical properties, and crystallinities of the paramylon esters over a wide range by adjusting the lengths of the side-chains. In particular, the paramylon esters with short side-chains had excellent thermal stabilities, with melting points above 190 °C and glass transition temperatures above 70 °C, most of which were higher than those of general petroleum-based plastics such as PE, PP, and PET. The crystal structures of PaPr, PaBu, and PaVa were determined by indexing the reflections of their 2-D X-ray fiber diagrams. All three paramylon esters were successfully packed in hexagonal crystal systems. The crystal lattice constants increased with increasing side-chain length, and the density of the crystal lattices decreased. Melt-spun fibers of paramylon esters were successfully manufactured at various take-up rates, without using any additives. By increasing the take-up rate and annealing, the tensile strengths of melt-spun paramylon ester fibers can be improved by the promotion of crystallinity.

Abbreviations

cellulose triacetate (CTA)
differential scanning calorimetry (DSC)
gel permeation chromatography (GPC)
polarized optical microscope (POM)
polyhydroxyalkanoates (PHAs)
polylactic acid (PLA)
polytetrafluoroethylene (PTFE)
tetramethylsilane (TMS)
thermogravimetric analysis (TGA)
trifluoroacetic anhydride (TFAA)
wide-angle X-ray diffraction (WAXD)

Acknowledgments

This work was performed as part of the "Innovative Synthesis of High-Performance Bioplastics from Polysaccharides" project supported by JST ALCA, Japan (grant number JPMJAL1502), and "Demonstration project for plastics resource circulation system for decarbonized society" in Ministry of the Environment. The synchrotron radiation experiments were mostly performed using the BL40B2 beamline at SPring-8 (2019A1213, 2019B1081). Some parts of the synchrotron experiments were performed using the BL03XU beamline at SPring-8 with the approval of the Japan Synchrotron Radiation Research Institute (JASRI) and the Advanced Soft Material Beamline Consortium (FSBL) (2017A1440, 2018A7232, 2019A7234, 2019B1081).

Appendix

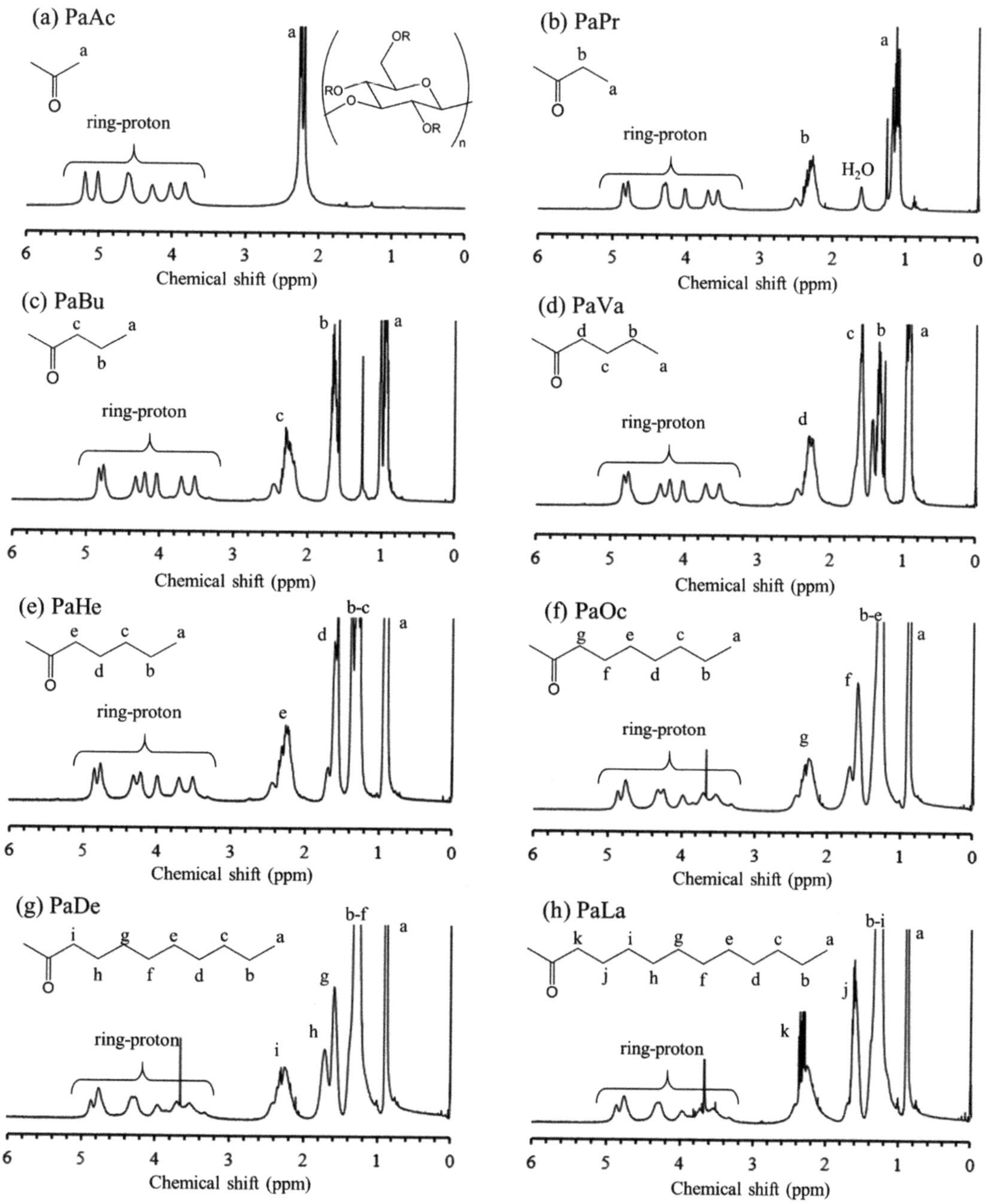

Figure A1. ^{1}H-NMR spectra of paramylon esters. DCDl$_3$ as solvent except for PaAc (TFA as solvent).

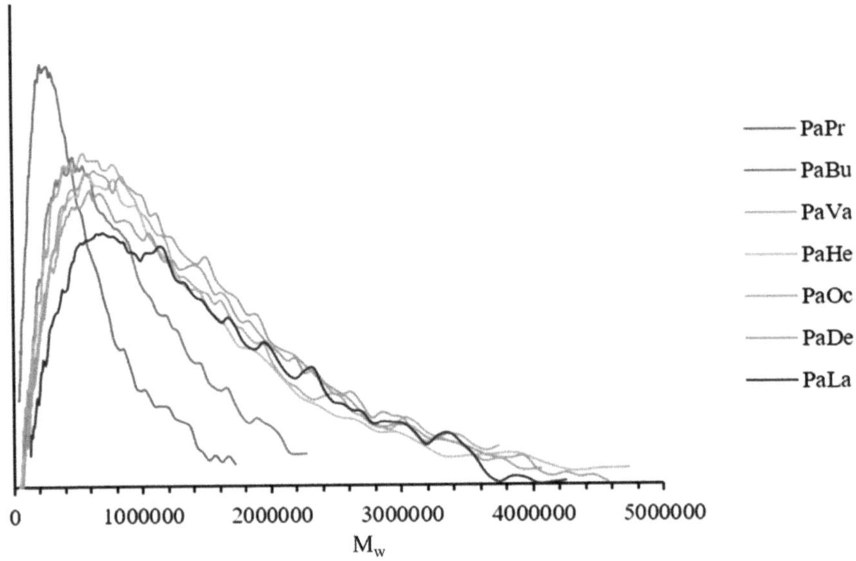

Figure A2. GPC curves from PaPr to PaLa (chloroform as solvent).

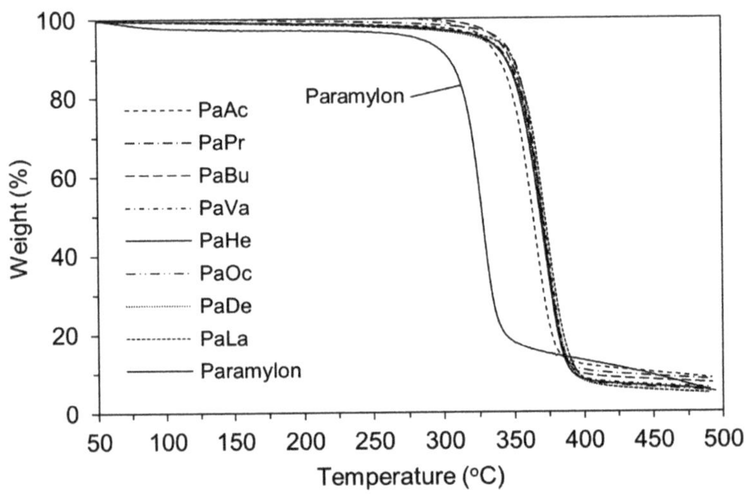

Figure A3. TGA curves of paramylon and paramylon eters.

References

1. Iwata, T. Biodegradable and Bio-Based Polymers: Future and Prospects of Eco-Friendly Plastics. *Angew. Chem., Int. Ed.* **2015**, *54*, 3210–3215.
2. Anders, S.; Stolt, M. Properties of lactic acid based polymers and their correlation with composition. *Prog. Polym. Sci.* **2002**, *27* (6), 1123–1163.
3. Lu, J.; Tappel, R. C.; Nomura, C. T. Mini-Review: Biosynthesis of Poly(hydroxyalkanoates). *Polym. Rev.* **2009**, *49* (3), 226–248.
4. Edgar, K. J.; Buchanan, C. M.; Debenham, J. S.; Rundquist, P. A.; Seiler, B. D.; Shelton, M. C.; Tindall, D. Advances in cellulose ester performance and application. *Prog. Polym. Sci.* **2001**, *26* (9), 1605–1688.

5. Fordyce, C. R.; Meyer, L. W. A. Plasticizers for cellulose acetate and cellulose acetate butyrate. *Ind. Eng. Chem.* **1940**, *32*, 1053–1060.
6. Harada, T.; Misaki, A.; Saito, H. Curdlan: A bacterial gel-forming β-1,3-glucan. *Arch. Biochem. Biophys.* **1968**, *124*, 292–298.
7. Clarke, A. E.; Stone, B. A. Structure of the paramylon from Euglena gracilis. *Biochim. Biophys. Acta.* **1960**, *44* (1), 161–163.
8. Marchessault, R. H.; Deslandes, Y.; Ogawa, K.; Sundararajan, P. R. X-ray diffraction data for β-(1-3)-D-glucan. *Can. J. Chem.* **1977**, *55* (2), 300–303.
9. Takeda, H.; Yasuoka, N.; Kasai, N.; Harada, T. X-ray Structural Studies of (1-3)-β-D-Glucan (Curdlan). *Can. J. Chem.* **1978**, *10*, 365–368.
10. Marchessault, R. H.; Deslandes, Y. Fine Structure of (1-3)-β-D-glucans: Curdlan and Paramylon. *Carbohydr. Res.* **1979**, *75*, 231–242.
11. Deslandes, Y.; Marchessault, R. H.; Sarko, A. Triple-helical structure of (1-3)-β-D-glucan. *Macromolecules* **1980**, *13*, 1466–1471.
12. Marubayashi, H.; Yukinaka, K.; Enomoto-Rogers, Y.; Takemura, A.; Iwata, T. Curdlan Ester Derivatives: Synthesis, Structure, and Properties. *Carbohydr. Polym.* **2014**, *103*, 427–433.
13. Marubayashi, H.; Yukinaka, K.; Enomoto-Rogers, Y.; Hikima, T.; Tanaka, M.; Iwata, T. Crystal polymorphism of curdlan propionate: 6-fold versus 5-fold helices. *ACS Macro Lett.* **2016**, *5*, 607–611.
14. Zhai, W.; Danjo, T.; Iwata, T. Synthesis and Physical properties of Curdlan Branched Esters. *J. Polym. Res.* **2018**, *25*, 181.
15. Zhai, W.; Iwata, T. Synthesis and properties of curdlan branched and linear mixed ester derivatives. *Polym. Dgrad. Stab.* **2019**, *161*, 50–56.
16. Shibakami, M.; Tsubouchi, G.; Hayashi, M. Thermoplasticization of euglenoid beta-1,3-glucans by mixed esterification. *Carbohydr. Polym.* **2014**, *105*, 90–96.
17. Shibakami, M.; Tsubouchi, G.; Sohma, M.; Hayashi, M. One-pot synthesisi of thermoplastic mixed paramylon esters using trifluoroacetic anhydride. *Carbohydr. Polym.* **2015**, *119*, 1–7.
18. Shibakami, M.; Sohma, M. Synthesis and thermal properties of paramylon mixed esters and optical, mechanical, and crystal properties of their hot-pressed films. *Carbohydr. Polym.* **2017**, *155*, 416–424.
19. Gan, H.; Enomoto, Y.; Kabe, T.; Ishii, D.; Hikima, T.; Takata, M.; Iwata, T. Synthesis, Properties and Molecular Conformation of Paramylon Ester Derivatives. *Polym. Degrad. Stab.* **2017**, *145*, 142–149.
20. Gan, H.; Kabe, T.; Kimura, S.; Hikima, T.; Takata, M.; Iwata, T. Crystal structures and crystalline elastic modulus of paramylon esters. *Polymer* **2019**, *172*, 7–12.
21. Gan, H.; Kabe, T.; Iwata, T. Manufacture, Characterization, and Structure Analysis of Melt-Spun Fibers Derived from Paramylon Esters. *J. Fiber Sci. Technol.* **2020**, *76* (5), 151–160.
22. Cochran, W.; Crick, F. H. C.; Vand, V. The structure of synthetic polypeptides I the transform of atoms on a helix. *Acta. Crystallogr.* **1952**, *5*, 581–586.

Chapter 10

Bioplastics from Vegetable Waste: A Versatile Platform for the Fabrication of Polymer Films

Roberto Simonutti,[*,1] Giovanni Perotto,[*,2] Laura Bertolacci,[2] and Athanassia Athanassiou[2]

[1]Dipartimento di Scienza dei Materiali, Università di Milano-Bicocca, Via Roberto Cozzi 55, 20125 Milano, Italy

[2]Smart Materials, Istituto Italiano di Tecnologia (IIT), Via Morego 30, 16163 Genoa, Italy

[*]Email: roberto.simonutti@unimib.it, giovanni.perotto@iit.it

The possibility to use the macromolecules present in vegetable biomass, with minimal processing, as alternative materials to produce new green bioplastics for packaging is very appealing. This strategy can help to mitigate environmental issues due to broad usage of conventional plastics. Since a large variety of vegetable waste, in terms of chemical composition and supramolecular organization, is produced by the food industry, many different bioplastics can be produced with the same protocol. Bioplastics films from carrot pomace, cocoa shells, spinach stems and orange peels will be characterized in terms of thermomechanical properties and biodegradability. The mechanical properties will be correlated with the composition and morphology as determined by means of a combination of techniques (solid-state NMR, SEM and confocal microscopy). The biodegradability of the films will be characterized in terms of Biochemical Oxygen Demand as function of time during 30 days of incubation in seawater.

Introduction

Nowadays thermoplastic polymers derived from oil are a big concern for public opinion, especially in western countries. Two main problems are envisaged: the carbon footprint of plastics manufacturing, that counts for almost the 50% of all the greenhouse gases emitted cradle to grave (*1*) even for products containing recycled material, and the limited, almost negligible, biodegradability of plastics. This aspect is quite relevant for all those applications that require single use products, like packaging and disposable items (e.g. tableware). In fact a significant portion of these objects, once used, end up in the environment and will eventually find their way in the seas, where fragmentation and mechanical degradation result in the presence of large quantities of microplastics and nanoplastics causing environmental alarms (2). In EU it is estimated that 40% of the plastic waste is produced by packaging (3). The 2016 report of the Ellen McArthur foundation estimated that less than 10% of plastic is recycled, while the majority of plastic waste (40%) is landfilled and around 32% is not collected and ends up dispersed in the environment generating plastic pollution (4).

With an estimated biodegradation time measured in hundreds of years, there is a large mismatch between the lifetime of the material and its actual use, that is usually within 48 months (5). On top of this, the shelf life of the food that is packaged is measured in the order of weeks. As consequence of that, polyolefin-based thermoplastics shouldn't be considered as optimal candidates for food packaging. On the other hand, food safety is relevant, especially after the pandemic of COVID-19, therefore food packaging and disposable items will be even more vital in the future. All these factors create a strong demand for materials with performances (mechanical, thermal, functional) similar to polyolefins, that are cheap, easily processed but with a much easier and safer biodegradation profile so that they could treated as organic waste and safely biodegraded within few weeks. Additionally, it is well known that the food industry generates a large amount of waste in all the phases of the supply chain and its disposal poses environmental and economic issues. Large part of this waste is constituted by fruits, vegetables, roots and tubers that do not meet the standards required by the large distribution, most of the times just for aesthetic reasons. In other cases, this waste is constituted by nonedible components like shells, kernels, hard leaves, stems, peels, hairs and feathers. Even considering a well organize and sustainable food chain, most of this waste is unavoidable and technologies for its valorization are needed. Considering the residual nutritional value and biodegradability of this waste, an obvious valorization strategy is its use as animal feed or as natural fertilizer. These options are always possible, but the added value, especially in the use as fertilizer, is extremely limited and safety issues might further reduce the attractiveness of these opportunities. From a materials' point of view, none of these strategies exploit the organization and intrinsic properties of the molecules comprising the food waste. In fact, vegetables are constituted by polymers organized in the space like polysaccharides (principally cellulose and hemicelluloses, but also pectins), proteins, aliphatic polyesters and lignin.

Pure cellulose and cellulose based materials like paper or cardboard are probably the most used polymeric materials, since the relevant mechanical properties of cellulose crystals. Besides them, also lignin and aliphatic polyesters can find application in the development of bio-based materials. The former, which plays a relevant structural role in wood, is used as filler in synthetic polymers, while the latter, can impart flexibility and significant elongations to blends, thanks to their low glass transition temperature. Nowadays, proteins are still little explored. Food waste provides almost infinite combinations of these elements that, in order to be harnessed, require technologies able to transform in a sustainable manner orange peels, parsley stem, cocoa shells, to name a few examples, into materials that can be easily processed as alternatives to plastics films or bags or food containers. Plastic materials obtained by direct conversion of vegetable biomass can be considered as a new concept of bioplastics.

For the last 20 years the term bioplastics was used to describe thermoplastic polymers obtained by the chemical polymerization of bio-sourced monomers (e.g. lactide) or natural occurring polymers obtained by fermentation such as polyhydroxyalcanoates. Starch based bioplastics were also commercialized since the '90s, taking advantage of the gelatinization of starch and its transformation in thermoplastic starch (6–8). These materials have the possibility to be reintegrated in the natural cycle that oil-based polymers lack and are also defined as biodegradable bioplastic. Bioplastics that are sourced from edible parts of vegetables (e.g. starch or sugars used to produce lactate and PHAs) pose ethical concerns, because the primary use of those resources should remain in the food value chain. In the quest to improve the sustainability of our plastic economy, the use of vegetable waste as a source of raw materials promises possibility to replace fossil oil with a renewable and more sustainable alternative with an improved biodegradability. Several materials based on cellulose, lignin or proteins were developed in a laboratory-scale showing promising results (*6–9*). Recently our groups developed an environmentally friendly, water-based method to directly convert waste and byproducts of the food industry into bioplastic films with interesting mechanical and functional properties suitable for packaging (*10–12*).

Experimental Section

Materials

Orange peel, cocoa shells, carrot pomace, spinach stems were obtained by local food processing companies as the byproduct or waste remaining from their processing. Trifluoroacetic acid (TFA), HCl (37% solution), formic acid were purchased from Sigma Aldrich. MilliQ grade water was used at all the times.

Processing

Typically, dried and micronized vegetable powders were used as received. In the case of spinach, the received dried stems were ground and then sieved with a 50 micron sieve. No further purification of the vegetable biomass was performed.

The vegetable powder was processed with one of the following methods, previously described in detail (*13–15*) and briefly summarized here:

i) TFA: vegetables were dispersed in anhydrous TFA and gently stirred at room temperature until complete dissolution of the powder. The time of dissolution varied for the different vegetables from 48 hours to two weeks.
ii) HCl: vegetable powder was dispersed in a 2 M water solution of HCl and vigorously stirred overnight at 40 °C.
iii) Formic Acid: vegetable powder was dispersed in a 2M water solution of formic acid and vigorously stirred overnight.

Bioplastic films could be obtained by casting of the vegetable solutions on a petri dish mold and allow for evaporation of the solvent.

Morphology

Scanning electron microscopy was used to image the surface of the bioplastic films. The samples were cut into pieces, fixed with carbon tape to a metallic stub and sputter coated with a 10 nm layer of gold before imaging. The samples were imaged with a JEOL JSM-6490LA Scanning Electron Microscope using a secondary electron detector and an acceleration potential of 5 kV.

Mechanical Properties

Mechanical properties of the films were measured by uniaxial tensile tests on a dual column Instron 3365 universal testing machine. Dog-bone shaped (Dumbbell) samples (width of 4 mm, useful length of 25 mm) of the bioplastic were stretched at a rate of 1 mm/min. At least three measurements were conducted for each sample and the results were averaged to obtain a mean value. The Young's modulus E, ultimate strength UTS and elongation at break values were evaluated from the stress-strain curves.

Biodegradation

Biodegradability of the samples was assessed by means of Biochemical Oxygen Demand (BOD), which can be easily determined by monitoring the oxygen consumption in a closed respirometer. In details, about 20 mg of sample were added to 432 ml of seawater as the single carbon source. The seawater was chosen to mimic real environmental conditions. Water already contains microbial consortia and the saline nutrients needed for their growth.

The experiment was conducted at room temperature inside dark glass bottles with a volume of 510 ml, hermetically closed with the OxiTop® measuring head. A CO_2 scavenger was added to sequestrate carbon dioxide produced during the biodegradation, and biotic consumption of the oxygen present in the free volume of the system was measured as a function of the decrease in pressure.

Samples were tested in triplicate. Raw data of oxygen consumption (mg O_2/L) were corrected subtracting the mean values of the blanks, obtained by measuring the oxygen consumption of the seawater in absence of any test material. After this subtraction, values were normalized on the mass of the individual samples and referred to 100 mg of material (mg O_2/100mg material). Finally, mean values of the triplicates were calculated and plotted vs time.

Theoretical Oxygen Demand (ThOD) was calculated starting from the elemental composition of the materials, according to ISO 14851:1999.

Solid-State NMR

^{13}C solid-state NMR spectra were obtained with an Avance 300 spectrometer (Bruker), using magic angle spinning (MAS) (spinning rate 10 kHz). Cross-Polarization (CP) (*16*) ^{13}C MAS NMR spectra exploit transfer of polarization under Hartmann-Hann conditions obtained by ramped amplitude variation of spin locking field (RAMP) (*17*) and two-pulse phase-modulated decoupling (TPPM) (*18*); the contact time of 1 ms and a recycling delay of 4 s were used, typically 2000 scans were acquired. Single pulse experiment (SPE) ^{13}C MAS NMR spectra were acquired with a recycle delay of 2 s and TPPM decoupling, typically with 1000 scans (*19*).

Results and Discussion

Preparation Methods

The last few years our groups have developed several methods for the preparation of a new generation of bioplastics films by direct conversion of several vegetables' residues with different chemical composition. All the methods are based on the partial dissolution of vegetables' molecules and all provide homogenous free-standing films upon casting into plastic dish molds and drying. As first step, wastes are chopped and pulverized, afterwards they are treated in acid conditions. These processes are also tested on vegetable wastes that were received as dried powder or pieces, from various food processing companies. Some of these wastes are the dried and powdered byproduct of the food processing (i.e. pomace obtained after extraction of nutrients or colorants; bran of grains), while some are chunks of the nonedible portions of vegetables (stems, leaves). This process is applied (data not shown here) also to fruit and vegetables that were noncommercialized because of aesthetic criteria or because they are too old (but not spoiled).

The first process to convert dried and micronized vegetable biomass, relies on mixing the powder in anhydrous TFA (3 wt% of vegetables). Vegetables are left for several days until visibly dissolved.

Each plant residue necessitates different time in TFA to completely dissolve and be suitable for solvent-casting. For instance, parsley stem powder can form a film from solution after 3 days in TFA, whereas spinach stems require up to 2 weeks in TFA. The difference can be mainly ascribed to the physiology of the plant tissue, moisture content and accessibility of bound cellulose by TFA. Other macromolecules such as protein, lignin, hemicellulose, pectin and waxes are also present with cellulose in the cell walls of plants. For instance, the presence of various types of fatty acids and triglycerides (cocoa butter) and of a substantial amount of pectin in cocoa shells significantly slows down the interaction with TFA.

Table 1. Main Characteristics of the Processes and Vegetables Wastes Processed

Method	*Pro*	*Cons*	*Applied*
TFA	Provides more flexible films Works also in presence of lignin	TFA is toxic, TFA residues in the material, It needs anhydrous conditions	Cocoa Shells Orange Peels Parsley stems Spinach leaves Cauliflower
HCl	Water based	Needs dialysis, Doesnt work with high lignin content	Carrot Pomace Parsley stems Cauliflower stems and leaves, Orange peel Radicchio stems and leaves Cocoa shells
Formic acid	Water based No residues Fast Higher yield	Higher content of Low MW components. More brittle because of lower degree of hydrolysis. Doesn't work with lignin	Carrot Pomace Parsley stems, Cauliflower stems and leaves, Orange peel, Radicchio stems and leaves, Cocoa shells

The second approach exploits a water solution of a strong acid, HCl, or HNO_3. In this acid environment water soluble polysaccharides like pectin can be easily dissolved and the amorphous part of cellulose fibers can be partially hydrolyzed. Films can be obtained by simple casting, still the introduction of an intermediate dialysis step allows the elimination of the acid residues in the material while retaining the macromolecules. Even though all the vegetable waste can be dispersed in solution during the HCl step, between 10% and 15% of the mass gets lost during the dialysis step. Overall, between 85% and 90% of the vegetable waste is finally converted in bioplastic.

The third process, based on a water solution of formic acid, allows the processing in the same range of vegetable concentration, time and temperature used for the HCl treatment and the bioplastic films showed a similar morphology with the ones obtained in our previous work. This method has an important advantage over the previous one: no dialysis step is necessary and no formic acid residues are found in the bioplastic after drying, eliminating the need for a dialysis step. This greatly simplify the process and significantly reduces water consumption. However, larger quantity of low MW components remains in the films, possibly leaching out during its use as packaging. The characteristics of the processes are summarized in Table 1.

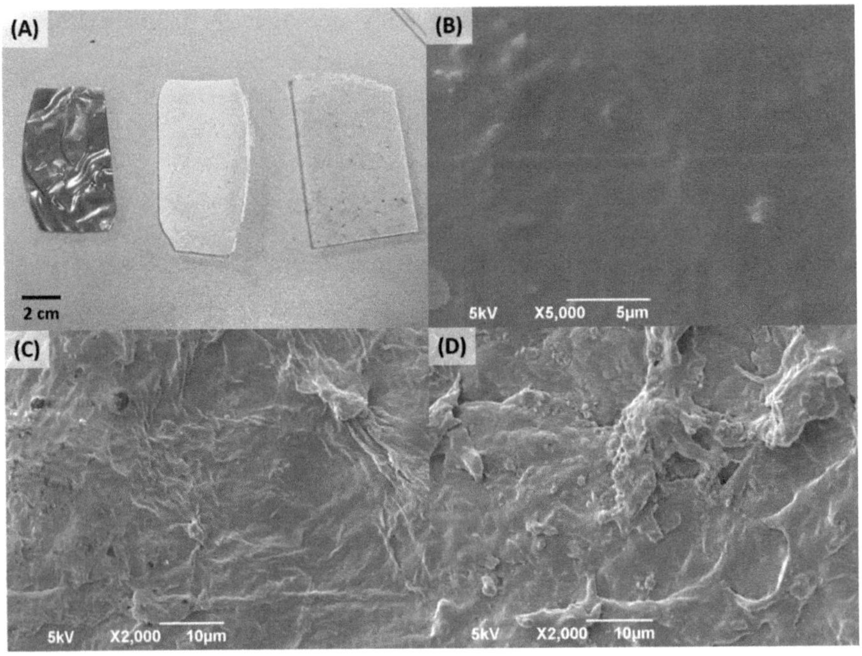

Figure 1. Bioplastic film made with the three processes described starting from carrot pomace. (A) picture of the bioplastic film made using TFA, HCl, and formic acid, respectively (starting from the left); (B) SEM micrograph of the carrot bioplastic synthesized with TFA; (C) SEM micrograph of the carrot bioplastic synthesized with HCl; (D) SEM micrograph of the carrot bioplastic synthesized with formic acid.

The final color, appearance and the microstructure of the bioplastics obtained, greatly depends on the processed used, as it is visible from Figure 1. Carrot bioplastic films made with TFA are more transparent and their microstructure, as showed by SEM, is very smooth. Films obtained with HCl were opaque because of the presence of cellulose crystals that causes scattering. The carrot bioplastic obtained with formic acid showed a structure similar to that of the HCl films, with some roughness and the presence of crystalline inclusions as shown by confocal microscopy (figures reported in ref (*14*) and (*15*)).

Table 2. Mechanical Properties of Bioplastic Films Obtained with Different Vegetable Biomass

	Youngs Modulus (MPa)	*UTS (MPa)*	*Elongation at Break (%)*
Cocoa shell	2400	12	1
Orange peel	1500	21	4
Spinach stem	1400	14	3.5
Carrot – TFA	300	13	17
Carrot – HCL	1300	38	6
Carrot – formic	3000	35	2.5

Bioplastic films showed mechanical properties that were dependent on the vegetable used but also on the different process that was applied for the conversion from powder to bioplastic (Table 2).

When vegetables with a high content of crystalline cellulose are used, such as carrot pomace, the resulting material are more rigid, with a higher elastic modulus and a lower elongation at break. The presence of other insoluble components such as lignin in the cocoa shell, made the material even more rigid and brittle.

The different chemicals used in the conversion process is another parameter that can be used to change the final properties of the materials: TFA has the ability to dissolve cellulose crystals, creating a final material in which all the polysaccharides are amorphous and thus softer and more stretchable. On the other side of the spectrum, formic acid dissolves only some of the polysaccharides (pectin and hemicellulose) creating a material that is more rigid. HCl can partially hydrolyze the cellulose, increasing the amount of low molecular weight and amorphous components and creating a somewhat plastic and stretchable material.

Another parameter that can be used to change the bioplastics' properties is the amount of moisture absorbed by the material. This was recently exploited to facilitate processing and allows fabrication of bioplastic films to produce complex objects (*15*).

Solid-State NMR Characterization

As reported above, a large part of edible vegetables, if not all, are constituted by the following systems: cellulose, hemicellulose, pectin, lignin, waxes and triglycerides (Figure 2).

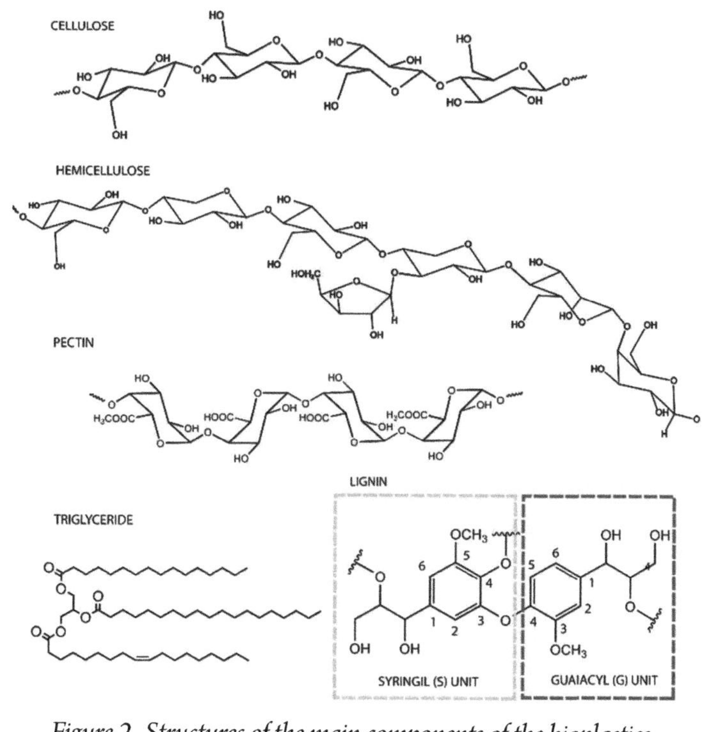

Figure 2. Structures of the main components of the bioplastics.

Also, starch can be important, but in the cases we studied starch was not present. Other components like flavonoids, antioxidants, vitamins, minerals are partially removed by the process of bioplastic formation and thus if present are minor components the bioplastics. They are relevant for the functional properties that they can impart to bioplastics (*14*, *15*, *20*) but not for the film formation and structural properties, so they are not considered in this context. The different balance among the polymeric components is fundamental in determining the mechanical properties of the bioplastics. Waxes, triglycerides and low molecular weight polysaccharides can behave as plasticizers, especially in the presence of moisture. Thus, it is extremely relevant to evaluate the composition of the bioplastic and that of the original vegetable waste. In fact, the developed processes allow to easily mix the vegetable "solutions" leading to a bioplastic with the intermediate desired final properties. Since some of the vegetable's components are hardly soluble or not soluble at all, ^{13}C MAS NMR is the preferred technique of characterization, providing either the identification of the chemical species present in the bioplastic or vegetable and the determination of the presence of crystalline o pseudo-crystalline structures.

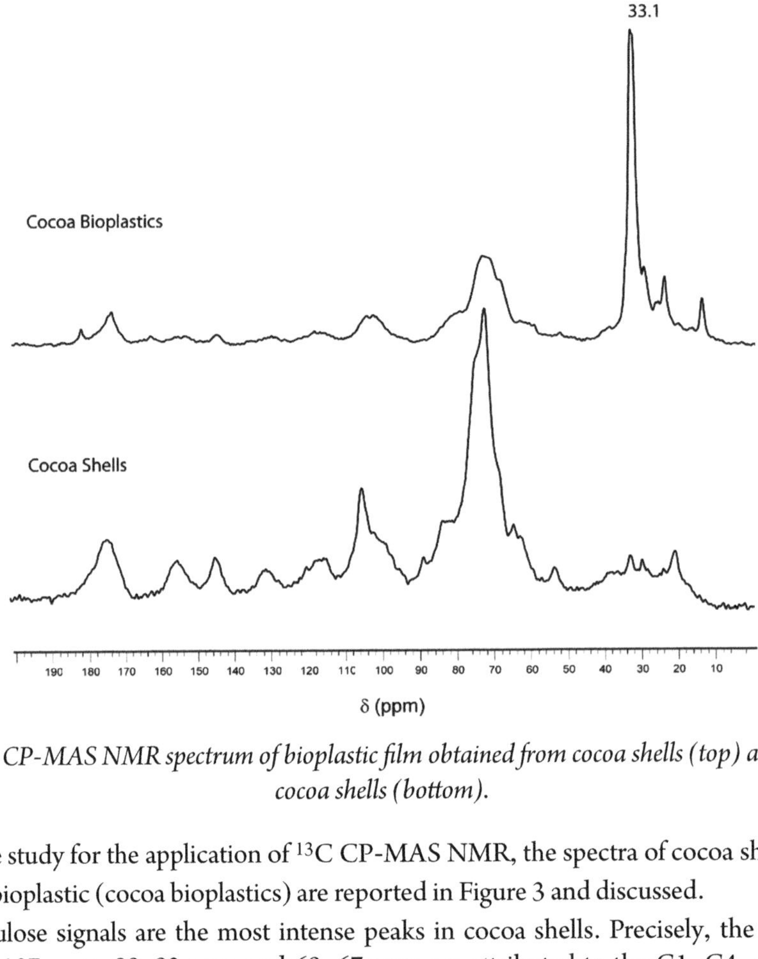

Figure 3. ^{13}C CP-MAS NMR spectrum of bioplastic film obtained from cocoa shells (top) and spectrum of cocoa shells (bottom).

As a case study for the application of ^{13}C CP-MAS NMR, the spectra of cocoa shells and cocoa shells based bioplastic (cocoa bioplastics) are reported in Figure 3 and discussed.

The cellulose signals are the most intense peaks in cocoa shells. Precisely, the chemical shift regions 102–107 ppm, 80–92 ppm and 60–67 ppm are attributed to the C1, C4, and C6 signals respectively of the glucose unit of the cellulose (Table 3) (*21*). The C4 signal in the carbon spectrum of microcrystalline cellulose is the most reliable resonance used to determine the degree of crystallinity. In fact, C4 cellulose amorphous carbons give rise to a fairly broad signal ranging from 80–85 ppm, while C4 cellulose crystalline carbons generate a sharper resonance in the range 85–92 ppm. In the case of cocoa shells a visible signal at 88.4 ppm is present indicating that cellulose is partially crystalline.

Table 3. ^{13}C Chemical Shifts Assignments

	Cocoa Shells		Cocoa Bioplastics	
	C.S. (ppm)	Intensity	C.S. (ppm)	Intensity
O-**C**OCH$_3$ R-**C**OO-R′ Pectin COO$^-$			181.0	0.05
Pectin COOH Pectin **C**OOCH$_3$	173.7	0.46	172.8	0.18
Lignin	154.3 144.2 130.1 116.0	0.17 0.11 0.08 0.20	162.1 155.5 152.4 144.9 131.1 117.7	0.02 0.05 0.01 0.03 0.03 0.12
C_1 Cellulose	104.8	0.42		
C_1 Hemicellulose (xylan) C_1 Pectin	101.8 98.4	0.13 0.10	103.7 101.4	0.98
C_4 Crystalline cellulose	88.4	0.06	88.6	0.13
C_4 Amorphous cellulose C_4 Pectin	82.3	0.23	83.4	0.74
$C_{2,3,5}$ Cellulose $C_{2,3,5}$ Hemicellulose (xyloglucan) $C_{2,3,4}$ hemicellulose (xylan)	71.9	1	80.9 73.2	0.12 0.51
$C_{2,3,5}$ pectin	69.2	0.28	69.2	0.41
C_6 Cellulose C_6 Hemicellulose (xyloglucan)	64.5	0.04		0.83
C_5 Hemicellulose (xilose)	62.7	0.09	61.6	0.08
COO**C**H$_3$ pectin	53.4	0.05	53.6	0.02
CH$_2$ fatty acids	37.9 33.1 29.9 25.2	0.15 0.09 0.05 0.1	32.9 29.6 25.9	1.00 0.10 0.07
O-CO**C**H$_3$			24.08	0.1
CH$_3$	21.1	0.16	16.6 14.0	0.05 0.07

A significant peak centered at 173.8 ppm is present; this resonance is typically associated with carboxylic carbons, either as free acid or as esterified group. In the range from 10 ppm to 50 ppm, due to aliphatic carbons, several peaks are detectable, however their overall intensity is significantly smaller compared to that of the carboxylic signal. This means that aliphatic esters (triglycerides) and polyesters are present and that there is also a relevant amount of pectin. In fact, the resonance at 175.8 ppm is associated with the non-esterified carboxylic group of galacturonan acid that appears as a shoulder of the main resonance. Close to the C1 signal of cellulose, at 101.8 ppm, C1 signal of hemicellulose is present. The other signals of hemicellulose are in part overlapped with cellulose signals and can be found as shoulders at lower chemical shift with respect to C6 signal of cellulose (*22–24*). However also the C1 of galacturonic acid, component of the pectin, resonates around 98 ppm (*25*).

Finally, the region between 125 and 160 ppm, is specific of aromatic carbons and carbons on double bonds. Lignin is characterized by a quite complex tridimensional structure, but it is well accepted that the main building block are syringil (S) and guaiacyl (G) units (Figure 2) (*26*). The signal at 154 ppm is assigned to C-3 and C-5 of S units that are etherified at C-4. The signal at 144 ppm is also assigned to C-3 and C-5 of S units, but for those with free phenolic groups at C-4. Additionally, the signal at 144 ppm is assigned to C-3 and C-4 of G units. The signal at 130 ppm is assigned to C-1 and C-4 of S and G units that are etherified at C-4. The protonated carbons of syringyl (C-2 and C-6) units and guaiacyl units (C-2, C-5 and C-6) resonate around 116 ppm.

At high fields, between 40 and 15 ppm we can find the resonances of aliphatic carbons nonbonded to heteroatoms, this is basically the region of waxes and fatty acids. Interestingly a significant signal is clearly detectable at 33.1 ppm, a value usually attributed to inner methylenes in all trans conformation.

This conformation is typical of alkanes crystallized in orthorhombic crystals (*27*), or crystallized fatty acids (*28*) or amines (*29*) Signal generation in CP-MAS experiments is based on polarization transfer from proton nuclei, a process that is more efficient in the presence of strong dipolar couplings typical of rigid phases.

Instead by using the SPE MAS sequence and short recycle it is possible to highlight the presence of mobile phases. In the case of cocoa shells, the most intense signals in the SPE spectrum, recorded with a recycle delay of 2 s (Figure 4), are those of the aliphatic chains, clearly indicating that the only mobile part is associated with waxes and fatty acids, but more interestingly the chemical shift of the inner methylenes resonates at 29.75 ppm: mobile chains presents a significant amount of gauche conformation typical of liquid alkanes. Thus, the waxes and fatty acids appearing in the CP-MAS spectrum belong to a well-defined crystalline phase, but do not represent all the triglycerides present in the system, also a liquid-like oily phase is present.

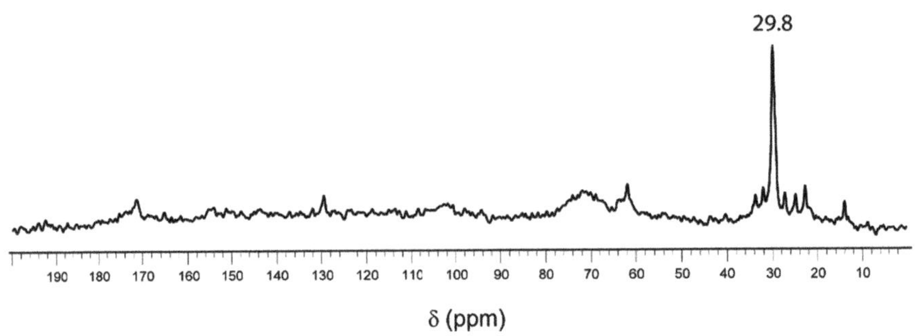

Figure 4. ^{13}C SPE MAS spectrum of the cocoa shells with short recycle delay (2s).

The ^{13}C CP-MAS spectrum of the cocoa bioplastics basically presents the same resonances of the cocoa shells and in Table 3 it is possible to appreciate the small variations. Conversely, the relative intensities of the peaks are dramatically different. In fact, in the cocoa shells the signals of cellulose dominate, instead for the cocoa bioplastics the fatty components are by far the more intense ones. This remarkable difference can be attributed to two main reasons: 1) the TFA treatment leads to a reduction of the degree of crystallinity of the cellulose and a consequence the cross-polarization efficiency decrease drastically, this phenomenon is confirmed by the relevant decrease of the Young modulus (Table 2); 2) the processing of the shells causes a larger crystallization of the fatty compounds that now are more easily detected.

The amorphization of the cellulose is supported by the strong decrease of the peak at 104.8 ppm due to crystalline anomeric carbon. This is a typical effect of the use of TFA (*14, 15*), already seen by ^{13}C CP-MAS NMR in the case of parsley and spinach bioplastics (*30*).

Biodegradation of Bioplastics

All the samples studied show a high level of biodegradability, as shown in Figure 5. The curves describing the oxygen consumption for the films and the cocoa shell powder represent typical bacterial growth curves: an initial linear behavior for the first 4-5 days is followed by a progressive decrease in the slope and, after one month, the BOD is approaching the plateau, achieving values between 73.0 and 79.5 mg O2/100mg material. The behavior of the two films seems superimposable, while biodegradability of the cocoa shell powder is slightly lower. Biodegradation of the reference material (microcrystalline cellulose), is slower and remains almost constant during the whole analysis showing a value of about 79.1 mg O_2/100mg material after 30 days of test in our experimental conditions.

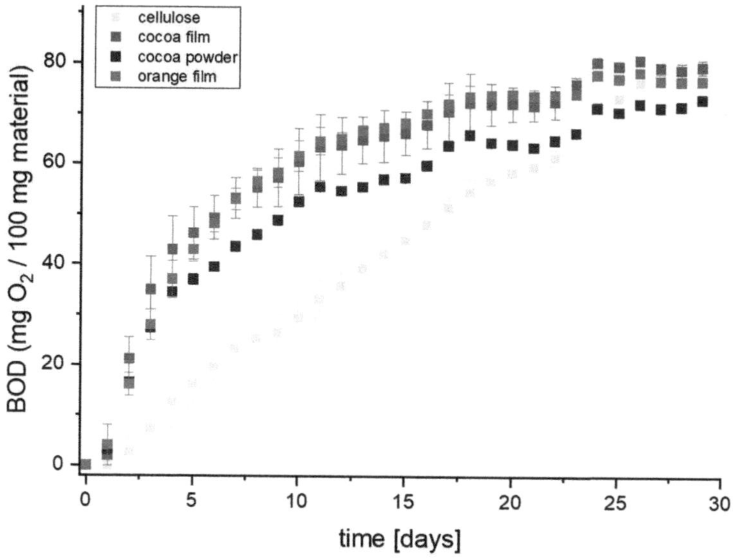

Figure 5. BOD_{30} [mg O_2/100mg material] measured for the bioplastic films derived from cocoa shell and orange peel for 3 weeks incubation in seawater. Comparison with cocoa shell powder. Microcrystalline cellulose was used as a reference biodegradable material.

It is noteworthy that, whereas the curves of BOD for the films and the cocoa shell powder are approaching the plateau, the slope of the curve describing cellulose BOD remains almost constant.

This indicates that during the first five days the films and the cocoa powder biodegrade with a faster rate than cellulose, during the time window between 5 and 10 days, the rates are comparable and, finally, the biodegradation rate of the films and cocoa shell powder becomes lower than the cellulose one. The relatively slow initial kinetics of cellulose biodegradation can be attributed to its crystallinity, a feature well-known to decrease biodegradation rate.

Biodegradation is finally defined by referring the experimental BOD to the ThOD and is expressed as a percentage, according to the official method ISO14851:1999 (*31*). The ThOD is the theoretical total amount of oxygen required to oxidize a chemical compound completely; it is calculated from the molecular formula and can be expressed as mg of oxygen required per 100 mg of test compound. As showed in Table 4, the films are characterized by a high level of biodegradability, more than 50% of the ThOD is consumed after only 30 days.

Table 4. Oxygen Consumption after 30 Days (BOD_{30}) and Percentage of Biodegradation

Sample	*Cocoa Film*	*Cocoa Shell Powder*	*Orange film*	*Microcrystalline Cellulose*
BOD_{21} [mg O_2/100 mg material]	79.5	73.0	76.7	79.1
Percentage of biodegradation	67%	62%	65%	67%

Conclusions

In this contribution we have summarized a new approach in the transformation of food waste in bioplastic films. By means of simple treatments with diluted inorganic or organic acids, dried and pulverized vegetable waste can be transformed in a coherent film with good and tunable mechanical properties, interesting optical properties (transparency, color) and safe biodegradation profile. This processing allow to retain the main structural components the vegetable (cellulose, hemicellulose, pectin, lignin and waxes) intact and exploit their molecular structure to provide the final properties to the bioplastic materials. Correlating the properties with composition obtained by solid-state NMR, it is possible to design new bioplastics with desired performance by mixing different precursors. Furthermore, the bioplastics are more quickly biodegradable than pure crystalline cellulose. Overall, these properties make bioplastics films from vegetable waste interesting for food packaging, especially considering that the films can be processed in the same plant where food is treated and packed. It can be envisaged that, in a near future, food factories could be self-sufficient in terms of packaging materials recycling their own byproducts to produce packaging that have a biodegradation time more fitting with its actual use.

Acknowledgments

This work has been supported by Fondazione Cariplo, grant n° 2018-1005

References

1. Dormer, A.; Finn, D. P.; Ward, P.; Cullen, J. Carbon footprint analysis in plastics manufacturing. *J. Clean Prod.* **2013**, *51*, 133–141.
2. Magri, D.; Sanchez-Moreno, P.; Caputo, G.; Gatto, F.; Veronesi, M.; Bardi, G.; Catelani, T.; Guarnieri, D.; Athanassiou, A.; Pompa, P. P.; Fragouli, D. Laser Ablation as a Versatile Tool

To Mimic Polyethylene Terephthalate Nanoplastic Pollutants: Characterization and Toxicology Assessment. *ACS Nano* **2018**, *12*, 7690–7700.

3. Eurostat; Packaging; Waste; Statistics, http://ec.europa.eu/eurostat/statistics-explained/index.php/Packaging_waste_statistics. 2017.
4. W.E.F. Ellen MacArthur Foundation, McKinsey and Company. *The New Plastics Economy: Rethinking the Future of Plastics*; 2016.
5. Geyer, R.; Jambeck, J. R.; Law, K. L. Production, use, and fate of all plastics ever made. *Sci. Adv.* **2017**, *3*, e1700782.
6. Doane, W. M. USDA Research on Starch-Based Biodegradable Plastics. *Starch - Stärke* **1992**, *44*, 293–295.
7. Shogren, R. L.; Fanta, G. F.; Doane, W. M. Development of Starch Based Plastics - A Reexamination of Selected Polymer Systems in Historical Perspective. *Starch - Stärke* **1993**, *45*, 276–280.
8. Jiménez, A.; Fabra, M. J.; Talens, P.; Chiralt, A. Edible and Biodegradable Starch Films: A Review. *Food Bioprocess Technol.* **2012**, *5*, 2058–2076.
9. Lenz, R. W.; Marchessault, R. H. Bacterial polyesters: Biosynthesis, biodegradable plastics and biotechnology. *Biomacromolecules* **2005**, *6*, 1–8.
10. Song, J. H.; Murphy, R. J.; Narayan, R.; Davies, G. B. H. Biodegradable and compostable alternatives to conventional plastics. *Philos. Trans. R. Soc. B-Biol. Sci.* **2009**, *364*, 2127–2139.
11. Snell, K. D.; Peoples, O. P. PHA bioplastic: A value-added coproduct for biomass biorefineries. *Biofuels Bioprod. Biorefin.* **2009**, 3 (4), 456–467.
12. Cataldi, P.; Condurache, O.; Spirito, D.; Krahne, R.; Bayer, I. S.; Athanassiou, A.; Perotto, G. Keratin-Graphene Nanocomposite: Transformation of Waste Wool in Electronic Devices. *ACS Sustainable Chem. Eng.* **2019**, *7*, 12544–12551.
13. Bayer, I. S.; Guzman-Puyol, S.; Heredia-Guerrero, J. A.; Ceseracciu, L.; Pignatelli, F.; Ruffilli, R.; Cingolani, R.; Athanassiou, A. Direct Transformation of Edible Vegetable Waste into Bioplastics. *Macromolecules* **2014**, *47*, 5135–5143.
14. Perotto, G.; Ceseracciu, L.; Simonutti, R.; Paul, U. C.; Guzman-Puyol, S.; Tran, T.-N.; Bayer, I. S.; Athanassiou, A. Bioplastics from vegetable waste via an eco-friendly water-based process. *Green Chem.* **2018**, *20*, 894–902.
15. Perotto, G.; Simonutti, R.; Ceseracciu, L.; Mauri, M.; Besghini, D.; Athanassiou, A. Water-induced plasticization in vegetable-based bioplastic films: A structural and thermo-mechanical study. *Polymer* **2020**, *200*, 122598.
16. Laws, D. D.; Bitter, H. M. L.; Jerschow, A. Solid-state NMR spectroscopic methods in chemistry. *Angew. Chem., Int. Ed.* **2002**, *41*, 3096–3129.
17. Metz, G.; Wu, X. L.; Smith, S. O. Ramped-Amplitude Cross-Polarization in Magic Angle Spinning NMR. *J. Magn. Reson. Ser. A* **1994**, *110*, 219–227.
18. Ernst, M.; Zimmermann, H.; Meier, B. H. A simple model for heteronuclear spin decoupling in solid-state NMR. *Chem. Phys. Lett.* **2000**, *317*, 581–588.
19. Bonetti, S.; Farina, M.; Mauri, M.; Koynov, K.; Butt, H. J.; Kappl, M.; Simonutti, R. Core@shell Poly(n-butylacrylate)@polystyrene Nanoparticles: Baroplastic Force-Responsiveness in Presence of Strong Phase Separation. *Macromol. Rapid Commun.* **2016**, *37*, 584–589.

20. Tran, T. N.; Athanassiou, A.; Basit, A.; Bayer, I. S. Starch-based bio-elastomers functionalized with red beetroot natural antioxidant. *Food Chem.* **2017**, *216*, 324–333.
21. Foston, M. Advances in solid-state NMR of cellulose. *Curr. Opin. Biotechnol.* **2014**, *27*, 176–184.
22. Wikberg, H.; Maunu, S. L. Characterisation of thermally modified hard- and softwoods by C-13 CPMAS NMR. *Carbohydr. Polym.* **2004**, *58*, 461–466.
23. Simmons, T. J.; Mortimer, J. C.; Bernardinelli, O. D.; Poppler, A. C.; Brown, S. P.; Deazevedo, E. R.; Dupree, R.; Dupree, P. Folding of xylan onto cellulose fibrils in plant cell walls revealed by solid-state NMR. *Nat. Commun.* **2016**, *7*, 13902.
24. Komatsu, T.; Kikuchi, J. Comprehensive Signal Assignment of C-13-Labeled Lignocellulose Using Multidimensional Solution NMR and C-13 Chemical Shift Comparison with Solid-State NMR. *Anal. Chem.* **2013**, *85*, 8857–8865.
25. Sinitsya, A.; Copikova, J.; Pavlikova, H. C-13 CP/MAS NMR spectroscopy in the analysis of pectins. *J. Carbohydr. Chem.* **1998**, *17*, 279–292.
26. Keeler, C.; Maciel, G. E. (13)C NMR spectral editing of humic material. *J. Mol. Struct.* **2000**, *550*, 297–305.
27. Ishikawa, S.; Kurosu, H.; Ando, I. Structural Studies of N-alkanes by variable-temperature solid-state high resolution C-13 NMR-spectroscopy. *J. Mol. Struct.* **1991**, *248*, 361–372.
28. Kameda, T.; Tamada, Y. Variable-temperature C-13 solid-state NMR study of the molecular structure of honeybee wax and silk. *Int. J. Biol. Macromol.* **2009**, *44*, 64–69.
29. Simonutti, R.; Comotti, A.; Bracco, S.; Sozzani, P. Surfactant organization in MCM-41 mesoporous materials as studied by C-13 and Si-29 solid-state NMR. *Chem. Mater.* **2001**, *13*, 771–777.
30. Simonutti, R.; Athanassiou, A. Unpublished results.
31. *ISO 14851:1999: Determination of the ultimate aerobic biodegradability of plastic materials in an aqueous medium — Method by measuring the oxygen demand in a closed respirometer.*

Chapter 11

Design and Evaluation of Agro-Based Food Packaging Films

H. N. Cheng,*[,1] Atanu Biswas,[2] Roselayne F. Furtado,[3] Carlucio R. Alves,[4] and Qinglin Wu[5]

[1]Southern Regional Research Center, USDA Agriculture Research Service, New Orleans, Louisiana 70124, United States

[2]National Center for Agricultural Utilization Research, USDA Agriculture Research Service, Peoria, Illinois 61604, United States

[3]Embrapa Agroindústria Tropical, Dra Sara Mesquita Street, 2270, 60.511-110 Fortaleza, Brazil

[4]State University of Ceará, Science and Technology Center, 60.714-903 Fortaleza, Brazil

[5]School of Renewable Natural Resources, Louisiana State University Agricultural Center, Baton Rouge, Louisiana 70803, United States

*Email: hn.cheng@usda.gov

There is current interest in using agro-based materials to produce food packaging that can increase the shelf lives of food items, minimize food spoilage, and reduce plastic waste. We have been conducting research in this area and have adopted several approaches in designing new packaging films: 1) Polymer blending in order to optimize the properties of the resulting films, with the advantage that no chemical synthesis is needed and the regulatory hurdle in product development is lower. 2) Incorporation of essential oils as antimicrobial component in films. 3) Modification of film surfaces to render them less hydrophilic. 4) Design of composites involving biobased materials and synthetic polymers to form semi-synthetic films. 5) Incorporation of nanomaterials to strengthen the film. Because the agro-based materials are structurally diverse, combinations and variations are possible, and these approaches offer many opportunities for customization and optimization. Examples of these approaches from the authors' work will be shown as illustrations in this review.

Introduction

In order to achieve a better and more sustainable future for all mankind, the United Nations in 2015 adopted 17 Sustainable Development Goals as the blueprint to address the global challenges we face, including those related to health, climate change, environmental degradation, industry innovation and infrastructure, responsible production and consumption (*1*). The use of biobased materials for product development and commercialization is fully compatible with these goals. In fact, in view of the increasing public awareness of the need to minimize fossil resource consumption, waste accumulation, and environmental pollution, biobased materials represent a growing opportunity for innovative scientists and engineers. A lot of research and development activities are ongoing, and several books and review articles have appeared covering different aspects of this field (*2–7*).

One promising market for biobased materials is the food packaging area. In general, synthetic polymers do a good job protecting food from contamination and spoilage, but polymer additives may migrate into food, and the packages after use may pose as waste materials causing disposal problems. In contrast, biobased materials are biodegradable, non-toxic and sustainable, and they are increasingly being engaged in packaging. The more common materials reported in the literature include starch, cellulosic derivatives, chitosan, food gums, and proteins, and they have been evaluated as packaging films, food wrappings, trays, containers, and edible coatings. The number of publications in this area are large, and quite a few reviews are available (*8–13*).

For many years, we have been doing research and development (R&D) in this area, and several approaches have been employed to produce new film materials. These include: 1) blend formation, 2) composite formation, 3) modification of film surface, 4) incorporation of essential oil in the film matrix, and 5) inclusion of nanomaterials in the film. In this chapter, these approaches are being reviewed, with the emphasis on the work done by us and our collaborators.

A summary of the approaches and the agro-based materials used by our teams is given in Table 1. We have utilized both natural and synthetic polymers as the continuous matrices for packaging films, e.g., polysaccharides, proteins, poly(lactic acid), poly(vinyl alcohol), ethylene-vinyl alcohol copolymer, polyethylene and polypropylene. Where appropriate, these materials can be selectively mixed to form polymer blends. For composite formation, several agricultural products or byproducts have been chosen as fillers, such as cotton gin trash, cottonseed hull, nutshells, and edible beans. These fillers are relatively inexpensive and have the added benefit of reducing the price of the composites. As for nanocomposites, various types of nanoparticles have been used.

Table 1. Types of Polymers and Selected R&D Approaches Towards Material Design as Illustrated in This Work

Type	*Polymers*	*Examples*	*Approaches*
Agro-based	Polysaccharides (and derivatives)	cellulose esters, carboxymethyl cellulose, hemicellulose, cashew gum, starch	1. Blend formation 2. Composite formation 3. Surface modification 4. Incorporation of essential oil 5. Use of nanomaterials
	Proteins	zein, gelatin, cottonseed protein, soy protein	
	Ag products or byproducts	edible beans, cotton gin trash, cottonseed hull, nut shells	
	Bio-derived materials	poly(lactic acid), polyhydroxyalkanoates	
Synthetic	Water soluble	poly(vinyl alcohol), ethylene-vinyl alcohol copolymer	
	Water insoluble	polyethylene, polypropylene	

Polymer Blends

Polymer blending is a well-known technique that can be utilized to formulate new products from two or more materials (*14*). Since blending does not involve chemical synthesis, this approach is potentially cheaper and poses lower regulatory and safety hurdles towards product development. This topic has been often reviewed in the literature. For example, an article has reviewed the use of polysaccharides (including some blends) as films and coatings for food packaging (*15*). A number of blends of anionic and cationic polysaccharides as given in the literature have been noted in a recent paper (*16*). The existing methods of preparing polymer blends of cellulose, chitin and chitosan with natural and synthetic polymers and their applications have been reviewed (*17*). Moreover, another review article has reported several blends involving polysaccharide/protein combinations (*18*).

In one of our publications, we reported the blends of two types of neutral hemicellulose (*19*). Previously, the reports of blends of two uncharged (electrically neutral) hemicelluloses are relatively rare in the literature. In this work, hemicelluloses were extracted from *Caesalpina pulcherrima* (25% yield) and *Tamarindus indica* (20% yield). The rheology data indicated that the solutions of films and blends formulated in this study had non-Newtonian behavior. Galactomannan film showed a higher permeability to moisture than xyloglucan and the blends. All films showed less than 20% opacity and were relatively transparent. Thus, the characterization and test results suggested that both of these hemicelluloses and their blends have satisfactory properties that are compatible with food-related and packaging applications. The use of the blends permits a wider range of properties to be accessible than the hemicelluloses alone.

Another promising blend system involved cashew gum and gelatin (*20*). These were formulated with 10% glycerol with the help of the central composite rotational design. The resulting films with different cashew gum/gelatin ratios were examined for water vapor permeability (WVP), thickness, solubility, mechanical and thermal properties, surface morphology, and biodegradability. The 1/1 cashew gum/gelatin blend showed the best WVP, but for elongation at break, the 2/1 cashew gum/gelatin provided the best results with over 100% elongation. The films were shown to be easily biodegradable, producing as much CO_2 as starch during the 22 days of an aerobic biodegradability test.

Another protein/polysaccharide combination was the zein/chitosan blend, which was previously shown to have good film properties (*21*). We studied the film preparation method and its effects on film thickness, color, water solubility, contact angle and tensile properties (*22*). The zein/chitosan blend was confirmed to have good film properties, and the films prepared with unheated chitosan showed better results for water solubility, contact angle, tensile strength and elastic modulus than heated chitosan.

In other studies, we have also made blends from galactomannan and cashew gum (*23*), cationic and anionic xylan (*16*), cottonseed protein and polycaprolactone (*24*), soy and cottonseed proteins (*25*), polysaccharides and cottonseed protein (*25*), anionic polymers and cottonseed protein (*26*), starch, poly(acrylic acid), and poly(N-vinyl pyrrolidone) (*27*).

Composites

By definition, a biocomposite has one of its components derived from biological origins (*28*). Frequently the discrete phase of the composite consists of agro-based materials, and the continuous phase is either a biopolymer or a synthetic polymer. This has been a popular approach to produce new materials, and many reviews have appeared on this topic (*10, 29–31*). A large amount of agro-based materials have been used as fillers, including starch, cellulose fibers, cellulose whiskers, microcrystalline cellulose, wood flour and fibers, flax fibers, hemp fibers, kenaf fibers, bamboo fibers, jute, corn husks and stover, apple solids and fibers, sugar beet pulp, oat husks, cocoa shells, wheat straw, soy stalks, abaca leaves, green coconut, cuphea and lasquerella seeds, agar, nutshells, cotton gin trash (CGT), and cottonseed hull (*32*).

In our studies, we have used PLA as the continuous matrix and pulverized cotton burr and hull (*33*), nutshells (*32*), and pinto bean flour (PBF) (*34*) as fillers. The addition of the fillers reduced the tensile strength, elongation at break, and Young's modulus, slightly decreased Tg and Tm, but increased the % crystallinity. The PLA-filler composites may be useful for lowering the cost of the materials in applications that can tolerate the reduced mechanical properties. In addition, the PLA-bean composite may be applied to situations where moisture uptake and water absorption properties are desirable.

In two studies, we made poly(vinyl alcohol) (PVOH) composites using fillers such as PBF and pinto bean high starch fraction (PBHSF) (*35*) and CGT (*36*). In general, the PBF composites performed better than the PBHSF composites. Films made from PVOH/PBF composites gave reduced tensile strength and Young's modulus, but enhanced elongation up to 40% PBF level (*35*). The PVOH-CGT composites gave a different behavior. For composites up to 60% CGT, the composites showed decreased tensile strength and elongation but enhanced Young's modulus (*36*). Thus, the addition of CGT fillers to PVOH might be beneficial in applications where some reductions in tensile strength are permissible, and reductions in cost, enhanced stiffness, and improved biodegradability are desirable.

In four papers, we used low-density polyethylene (LDPE) as the continuous matrix for composite formation, e.g., LDPE-PBF (*34*), LDPE-nutshell (*32*), LDPE-cotton burr and hull (*33*), and LDPE-CGT (*37*). For LDPE-cotton burr and LDPE-CGT, addition of the fillers slightly lowered the composite's Tm, increased the composite's Young's modulus, but decreased the composite's tensile strength and elongation at break (*33, 37*). Incorporation of maleic anhydride and peroxide only slightly improved the tensile properties (*33*). For LDPE-nutshell (*32*) and LDPE-PBF (*34*), the addition of fillers caused reductions in tensile strength and elongation, mixed results for Young's modulus, and slightly reduced Tm. The use of maleic anhydride and peroxide with the fillers had a slightly positive effect on the stiffness of polyethylene. Again, the composites may be helpful in lowering the cost of the materials in applications that can tolerate the decreased tensile properties.

A recent paper (*38*) looked at the composites of polypropylene (PP) and CGT that was modified by two methods: toluene washed only, and toluene washed and treated with maleic anhydride-grafted polypropylene (MAPP). The PP composites displayed tensile strengths similar to that of the neat PP, and exhibited higher values of Young's modulus and lower elongation at break than the neat PP. These results were attributed to the better matrix-filler adhesion for the treated versus untreated fillers.

Surface Modifications

Many of the agro-based materials are hydrophilic, and when they are used as food packaging, sometimes undue water vapor transmission can accelerate food spoilage and reduced shelf lives (*39, 40*). Moreover, if the package gets wet, the water may alter the properties of the packaging material and render it less effective. There are many approaches to solving this problem, such as the incorporation of nanomaterials, multilayer coatings, coating with a hydrophobic material, polymer crosslinking, and surface modification reactions. These approaches have been reviewed in the literature (*39–41*). In our work, we have placed more emphasis on hydrophobic surface modification reactions.

In an earlier work, hydroxyethyl cellulose film was grafted with polycaprolactone (*42*) using a lipase-catalyzed ring-opening graft reaction. The degree of substitution (DS) was 0.10-0.32 per anhydroglucose unit. Another surface-modified polysaccharide film was cashew gum, which is obtained from the bark of the cashew tree and contains β-D-galactose (72%) and α-D-glucose (14%), with less than 5% of each of α-L-arabinose (4.6%), α-L-rhamnose (3.2%) and β-D-glucuronic acid (4.5%). For example, in a recent work, antibodies were immobilized on carboxymethylated cashew gum film, which was used as a biosensor for the rapid and efficient detection of *Salmonella Typhimurium* bacteria in food (*43*).

In two papers, cashew gum was modified with hydrophobic functional groups (*44, 45*). In the first paper (*44*), hydrophobic modification was carried out with alkyl ketene dimer (AKD) at 90 °C, using dimethyl sulfoxide as a solvent and 4-dimethylaminopyridine as a promoter. Samples with DS up to 0.016 have been made; the higher DS samples tend to form insoluble dispersions. In the second paper (*45*), cashew gum was modified with alkenyl succinic anhydride (ASA) using two reagents: octenyl succinic anhydride and tetrapropenyl succinic anhydride. Reactions were conducted at 120 °C using dimethyl sulfoxide as a solvent, with conversions better than 88%. Samples with DS between 0.02 and 0.20 were produced. These new polymers showed some of the properties of other ASA- and AKD-derivatized polysaccharides and may be helpful additions to the family of hydrophobically modified polysaccharides used as polymeric surfactants and as thickeners in water-based formulations.

Another biopolymer that is very promising for food packaging applications is zein. It is a protein and a product of industrial corn processing (*46, 47*). Zein has been subjected to derivatization reactions, and these reactions have been previously reviewed (*46, 48*). For food packaging applications, there is a need to reduce its water absorption and surface wetting behavior. A new method to carry out surface modification of zein film was reported by one of us (*48*), where octenyl succinic anhydride or AKD was placed on zein film surface and the assembly baked at 100 °C overnight. The resulting films exhibited decreased water absorption and increased contact angle with water.

A different reaction involving isocyanates and diisocyanates was reported to reduce the surface wetting behavior of zein (*49*). The moisture uptake decreased with isocyanate and diisocyanate modifications, and the surface free energy of the modified zein decreased, indicating reduced hydrophilicity. Based on results from NMR and FTIR, we identified the major pathways of this reaction. Our findings suggested that the reaction of zein with diisocyanates might find further applications in bioplastics.

Incorporation of Essential Oils

Antimicrobial packaging is a term that refers to the different methods whereby antimicrobial activity is introduced into the packaging film material in order to kill microorganisms and extend the shelf life of food (*50–52*). As noted in the reviews (*52, 53*), antimicrobial packaging can take several forms, such as: 1) inclusion of sachets containing antimicrobial agents in packages; 2) adding antimicrobial compounds directly into polymers; 3) placing an antimicrobial coating onto polymer surfaces; 4) attaching antimicrobial compounds to polymers via ionic or covalent linkages; 5) using inherently antimicrobial polymers.

Many antimicrobial compounds have been added directly into polymers for food packaging. Among them, essential oils are notable for being eco-friendly and natural compounds (*54*). Some essential oils like rosemary, clove, cinnamon, oregano, nutmeg and basil, have stood out because of their strong antimicrobial efficacy. In each case, it is necessary to evaluate the effects of these compounds and their interactions with the polymer matrix, together with the mechanical and thermal properties of the packaging. A review article on the use of essential oils in active food packaging is available (*55*).

In a collaborative publication between Embrapa and USDA scientists (*56*), cellulose ester-based films were made that incorporated essential oils (EO's) from lemongrass (*Cybopogon citratus*), rosemary pepper (*Lippia sidoides*) and basil (*Ocimum gratissimum*) at concentrations of 10 and 20% (v/w). The addition of the essential oil caused a decrease in the WVP due to the hydrophobic nature of the oil. Moreover, the presence of EO's reduced Young's modulus and tensile strength and increased elongation at break of the cellulose ester films, suggesting that the essential oils acted as plasticizers.

In a follow-up expanded study, nine EO's were incorporated into cellulose acetate, cellulose acetate propionate and cellulose acetate butyrate (*57*). The EO's were derived from lime, nutmeg, eugenol, pimenta berry, rosemary, petitgrain, coffee, anise, and trans-cinnamaldehyde. In almost all cases, the addition of EO's to cellulose ester reduced tensile strength and Young's modulus but increased elongation at break. Thus, it was confirmed that an EO acted like a plasticizer that enhanced the flexibility of the polymer. The water barrier property was improved the most when EO's were added to cellulose acetate; however, different cellulose ester/EO combinations showed somewhat different effects. Whereas most of the EO's decreased the transparency of the films, eugenol, pimento berry, and anise were notable exceptions. Thus, depending on a specific application, a particular cellulose ester/EO combination could be used to give the optimal performance.

In an effort to produce an edible food packaging, two carboxymethyl cellulose (CMC) materials with different DS were compounded with four different EO's (eugenol, rosemary oil, coriander oil, and nutmeg oil) (*58*). The tensile strength, elongation at break, Young's modulus, opacity, and water vapor permeation were evaluated. In general, the EO-embedded CMC with the two DS values behaved rather differently. In particular, the EO-embedded CMC with 0.7 DS gave stronger and more flexible films than EO-embedded CMC with 1.2 DS and may be more suited for use in food packaging.

Use of Nanomaterials

Currently there is a lot of interest in nanotechnology (*59, 60*), and a good approach to strengthening a food packaging polymer is to add an appropriate nanomaterial to it. Numerous papers have been published on this topic, including several reviews (*61–65*). Some commonly used materials include nanoclays (*61, 62, 65*), carbon nanotubes (*62*), graphene nanoplates (*62*), silver nanoparticles (*65*), and nanocellulose (*63, 64*).

In a recent collaborative work (*66*), a nanocomposite packaging system was designed that reduced the oxygen content in the package in order to minimize product oxidation and spoilage, and prolong product shelf life. This system comprised cellulose nanocrystals and Pd nanoparticles embedded in an ethylene-vinyl alcohol copolymer (EVOH). The nanocellulose reduced Pd precursor to Pd metal and also enhanced the physical properties of the EVOH film. Pd nanoparticles reacted with oxygen to serve as oxygen scavenger. The cellulose nanocrystals were also optionally oxidized, which provided improved oxygen absorption. These features made these nanocomposites promising candidates as active packaging materials.

A different approach was taken (*67*) to isolate cellulose nanoparticles (CN) with different morphologies from cottonseed hull and to observe their influence on rheological properties of PVOH suspensions. Three types of CN were obtained, viz., fibrous cellulose nanofibers, rod-like cellulose nanocrystals and spherical CN. For PVOH/CN suspensions, the PVOH/CN concentration, CN morphology, and CN/PVOH weight ratio were the main factors that influenced their rheology. This information will be useful in the preparation of films, fibers, and hydrogels based on PVOH/CN suspensions.

In another work (*68*), partially deacetylated chitin nanofibers were added as reinforcing and antimicrobial agents in CMC films. Up to 10 wt % nanofibers were incorporated in the films, and their mechanical properties and antimicrobial activities were evaluated. These nanocomposite films gave very good tensile strength and Young's modulus, and strong antimicrobial activities against both *E. coli* and *S. aureus*, suggesting that they might be used for food packaging applications.

A different strategy was adopted (*69*) to graft poly(methyl methacrylate) (PMMA) and poly(butyl acrylate) (PBA) onto cellulose nanocrystals (CNC), thereby imparting hydrophobic traits to CNC. The nanocomposites with 7 wt % of this modified CNC in PBA-PMMA copolymer showed an increase of >25-fold in Young's modulus and 3-fold in tensile strength compared to those of neat PBA-co-PMMA. Therefore, the modified CNCs are effective reinforcing agents for the PBA-co-PMMA copolymers. Another application of this strategy was to graft PMMA onto CNC (*70*). This modified CNC provided increased thermal stability, improved tensile strength and Young's modulus, but reduced elongation at break to PBA-co-PMMA nanocomposites. It was also shown to play a crucial role in controlling the interfacial miscibility and tuning the phase morphology of the nanocomposites. The nanocomposite materials containing these modified CNC's are likely to be useful in packaging applications.

Opportunities and Challenges

Food packaging is needed to aid handling, facilitate transport, and minimize food spoilage. Because most of the commonly used synthetic packaging materials are petroleum-based, which are not sustainable and degradable, leading to increasing volumes of plastic waste in landfills, there is ongoing interest in developing biopolymers as alternative packaging materials. In this work, polysaccharides and their derivatives, proteins and their derivatives, agricultural products or byproducts, and bio-derived materials have all been shown to be viable materials for the production of food packaging films. Five approaches are especially highlighted, including blend and composite formation, surface modification, incorporation of essential oil, and use of nanomaterials in the films. For composites and blends, sometimes synthetic polymers are included in order to achieve the desired properties. It is apparent that a large number of new materials can be generated through the application of these approaches, and the examples shown in this review represent only a small fraction of all the possibilities. Thus, this area has the potential for future innovation and growth.

Whereas there are ample opportunities for creative research and development in this area, it is more challenging to transition research findings to commercial products (*71*, *72*). A successful development of a commercial product for a targeted application often requires formulation studies with different compositions, process variations to optimize the product quality and properties, and design and acquisition of application data. This development process can be time-consuming and is often done in collaboration with a company that has the capability to manufacture and market the product. Of course, there are many other considerations as well, such as profit/cost analysis, regulatory compliance, environmental and safety considerations, and protection of proprietary information. Yet, despite the challenges, many biobased films have been successfully commercialized (73–75). In view of continued activities in both R&D and commercial development, we expect more biobased film products to appear in the future.

Acknowledgments

The authors thank their collaborators over the years for the productive partnerships and the encouragements of the managements of their respective organizations. Thanks are also due to Dr. K. T. Klasson for his helpful input. Mention of trade names or commercial products in this publication is solely for the purpose of providing specific information and does not imply recommendation or endorsement by the U.S. Department of Agriculture (USDA). USDA is an equal opportunity provider and employer.

References

1. United Nations. https://www.un.org/sustainabledevelopment/sustainable-development-goals/ (accessed 2020-06-06).
2. *Biobased Products and Industries*; Galanakis, C. M., Ed.; Elsevier: Amsterdam, 2020.
3. *Green Polymer Chemistry: New Products, Processes, and Applications*; Cheng, H. N., Gross, R. A., Smith, P. B. , Eds.; ACS Symposium Series 1310; American Chemical Society: Washington, DC, 2018.
4. *Green Polymer Chemistry: Biobased Materials and Biocatalysis*; Cheng, H. N., Gross, R. A., Smith, P. B., Eds.; ACS Symposium Series 1191; American Chemical Society: Washington, DC, 2015.
5. Valdés, A.; Mellinas, A. C.; Ramos, M. Natural additives and agricultural wastes in biopolymer formulations for food packaging. *Front. Chem.* **2014**, 2, 1–10.
6. *Bio-Based Plastics: Materials and Applications*; Kabasci, S., Ed.; John Wiley & Sons: Chichester, UK, 2014.
7. Babu, R. P.; O'Connor, K.; Seeram, R. Current progress on bio-based polymers and their future trends. *Prog. Biomater.* **2013**, *2*, 8.
8. Halonen, N.; Pálvölgyi, P. S.; Bassani, A.; Fiorentini, C.; Nair, R.; Spigno, G.; Kordas, K. Bio-Based Smart Materials for Food Packaging and Sensors – A Review. *Front. Mater.* **2020**, 7, 82.
9. *Bio-based Materials for Food Packaging*; Ahmed, S., Ed.; Springer Nature: Singapore, 2018.
10. Johansson, C.; Bras, J.; Mondragon, I.; Nechita, P.; Plackett, D.; Simon, P.; Svetec, D. G.; Virtanen, S.; Baschetti, M. G.; Breen, C.; Aucejo, S. Renewable fibers and bio-based materials for packaging applications – a review of recent developments. *BioResources* **2012**, 7, 2506–2552.
11. Weber, C. J.; Haugaard, V.; Festersen, R.; Bertelsen, G. Production and applications of biobased packaging materials for the food industry. *Food Addit. Contam.* **2002**, *19* (Sup.1), 172–177.
12. Petersen, K.; Nielsen, P. V.; Bertelsen, G.; Lawther, M.; Olsen, M. B.; Nilssonk, N. H.; Mortenseny, G. Potential of biobased materials for food packaging. *Trends Food Sci. Technol.* **1999**, *10*, 52–68.

13. Helanto, K.; Matikainen, L.; Talja, R.; Rojas, O. J. Bio-based polymers for sustainable packaging and biobarriers: A critical review. *BioResources* **2019**, *14*, 4902–4951.
14. Paul, D. R.; Bucknall, C. B. *Polymer Blends*; Wiley: New York, 2000.
15. Cazón, P.; Velazquez, G.; Ramírez, J. A.; Vázquez, M. Polysaccharide-based films and coatings for food packaging: A review. *Food Hydrocolloids* **2017**, *68*, 136–148.
16. Cheng, H. N.; Ford, C. V.; Kolpak, F. J.; Wu, Q. Preparation and characterization of xylan derivatives and blends. *J. Polym. Environ.* **2018**, *26*, 4114–4123.
17. Rogovina, S. Z.; Vikhoreva, G. A. Polysaccharide-based polymer blends: Methods of their production. *Glycoconjugate J.* **2006**, *23*, 611.
18. Benbettaieb, N.; Gay, J.-P.; Karbowiak, T.; Debeaufort, F. Tuning the functional properties of polysaccharide–protein bio-based edible films by chemical, enzymatic, and physical cross-linking. *Compr. Rev. Food Sci. Food Saf.* **2016**, *15*, 739–752.
19. Mendes, F. R. S.; Bastos, M. S. R.; Mendes, L. G.; Sousa, F. D.; Silva, A. R. A.; Monteiro-Moreira, A. C. O.; Cheng, H. N.; Biswas, A.; Moreira, R. A. Preparation and evaluation of hemicellulose films and their blends. *Food Hydrocolloids* **2017**, *70*, 181–190.
20. Oliveira, M. A.; Furtado, R. F.; Bastos, M. S. R.; Benevides, S.; Leitão, R. C.; Muniz, C. R.; Biswas, A.; Cheng, H. N. Performance evaluation of cashew gum and gelatin blend for food packaging. *Food Packaging Shelf Life* **2018**, *17*, 57–64.
21. Garcia, M. E.; Dominguez, G. C.; Perez, J. J. C.; Rebollo, R. R. F.; Adame, J. A. A.; Vazquez, I. A.; Mendez, J. V.; Ruiz, L. A. M. Physical and structural characterization of zein and chitosan edible films using nanotechnology tools. *Int. J. Biol. Macromol.* **2013**, *61*, 196–203.
22. Pinho, L. X.; Oliveira, M.; Oliveira, W. Q.; Furtado, R. F.; Bastos, M. S. R.; Muniz, C.; Biswas, A.; Cheng, H. N. Preparation and characterization of zein and chitosan edible film. *Proc. 25th Food Sci. Technol. Brazilian Congress (CBCTA) and the 10th CIGR Section IV Int. Technical Symp.*, October 24–27, 2016, Gramado/RS, Brazil.
23. Mendes, L. G.; Mendes, F. R.; Furtado, R. F.; Bastos, M. S.; da Costa, J. M.; Monteiro-Moreira, A. C.; Biswas, A.; Cheng, H. N.; Moreira, R. A. Plant polysaccharide blends for rosemary essential oil encapsulation, manuscript in preparation.
24. Cheng, H. N.; Ford, C.; He, Z. Evaluation of polyblends of cottonseed protein and polycaprolactone plasticized by cottonseed oil. *Int. J. Polym. Anal. Charact.* **2019**, *24*, 389–398.
25. Cheng, H. N.; Ford, C.; Dowd, M. K.; He, Z. Soy and cottonseed protein blends as wood adhesives. *Ind. Crops Prod.* **2016**, *85*, 324–330.
26. Cheng, H. N.; Wyckoff, W.; Dowd, M. K.; He, Z. Evaluation of adhesion properties of blends of cottonseed protein and anionic water-soluble polymers. *J. Adhes. Sci. Technol.* **2019**, *33*, 66–78.
27. Biswas, A.; Willett, J. L.; Gordon, S. H.; Finkenstadt, V. L.; Cheng, H. N. Complexation and blending of starch, poly(acrylic acid), and poly(n-vinyl pyrrolidone). *Carbohydr. Polym.* **2006**, *65*, 397–403.
28. *Biocomposites: Biomedical and Environmental Applications*; Ahmed, S., Ikram, S., Kanchi, S., Bisetty, K., Eds.; Pan Stanford Publishing: Singapore, 2018.
29. Mohanty, A. K.; Misra, M.; Drzal, L. T. Sustainable bio-composites from renewable resources: opportunities and challenges in the green materials world. *J. Polym. Environ.* **2002**, *10*, 19–26.

30. Šimkovic, I. Unexplored possibilities of all-polysaccharide composites. *Carbohydr. Polym.* **2013**, *95*, 697–715.
31. Väisänen, T.; Das, O.; Tomppo, L. A review on new bio-based constituents for natural fiber-polymer composites. *J. Cleaner Prod.* **2017**, *149*, 582–596.
32. Sutivisedsak, N.; Cheng, H. N.; Burks, C. S.; Johnson, J. A.; Siegel, J. P.; Civerolo, E. L.; Biswas, A. Use of nut shells as fillers in polymer composites. *J. Polym. Environ.* **2012**, *20*, 305–314.
33. Sutivisedsak, N.; Cheng, H. N.; Dowd, M. K.; Selling, G. W.; Biswas, A. Evaluation of cotton byproducts as fillers for poly(lactic acid) and low density polyethylene. *Ind. Crops Prod.* **2012**, *36*, 127–134.
34. Sutivisedsak, N.; Cheng, H. N.; Liu, S.; Lesch, W. C.; Finkenstadt, V. L.; Biswas, A. Mechanical, thermal, and moisture properties of plastics with bean as filler. *J. Biobased Mater. Bioenergy* **2012**, *6* (1), 59–68.
35. Sutivisedsak, N.; Biswas, A.; Finkenstadt, V. L.; Cheng, H. N. Evaluation of composite films containing poly(vinyl alcohol) and edible bean. *ACS Polym. Prepr.* **2011**, *52* (2), 256–257.
36. Biswas, A.; Cheng, H. N.; Evangelista, R.; Evangelista, M.; Boddu, V.; Kim, S. Evaluation of composite films containing poly(vinyl alcohol) and cotton gin trash. *J. Polym. Environ.* **2020**, *28*, 1998–2007.
37. Cheng, H. N.; Dowd, M. K.; Finkenstadt, V. L.; Selling, G. W.; Evangelista, R. L.; Biswas, A. Use of cotton gin trash and compatibilizers in polyethylene composites. *ACS Symp. Ser.* **2013**, *1144*, 423–431.
38. Ge, C.; Cheng, H. N.; Miri, M. J.; Hailstone, R. K.; Francis, J. B.; Demyttenaere, S. M.; Alharbi, N. A. Preparation and evaluation of composites comprising polypropylene and cotton gin trash. *J. Appl. Polym. Sci.* **2020**, *137*, e49151.
39. Vartiainen, J.; Vähä-Nissi, M.; Harlin, A. Biopolymer films and coatings in packaging applications—a review of recent developments. *Mater. Sci. Appl.* **2014**, *5* (10), 48968. DOI: 10.4236/msa.2014.510072.
40. Ahmed, I.; Lin, H.; Zou, L.; Brody, A. L.; Li, Z.; Qazi, I. M.; Pavase, T. R.; Lv, L. A comprehensive review on the application of active packaging technologies to muscle foods. *Food Control* **2017**, *82*, 163–178.
41. Garavand, F.; Rouhi, M.; Razavi, S. H.; Cacciotti, I.; Mohammadi, R. Improving the integrity of natural biopolymer films used in food packaging by crosslinking approach: A review. *Int. J. Biol. Macromol.* **2017**, *104*, 687–707.
42. Li, J.; Xie, W.; Cheng, H. N.; Nickol, R. G.; Wang, P. G. Polycaprolactone-modified hydroxyethylcellulose films prepared by lipase-catalyzed ring-opening polymerization. *Macromolecules* **1999**, *32*, 2789–2792.
43. Melo, A. M. A.; Oliveira, M. R. F.; Furtado, R. F.; Borges, M. F.; Biswas, A.; Cheng, H. N.; Alves, C. R. Preparation and characterization of carboxymethyl cashew gum grafted with immobilized antibody for potential biosensor application. *Carbohydr. Polym.* **2020**, *228*, 115408.
44. Biswas, A.; Kim, S.; Buttrum, M.; Furtado, R. F.; Alves, C. R.; Cheng, H. N. Preparation of hydrophobically modified cashew gum through reaction with alkyl ketene dimer. In Green Polymer Chemistry: New Products, Processes, and Applications; Cheng, H. N., Gross, R. A.,

Smith, P. B., Eds.; American Chemical Society: Washington, DC, 2018. *ACS Symp. Ser.* **2018**, *1310*, 137–146.

45. Biswas, A.; Cheng, H. N.; Kim, S.; Alves, C. R.; Furtado, R. F. Hydrophobic modification of cashew gum with alkenyl succinic anhydride. *Polymers* **2020**, *12*, 514.
46. Lawton, J. W. Zein: a history of processing and use. *Cereal Chem.* **2002**, *79*, 1–18.
47. Anderson, T. J.; Lamsal, B. P. Zein extraction from corn, corn products, and coproducts and modifications for various applications: A review. *Cereal Chem.* **2011**, *88*, 159–173.
48. Biswas, A.; Selling, G. W.; Woods, K. K.; Evans, K. O. Surface modification of zein films. *Ind. Crops Prod.* **2009**, *30*, 168–171.
49. Sessa, D. J.; Cheng, H. N.; Kim, S.; Selling, G. W.; Biswas, A. Zein-based polymers formed by modifications with isocyanates. *Ind. Crops Prod.* **2013**, *43*, 106–113.
50. Malhotra, B.; Keshwani, A.; Kharkwal, H. Antimicrobial food packaging: potential and pitfalls. *Front. Microbiol.* **2015**, *6*, 611.
51. Cha, D. S.; Chinnan, M. S. Biopolymer-based antimicrobial packaging: A review. *Crit. Rev. Food Sci. Nutr.* **2004**, *44*, 223–237.
52. Appendini, P.; Hotchkiss, J. H. Review of antimicrobial food packaging. *Innov. Food Sci. Emerging Technol.* **2002**, *3*, 113–126.
53. Coma, V. Bioactive packaging technologies for extended shelf life of meat-based products. *Meat Sci.* **2008**, *78*, 90–103.
54. Raut, J. S.; Karuppayil, S. M. A status review on the medicinal properties of essential oils. *Ind. Crops Prod.* **2014**, *62*, 250–264.
55. Ribeiro-Santos, R.; Andrade, M.; Melo, N. R.; Sanches-Silva, A. Use of essential oils in active food packaging: Recent advances and future trends. *Trends Food Sci. Technol.* **2017**, *61*, 132–140.
56. Bastos, M. S. R.; Laurentino, L. S.; Canuto, K. M.; Mendes, L. G.; Martins, C. M.; Silva, S. M.; Furtado, R. F.; Kim, S.; Biswas, A.; Cheng, H. N. Physical and mechanical testing of essential oil-embedded cellulose ester films. *Polym. Test.* **2016**, *49*, 156–161.
57. Biswas, A.; Bastos, M. S. R.; Furtado, F. F.; Kuzniar, G.; Boddu, V.; Cheng, H. N. Evaluation of the properties of cellulose ester films that incorporate essential oils. *Int. J. Polym. Sci.* **2020**, 4620868.
58. Biswas, A.; Furtado, R. F.; Bastos, M. S. R.; Benevides, S. D.; Oliveira, M. A.; Boddu, V.; Cheng, H. N. Preparation and characterization of carboxymethyl cellulose films with embedded essential oils. *J. Mater. Sci. Res.* **2018**, *7*, 16–25.
59. *Nanotechnology: Delivering on the Promise; Volume* 1; Cheng, H. N., Doemeny, L., Geraci, C. L., Schmidt, D. G., Eds.; ACS Symposium Series 1220; American Chemical Society: Washington, DC, 2016.
60. Cheng, H. N.; Klasson, K. T.; Asakura, T.; Wu, Q. Nanotechnology in agriculture. *ACS Symp. Ser.* **2016**, *1224*, 233–242.
61. Attaran, S. A.; Hassan, A.; Wahit, M. U. Materials for food packaging applications based on bio-based polymer nanocomposites: A review. *J. Thermoplast. Compos. Mater.* **2017**, *30*, 143–173.
62. Müller, K.; Bugnicourt, E.; Latorre, M.; Jorda, M.; Sanz, Y. E.; Lagaron, J. M.; Miesbauer, O.; Bianchin, A.; Hankin, S.; Bölz, U.; Pérez, G.; Jesdinszki, M.; Lindner, M.; Scheuerer, Z.; Castelló, S.; Schmid, M. Review on the processing and properties of polymer nanocomposites

and nanocoatings and their applications in the packaging, automotive and solar energy fields. *Nanomaterials* **2017**, 7 (4), 74.

63. Stark, N. M. Opportunities for cellulose nanomaterials in packaging films: a review and future trends. *J. Renew. Mater.* **2016**, *4*, 313–326.
64. Hubbe, M. A.; Ferrer, A.; Tyagi, P.; Yin, Y.; Salas, C.; Pal, L.; Rojas, O. J. Nanocellulose in thin films, coatings, and plies for packaging applications: A review. *BioResources* **2017**, *12*, 2143–2233.
65. Silvestre, C.; Duraccio, D.; Cimmino, S. Food packaging based on polymer nanomaterials. *Prog. Polym. Sci.* **2011**, *36*, 1766–1782.
66. Cherpinski, A.; Biswas, A.; Lagaron, J. M.; Dufresne, A.; Kim, S.; Buttrum, M.; Espinosa, E.; Cheng, H. N. Preparation and evaluation of oxygen scavenging nanocomposite films incorporating cellulose nanocrystals and Pd nanoparticles in poly(ethylene-co-vinyl alcohol). *Cellulose* **2019**, *26*, 7237–7251.
67. Zhou, L.; He, H.; Li, M.-C.; Song, K.; Cheng, H. N.; Wu, Q. Morphological influence of CNs from cottonseed hulls on rheological properties of polyvinyl alcohol/CN suspensions. *Carbohydr. Polym.* **2016**, *153*, 445–454.
68. Li, M.; Wu, Q.; Song, K.; Cheng, H. N.; Suzuki, S.; Lei, T. Chitin nanofibers as reinforcing and antimicrobial agents in carboxymethyl cellulose films: influence of partial deacetylation. *ACS Sustainable Chem. Eng.* **2016**, *4*, 4385–4395.
69. Zhang, J.; Li, M.-C.; Zhang, X.; Dong, J.; Lee, S.; Cheng, H. N.; Lei, T.; Wu, Q. Cellulose nanocrystal driven microphase separated nanocomposites: enhanced mechanical performance and nanostructured morphology. *Int. J. Biol. Macromol.* **2019**, *130*, 685–694.
70. Zhang, J.; Li, M.-C.; Zhang, X.; Ren, S.; Dong, L.; Lee, S.; Cheng, H. N.; Lei, T.; Wu, Q. Surface modified cellulose nanocrystals for tailoring interfacial miscibility and microphase separation of polymer nanocomposites. *Cellulose* **2019**, *26*, 4301–4312.
71. Datta, A.; Mukherjee, D.; Jessup, L. Understanding commercialization of technological innovation: taking stock and moving forward. *R&D Management* **2015**, *45*, 215–249.
72. Hall, J.; Matos, S.; Bachor, V.; Downey, R. Commercializing university research in diverse settings: moving beyond standardized intellectual property management. *Res.-Technol. Manag.* **2014**, *57*, 26–34.
73. Anon, Bio-based multilayer transparent barrier films are now reality. *Bioplastics Magazine*, 6/19/2018. https://www.bioplasticsmagazine.com/en/news/meldungen/20180619Bio-based-multilayer-transparent-barrier-films-are-now-reality.php (accessed, 2020-06-06).
74. Anon, Will we shift from fossil-based to biobased plastic packaging? *Sustainability Times*, 2020, Volume 10, Issue 2. https://www.tricorbraun.com/sustainability-times-newsletter/volume-10/issue-2/#will-we-shift-from-fossil-based-to-biobased-plasti (accessed 2020-06-06).
75. Briassoulis, D.; Giannoulis, A. Evaluation of the functionality of bio-based food packaging films. *Polym. Test.* **2018**, *69*, 39–51.

Editors' Biographies

H. N. Cheng

H. N. Cheng (Ph.D., University of Illinois) is currently a research chemist at Southern Regional Research Center of the U.S. Department of Agriculture in New Orleans, where he works on projects involving improved utilization of commodity agricultural materials, green chemistry, and polymer reactions. Prior to 2009 he worked for Hercules Incorporated where he was involved at various times with new product development, team and project leadership, new business evaluation, pioneering research, and supervision of analytical research. Over the years, his research interests have included green polymer chemistry, agro-based materials, biocatalysis, pulp and paper chemistry, functional foods, polymer characterization, and NMR spectroscopy. An ACS Fellow (2009), a POLY Fellow (2010), and an AGFD Fellow (2018), he has authored or co-authored over 280 papers, 26 patent publications, co-edited 21 books, and organized or co-organized 40 symposia since 2000. He is currently (2020) the President-Elect of ACS and will serve as ACS President in 2021.

Richard A. Gross

Richard A. Gross (Ph.D., Polytechnic University) holds Constellation Chaired Professorship at Rensselaer Polytechnic Institute (RPI) and is also a member of RPI's Departments of Chemistry and Biology as well as Biomedical Engineering. Previously he was on the faculty of University of Massachusetts (Lowell) (1988–1998) and occupied the Herman F. Mark Chair Professorship at Polytechnic University (1998-2013). His research is focused on developing biocatalytic routes to biobased materials, combining chemical methods with cell-free and whole-cell biocatalytic systems to produce a large number of new biobased materials. He has over 400 publications in peer-reviewed journals with approximately 26,000 citations, has edited 7 books, and has been granted or filed 26 patents. He has received numerous awards, including Presidential Green Chemistry Award in the academic category (2003), induction into the American Institute for Medical and Biological Engineering (2007), Turner Alfrey Visiting Professor (2010), POLY Fellow (2015), Lifetime Achievement Award from the Bio-Environmental Polymer Society (2017) and the 2018 recipient of the Affordable Green Chemistry Award from the ACS.

Indexes

Author Index

Subject Index

H

N

P

S

V

Printed in the USA/Agawam, MA
December 30, 2021

787059.002